Empire Biota

Taxonomy and Evolution

2nd Edition

Bernard Pelletier

978-1-329-87400-8

Acknowledgement

I am extremely grateful to Dr. James Carpenter, of the Am. Mus. Nat. Hist., former editor-in-chief of Cladistics and now administrative editor, and administrator with the Willi Hennig Society, for his invaluable assistance in doing the computer calculations for the 3 TNT analyses.

Table of Contents

Introduction	5
Part 1	
Chapter 1-Historical Overview	9
Chapter 2-Some Words on Nomenclature	15
Chapter 3-Taxonomy-Meaning and Methods	17
Chapter 4-Prokaryotes-TNT Analyses	28
Chapter 5-Prokaryotes-PAUP Analyses	35
Chapter 6-Eukaryotes-TNT Analysis	51
Chapter 7-Eukaryotes-PAUP Analysis	58
Chapter 8-A Convenience Classification	78
Chapter 9-Conclusions	83
References	85
Part 2	
Chapter 10-Bacteria	100
Chapter 11-Red Algae	129
Chapter 12-Plantae	
Historical Overview	135
Green Algae	153
Bryophytes	159
Pteridophytes	168
Spermatophytes	175
Chapter 13-Chromophytes	224
Chapter 14-Fungi	235
Chapter 15-Animalia	254
Chapter 16-Protozoans	273
Chapter 17- Statistics	284
Chapter 18-Ecology and Biogeography	288
Chapter 19-Geological Time	302
Diagnoses	314
Appendices	
Data Set for Bacteria	315
Data Matrix for Bacteria	323
Data Set for Eukaryotes	326
Data Matrix for Eukaryotes	337
Data Set for Eubacteria	342
Data Matrix for Eubacteria	343
Figures	345
Author Profile	352

Introduction

Phylogenetic classification systems of bacteria (=prokaryotes) and eukaryotes are presented, based on a cladistic analysis of morphology, chemistry, physiology, and molecular biology. Six large-scale cladistic analyses were done, along with 2 small scale ones for Eubacteria alone, using TNT, PAUP, and Phylip's Penny, and are designed to more accurately reflect evolutionary kinship. The bacterial analyses are the first comprehensive high-level phylogenies of prokaryotes based on classical evidence. The results are in basic agreement with Gupta's protein phylogenies, i.e., Gram negatives as monophyletic and Metabacteria (Mendosicutes, Archeota) as directly related to Gram positives. However, Gram negatives have very weak resampling support. Monodermata, that is, Mollifirmicutes plus Metabacteria, also have low confidence.

Prokaryotes in the first TNT analysis, including eukaryotes as a single taxon, had 10 kingdoms which are: Thermosiphia, Fervidobacteria, Thermotogae, Gracilicutes (Didermata)(Gram negatives), Mollifirmicutes, and the 5 Metabacteria groups. The major supergroup was Aerobia comprising Fervidobacteria, Thermotogae, Gracilicutes, and Monodermata.

Another analysis was done with TNT but for bacteria only, which resulted in the same major monophyletic groups and there are 5 kingdoms: Togabacteria, Heliobacteria, Gracilicutes, Mollifirmicutes, and Metabacteria. Again, Monodermata and Didermata have weak support.

Analyses were also done with PAUP, one for prokaryotes, with eukaryotes as a single taxon and another for bacteria only, resulting in 6 (Togabacteria, Protometaerobia, Saproprotei, Metabacteria, and Heliobacteria) and 8 (Togabacteria, Subchlamydiae, Chlamydiae, Planctobacteria, Spirochetes, Chlorobacteria, Metabacteria, and Heliobacteria) kingdoms, respectively.

There were 14 kingdoms in the 1st eukaryote classification (using TNT) and 16 in the 2nd (using PAUP). Combined there are 15 and they are: Cyanidioschyza, Cyanidibiota, Glaucobiota, Rhodobiota, Plasmodiophorae, Spongomonada, Plantae, Chromista, Cercomyxa (Cercomonada and Myxobiota), Animalia, Fungi, Retaria (Actinopoda and Foraminifera), Dinaria (alveolates and excavates), Lobosa (Pelomyxida and Eulobosa), and Filosa. So there are 20 to 25 kingdoms in all, but if Eubacteria and Metabacteria are domains, as may well be the case, then there are 17, since the prokaryotic domains, which should be treated as subempires for ranking, should have only 1 kingdom each as they do not require more.

Archeoplastida, Rhizaria, Cercozoa, Alveolata, Unikonta, and Bikonta are not recovered. Chromistans and excavates are recovered but with no support, and Dinaria has no support either. Cercomyxa, Lobosa, and Filosa have weak support. Histonia (Plantae, Animalia, and Fungi) and Histonia-Chromista are also recovered but with no support. The only eukaryotic supergroups strongly supported by both types of evidence and confidence measures are Opisthokonta and possibly Retaria.

The average confidence value for the 1st bacterial analysis (including Eukaryota) was 32, for the 2nd bacterial analysis it was 32.4, and for the 1st eukaryotic analysis it was 28.8. The average CI was .51 and the average RI was .60.

The following table gives a summary of the major statistics for each analysis.

	phylogeny program	parsimony criterion	no. of taxons	no.of chrs.	CI	RI	no. of steps	MPTs	no. of rearrgmts.	year
Bacteria+Eukaryota	TNT	Wagner	23	263	.67	.76	315	8	176 mln.	2006
Bacteria	TNT	Wagner	23	277	.58	.67	347	64	15 mln.	2013
Bacteria	PAUP	Fitch	31	281	.50	.71	522	1	1.8 mln.	2015
Bacteria+Eukaryota	PAUP	Fitch	32	281	.41	.61	654	10	4.3 mln.	2015

Eubacteria	Penny	Wagner	29	78	.48		163	27		2013
Eubacteria	PAUP	Fitch	30	78	.43	.58	194	1	6 mln.	2014
Eukaryota	TNT	Wagner	27	283	.58	.45	448	13	208 mln.	2007
Eukaryota	PAUP	Fitch	41	320	.40	.44	1050	3	4.3 mln.	2014

Some of the new taxons are Mureinimurus, Clatobacteria, Metakaryota, Neokaryota, Anakaryota, Cyanidiobiota, Dinaria, Cercobiota, and Neochromista.

Some of the new names are Proteinomurus, Heterogracilicutes, Autogracilicutes, Negaerobia, Neganaerobia, Mollifirmicutes, Celestina, Anisokonta, Euglenaria, Cenophyta, Anaphyta, Apicophyta, Stomophyta, Coscinophyta, and Dinociliata (or Dinociliophora).

The infallibility or superior reliability of genotypic evidence, gradism as phylogenetic, eukaryote-first hypothesis, species as individuals, and falsificationism are rejected and refuted.

A modified Farris system is used for rank prefixes. Also included are comprehensive historical overviews, taxon tally tables (one for prokaryotes in Chp. 10, one for the largest flowering familes in Chp. 12, and a general one in Chp. 18), 4 standardized suffix systems, and the MRT (most representative tree) (a solution to the problem of multiple MPTs (equally parsimonious trees)), and convenience classifications, the general one containing 4 kingdoms: Bacteria, Phyta, Mycota, and Zoa. In Part 1 there are also discussions of various theories of vertical and horizontal descent in the origin of eukaryotes and LGT (lateral gene transfer).

Most of Part 1 was originally a condensed article submitted to about 10 peer-reviewed scientific periodicals, where it was sent to review but was not accepted for various reasons, none very convincing, or because it had already been posted on the Internet (empirebiota.info, which I deleted but recently brought back with another web host but only for Part 1 but which is still not visible as of this writing and might not be renewed), but for one there was a technical problem. At any rate, they are not in the habit of publishing independent theorists/researchers. I recently tried again with a revised version, and separated it into bacterial and eukaryotic articles, with several magazines, but the submission has to be electronic and there were technical problems for many so they were unsuccessful with those (a couple of the specialized, bacteriological magazines, where it may have stood a better chance, did try hard to get the bacterial article successsfully submitted but to no avail), and with others they were rejected, but one, although it had trouble finding reviewers because of molecular extremism, did accept the eukaryotic article, with a major revision, based on 1 reviewer, but did a 180 based on a subsequent reviewer's critique, which had errors and bias, as with most of the reviews. The bacterial article was resubmitted according to the recommendation of a reviewer but rejected in the end. A reason given by the magazine for declining the articles was that the readership would want a combined genotypic-phenotypic analysis, which is not the purpose of the articles, and I don't have the expertise for it in any case.

I did not submit the ms. to any literary agents nor conventional book publishers prior to the 1st edition, but in the meantime I submitted a book proposal for Part 2 to a conventional publishing company, but, although they found it interesting, they did not find it right for them.

Part 2 covers taxonomy at lower levels, includes historical overviews (which incorporate profiles of important taxonomists), provides descriptions of the various taxons, goes into evolution in angiosperms (foliar theory, floral evolution, and origin) and fossils (including the Cambrian Explosion, Burgess Shale, Ediacara Biota, Tully Monster in animals), and statistics. There are also discussions of bacterial photosynthesis, moonmilk, the Gram stain, wind pollination in flowers, night bloomers, luminous mushrooms, and the *Armillaria* clonal colony.

For the 2nd edition I have made edits in most of the chapters, including adding new material--2 new chapters (one on ecology and biogeography and one on geological time), updates for Fungi, material on hornworts, expansion of the bryophyte and pteridophyte sections, information on cercomonads and excavates, extra information on prerhodophyceans and glaucophyceans, an extra figure and a synoptic key to Chapter 8, author names for many genera, some families, and some orders, more diagnoses, and quotes at the head of chapters; relocating the material on plasmodiophorans and spongomonads from Chapter 7 to the protozoan chapter, some of the material on prerhodophyceans and glaucophyceans from Chapter 7 to the red algae chapter, and the Tally Table from Chapter 7 as a separate chapter; addition of a chapter on geological time; as well as many relatively minor corrections.

Part 1

Chapter 1 Historical Overview

The living world cannot be neatly divided into a small number of monophyletic kingdoms because of the tremendous diversity of organisms. The major taxonomic objective of establishing monophyletic assemblages of organisms should not be compromised for the sake of convenience as has occurred with the kingdoms Protista and Protoctista in the Whittaker and Copeland schemes, respectively. - Peter Edwards, 1976.

Plants and animals were recognized since ancient times, the concept of the plant-animal dichotomy being introduced by Aristotle. Alexandrian philosopher Ammonius Hermiae, in the 400s AD, recognized animals, plants, and zoophytes and may have been the first to use the term "zoophytes" (if not, the honour would go to Sextus Empiricus or Iamblichus, both in the late 200s AD). The group was first formally established as part of Animalia by Edward Wotton in 1552.

The Linnean system (Linneus, 1735) contained 24 plant classes, the first 23 for phanerogams and the 24th for algae, funguses, mosses, and ferns; and 6 animal classes: Vermes, Insecta, Pisces, Amphibia, Aves, and Quadripedia, in the 2 traditional kingdoms. Linneus (1767) conceived the chaotic kingdom (Regnum Chaoticum) which had only 2 brief life-spans. Regnum Neutrum was established by von Munchhausen (1765-66) for polyps, corals, funguses, and lichens.

German medic and mathematician Gottfried Reinhold Treviranus (1802-22), who coined the word "biology" in its current meaning, and, like de Monet (Lamarck), was a proponent of species transmutation ("descent with modification"), recognized kingdoms Plantae, Amphorganicum (divided as animal-plants: zoophytes and infusorians (variously circumscribed by different authors and recognized as a kingdom by Nees Esenbeck)(infusorians [Animalcula Infusoria Ledermüller 1760-1763] being life forms able to produce dessication-resistant stages and which can be reactivated by an infusion of water to hay or pepper contaminated with such resting stages); and plant-animals: fungi, confervae, fuci, bryophytes, ferns, and Najadales), and Animalia.

French naturalist, geographer, and explorer Jean-Baptiste Bory de Saint-Vincent (1824) established the Règne Psychodiaire (2-souled kingdom) for zoophytes, vorticellids, and diatoms, which contained 3 classes, Ichnozoaires, Phytozoaires, and Lithozoaires (corals). In his animal kingdom were included the Microscopiques, comprising 5 orders: Gymnodes, Trichodes, Stomoblephardes, Rotifères, and Crustodes. It was the first order which contained bacteria, along with green algae, monads, amebas, ciliates, and cercaria. Still in France and shortly after, there was also a Règne des Némazoaires by Gaillon (1833).

Russian botanist Paulus Horaninoff (Pavel Gorianinov)(1834) proposed 2 kingdoms of nature, the inorganic and the organic, each with 4 unranked divisions: for the former, fire, water, air, and circulum corporum, and for the latter, Vegetabilia (4 classes: Plantae sporophorae, Pseudospermae, Coccophorae, and Plantae spermophorae), Phytozoa (4 classes: Fungi, Algae, Polyparii [polyps], Acelaphae [sponges and cnidarians]), Animalia (12 classes), and *Homo sapiens*. In 1843 he elevated his kingdoms to worlds or orbits (Orbis Anorganicum and Orbis Organicum), the 8 unranked groups becoming kingdoms (Ethereum, Aqueum, Aereum, and Minerale; Vegetabile, Amphorganicum, Animale, and Hominis). The various classes were arranged in concentric rings.

In the late 1800s, with the advent of Darwin's theory of evolution, a re-evaluation of the Linnean system produced several attempts at improved kingdom-level taxonomies, which were by the Englishman Richard Owen (1859, 1861), the Englishman John Hogg (1860), the Americans Wilson and Cassin (1863), and the German Ernst Haeckel (1866, 1878, 1894), all with 3 organic kingdoms, the 3rd respectively named Protozoa (Acrita in 1861), Primigenum (or Protoctista), Primalia, and Protista, the most popular of these being the Haeckel system. In the 3rd kingdom Owen included Amorphozoa or sponges, Foraminifera or rhizopods, Polycistineae, Diatomaceae, Desmideae, Gregarinae, and most of Ehrenberg's Polygastria or infusorial animalcules, which means the green, red, and brown algae and funguses were in Plantae (Hogg, 1860). Wilson and Cassin identified 5 subkingdoms: Algae, Lichenes, Fungi, Spongiae, and Conjugata. Presumably, Conjugata contained the protozoans. Hogg divided it into Protophyta, Protozoa, and Amorphoctista, but did not specify their contents. Haeckel included Moneres, Diatomeae, Protoplasta (some protozoans), Flagellata, Myxocystoda (Noctilucae), Rhizopoda (mostly actinopods), Myxomycetes, and Spongiae. In Plantae were Archephyta (most green algae and blue bacteria), Inophyta (funguses), Characeae, Florideae, Fucoideae, and embryophytes. And in Animalia were the infusorians, which by that time had essentially become the ciliates. So it was arranged much like Owen's. In 1878 he relocated Fungi, Myxomycetes, and Infusoria, renamed Ciliata, to Protista. In 1892 he divided Protista into Protophyta, Protozoa, and Protista Neutralia, similarly to Hogg.

In the 20th century, Haeckel's final version, appearing in 1904, included kingdom Histonia, for the higher (tissue-bearing) organisms. But a major step occurred with recognition of the eukaryote-prokaryote division (Chatton, 1925, 1938), classed as superkingdoms Akaryonta and Karyonta (viruses being named Aphanobionta) by Novak (1930). It was Ellsworth Dougherty (1957) who formally named them. Neushul (1974) established subkingdoms Prokaryota and Eukaryota.

Edouard Chatton, in the whole text of his 1925 article never mentions the two categories, but on pages 76 and 77, there are diagrams, and the two main branches of the tree are named Procaryotes and Eucaryotes. The 3 twigs of the Procaryotes are Cyanophycées, Bacteriacées, and Spirochaetacées. The Eucaryotes have only protistan twigs, Mastigiés, Ciliés, and Cnidiés, which means that at the time he did not think of plants and animals as eucaryotes. (History of the Terms Prokaryotes and Eukaryotes-Red Orbit.Com).

American pteridologist E.B. Copeland (1928) brought attention to the inadequacies of the 2-kingdom system and proposed the possible utility of multiple kingdoms. The first modern 4-kingdom system was by his son H.F. Copeland, with subsequent revisions (Copeland 1938, 1947, 1956), the final version being: Mychota (blue algae and bacteria), Protoctista (other eukarytic algae, fungi, slime molds, and protozoans), Plantae (including only embryophytes and green algae), and Animalia (including sponges).

Other 4-kingdom schemes followed: Barkley (1939, 1949), Rothmaler (1948), both similar to Copeland's but including green algae in Protista, Whittaker (1957, 1959), Takhtajan (1973), and Leedale's pteropod scheme (Leedale, 1974), all including a protistan kingdom except the last 2, which had Monera, Plantae, Fungi, and Animalia. Bold et al (1987) also recognized 4 kingdoms (the 4th being Animalia, which was, however, not included but was implicit): Monera with 2 phylums, Bacteria and Cyanophyta; Myceteae (Fungi), with 3 phylums, Gymnomycota, Mastigomycota, and Amastigomycota; and Phyta (Plantae), arranged in some 20 consecutive phylums comprising cormophytes and algae. Parker's (1982) 4 kingdoms were Virus and Monera in superkingdom Prokaryotae and Plantae and Animalia in Eukaryotae, the former with subkingdoms Thallobionta (ncluding funguses as well as algae) and Embryobionta, the latter with subkingdoms Protozoa, Placaozoa, Parazoa, and Eumetazoa.

American botanist Henry Conard (1939), Russian botanist A. Y. Vada (1952), and American ecologist and botanist Robert Whittaker (1957) proposed a 3-kingdom scheme, which were fungi-bacteria, plants (algae and cormophytes), and animals (protozoans and metazoans), corresponding to the 3 nutrional modes and the ecologist's functional communities. The 1st of these authors used the names Mycetalia, Phytalia, and Animalia, the 2nd did not use an explicit rank nor names (blue algae were included in Plantae), and the last only informally suggested them. Canadian biologist Edward

Dodson's 3 kingdoms (1971) were somewhat different: Mychota (blue algae, bacteria, and viruses), Plantae (including all eukaryotic algae plus all fungi), and Animalia (including Protozoa).

Jahn and Jahn (1949) presented 6 kingdoms: Archetista (viruses), Monera (bacteria), Protista, Metaphyta, Metazoa, and Fungi.

American evolutionist Verne Grant (1963) devised the first 5-kingdom system. This comprised Monera (blue algae, bacteria, and viruses), Protista (protozoans, diatoms, and phytoflagellates), Fungi, Plantae (embryophytes, red, green, and brown algae), and Animalia. Whittaker, 6 years later, published a 5-kingdom arrangement (Whittaker, 1969), inspired by Grant's, containing Monera (excluding viruses), Protista (comprising other eukaryotic algae, protozoans, chytrids, hyphochytrids, and plasmodiophorans), Fungi (including oomycotes, slime molds, and slimes nets), Plantae (same as Grant), and Animalia. Margulis (1974, 1998) presented a 5-kingdom scheme as well, based on Whittaker's, these 2 authors teaming up for a 5-kingdom article (Whittaker and Margulis, 1978).

Others have proposed alternative kingdom-level arrangements. Walton (1930) and Dillon (1963) presented one kingdom systems, the former with 3 subkingdoms Protistodeae, Metaphytodeae (multicellular plants), and Zoodeae (multicellular animals) and called Bionta, the latter with 14 subkingdoms and called Plantae.

Stewart and Mattox (1980) proposed 2 eukaryotic kingdoms Bodonobiota (with flat cristae) and Dinobiota (with tubular cristae). They, like Taylor (1978), emphasized the evolutionary importance of cristal structure, recognizing a flat-tubular split, and the latter proposed a link between this split and plastid symbiogenesis--flat corresponding to chlorophyll b and tubular to chlorophyll c. This is generally the case, as plants and euglenoids have flat cristas and chlorophyll b, and the chromalveolates have chlorophyll c and tubular cristas, however, chlorarachnians have tubular cristas and chlorophyll b, and cryptomonads have chlorophyll c and flattened (albeit tubular) cristas.

Gordon Leedale (1974), botanist at Leeds, as well as the pteropod scheme, also proposed a 19-kingdom fan scheme (Monera, red algae, Plantae, heterokonts, eustigs, haptophytes, cryptomonads, dinoflagellates, chytrids, Fungi, euglenoids, zooflagellates, myxomycetes, sarcodines, ciliates, sporozoans, sponges, Animalia, and mesozoans).

Edwin Möhn's (1984) also had 19 kingdoms, which were distributed into the usual 2 superkingdoms with 6 suprakingdoms being Archeobacteria comprising the single kingdom Archeobacteriobionta; Neobacteria comprising kingdoms Bacteriobionta and Cyanobionta, and Aconta (with Erythrobionta and Rhodecyanobionta); Contophora (containing 8 kingdoms: Chlorobionta, Flagelloopalinida, Euglenophytobionta, Eumycota, Dinophytobionta, Cryptophytobionta, Colponemata, and Chloromonadaphytobionta); Cormobionta (with a single kingdom bearing the same name); Animalia, containing middle-kingdoms Parazoa (with kingdoms Porifera, Archeata, and Placozoomorpha), and Eumetazoa (with kingdoms Bilateria and Radiata).

Charles Jeffrey's 7 kingdoms (1971) were in 3 superkingdoms Acytota (viruses); Procytota with kingdoms Bacteriobiota and Cyanobiota; and Eucytota with Kingdoms Rhodobiota, Chromobiota (algae with chlorophyll c plus oomycotes), Zoobiota (animals and protozoans), Mycobiota (true funguses, apparently including chytrids but not stated), Chlorobiota (embryophytes, green algae, and euglenophytes). His 1982 5-kingdom system arranged as Prokaryota with kingdoms Bacteriobiota and Arcbeobacteriobiota and superkingdom Eukaryota with kingdoms Phytobiota (4 subkingdoms: Protistobionta, comprising dinoflagellates, euglenoids, cryptophytes, protozoans, and sponges; most chromobiotes; plants (but including haptophytes); and red algae), Mycobiota (true fungi, including chytrids), and Zoobiota (Placozoa, Mesozoa, and Metazoa).

Peter Edwards (1976) identified 9 kingdoms, with Prokaryota contaning Bacteria and Cyanophyta, and Eucaryota construed as 4 groups, the 1st containing Erythrobionta (red algae), the 2nd containing Myxobionta (slime molds), the 3rd animals, and the 4th Fungi 1 (true funguses, including chytrids)

+Chlorobionta (Tracheo-, Bryo-, Chloro-, and Euglenophyta), and Ochrobionta (Pheo-, Chryso-, Pyrrho-, and Cryptophyta)+Fungi 2 (water molds and slime nets).

Yaroslav Starobogatoff (1986), of the Zoological Institute of the Russian Academy of Sciences, also presented 9 kingdoms but in 3 eukaryotic superkingdoms: Aconta comprising Rhodymeniontes (red algae), Mychota (fungi including Microsporidia), Lamellicristata comprising Cryptomonadontes (cryptophytes, glaucophytes, centrohelians, and pseudociliates), Euglenontes, Plantae (including green algae), and Animalia (including choanoflagellates), and Tubulicristata made up of Ellipsoidiontes (vacuolarians, ellipsoids, sporozoans, trichomonads, entamebeans, opalinates, and ciliates), Peridiniontes (peridiniophytes, syndineans, ellobiophytes, spheriparaians, eberideans, sticholoncheans, and radiolarians), Chromulinontes (heterokonts, haptophytes, heliozoans, haplo- and myxosporidians, myxomycetes, forams, acantharians, and archeocyathans).

Kussakin and Starobogatov (1973) and Kussakin and Drozdoff (1994, 1998) also published kingdom level taxonomies. The latter divided Empire Cellulata Vorontsov 1965 into 22 kingdoms, recognizing Dominion Archaebacteria, containing kingdoms Thermoacidobacteriobiontes, Archetenericutobacteriobiontes, Halobacteriobiontes, and Methanobacteriobiontes; Dominion Eubacteria, containing Superkingdom Gracilicutes, with kingdoms Cyanobiontes, Anoxyphotobacteriobiontes, Scotobacteriobiontes, and Spirochetobacteriobiontes, and Superkingdom Firmicutobiontoi, with kingdoms Actinobacteriobiontes, Eufirmicutobiontes, and Tenericutobiontes; and Dominion Eukaryota, containing Kingdoms Microsporobiontes, Archemonadobiontes, Euglenobiontes, Myxobiontes, Rhodobiontes, Alveolates, Heterokontes, Chlorobionta, Mycobionta, Inferiobionta (Parazoa), and Metazoa. Viruses were named Empire Noncellulata. H. H. Vorontsov probably also presented a similar system in 1987 in "System of the Organic World" (cited in Kussakin and Drozdoff, 1998).

Cavalier-Smith had several taxonomies (1978, 1981, 1983, 1986), his first being a eukaryote scheme consisting of 7 kingdoms: Aconta (red algae and fungi), Haptophyta, Cryptophyta, Heterokonta, Corticoflagellata (most protozoans plus animals), Euglenoida, and Chlorophyta. In 1981 he proposed 9 eukaryotic kningdoms and a 2nd scheme that had 6 over-all, which were Bacteria (including Archeobacteria) in Prokaryota, and Protozoa (including dinoflagellates and euglenoids), Animalia, Fungi, Plantae, and Chromista in Eukaryota, which he presented also in 1983, 1986, 1998, and 2004. He had 8 kingdoms in 1991, comprising Empire Prokaryota composed of Archeobacteria and Eubacteria, and Empire Eukaryota with superkingdom Archeozoa including a single kingdom by the same name, and superkingdom Metakaryota made up of Protozoa, Plantae, Animalia, Fungi, and Chromista. The empires and the 8-kingdom system he later abandoned.

Mayr's system (1990) also had 6, comprising Domain Prokaryota with subdomains Eubacteria (with a single kingdom) with a subsequent version the following year (1991), and Archeobacteria (containing kingdoms Euryarcheota and Crenarcheota); Domain Eukaryota with subdomains Protista (with a single kingdom) and Metabionta (containing Kingdoms Metaphyta, Fungi, and Metazoa).

Protistologist John Corliss recognized the 4 eukaryotic kingdoms (1984) of the 5-kingdom system, but then recognized 6 eukaryotic kingdoms (1994, 1995), which were similar to Cavalier-Smith's and included Archeozoa, Protozoa, Chromista, Plantae, Fungi, and Animalia.

Blackwell and Powell presented 7 kingdoms in 1995: Monera, Protozoa, Stramenopila (Heterokonta), Biliphyta, Plantae, Fungi, Animalia; and 8 kingdoms in 1999 and 2001 with the separation of Monera into Archeobacteria and Eubacteria (Blackwell and Powell, 2004).

All these were largely artificial, so Diana Lipscomb's eukaryotic system (1985, 1989, 1991), was the first based on cladistic analysis of classical evidence.

Molecular phylogenies were published by Woese and Fox (1977), Woese, Kandler, and Wheelis (1990), which identified 3 domains or primary kingdoms, Archeota, Bacteria, and Eucaryota, and by

James Lake (1988), which was a 2-primary-kingdom arrangement (Parkaryotae + eukaryotes and eocytes + Karyotae). This latter author (Lake. 1986; Lake et al., 1986) has also suggested a 5-primary-kingdom scheme (Eukaryota, Eocyta, Methanobacteria. Halobacteria, and Eubacteria) based on ribosomal structure and a 4-primary-kingdom scheme (Eukaryota, Eocyta, Methanobacteria, and Photocyta), bacteria being classified according to 3 major biochemical innovations: photosynthesis (Photocyta), methanogenesis (Methanobacteria), and sulfur respiration (Eocyta).

Sandra Baldauf et al (2000) presented an 8-kingdom taxonomy for eukaryotes based on a synthesis of molecular phylogenies. These were Polymastigota, Chromalveolata (alveolates+heterokonts)+Excavata, Plantae, Rhodophyceae, Glaucophyceae, Lobosa-Mycetozoa, Animalia, and Fungi.

Adl et al (2005) published a taxonomy which was a revision of the classification of unicellular eukaryotes that updates Levine et al (1980) for the protozoans and expands it to include other protistans and incorporates the results of fine structural studies as well as molecular studies, whereas the previous revision included primarily the former. There are 7 kingdoms (the authors acknowledge 6 clusters they say correspond to traditional kingdoms) arranged as follows: Amorphea (Amebozoa+Opisthokonta), Diaphoretikes (SAR (stramenopiles, alveolates, rhizarians)+Archeoplastida (red algae, glaucophyceans, and plants), and Excavata.

Jack Holt and Carlos Iudica (Taxa of Life website, 2007) presented an arrangement which comprises 22 kingdoms in 3 domains: Eubacteria (Proteobacteriae, Spirochetae, Oxyphotobacteriae, Saprospirae, Chloroflexae, Chlorosulfatae, Pirellae, Firmicutae, and Thermotogae), Archeota (Euryarcheota, Crenarcheota), Eukaryota (Rhodophytae and Viridiplantae in supergroup Planta; Cercozoae in supergroup Rhizaria; Alveolatae, Heterokontae, Eukaryomonadae in supergroup Chromalveolata; Discicristatae and Euexcavatae in supergroup Excavata; and Amebozoae, Fungi, and Animalia in supergroup Unikonta).

The following is a tally of the various kingdoms proposed throughout history.

Table 1-1.

Aristotle	2	300s BC
Ammonius Hermiae	3	400s AD
Linneus	2	1735
Linneus	3	1767
Munchausen	3	1765-66
Treviranus	3	1802-22
Gaillon	3	1833
Bory de St. Vincent	3	1824
Horaninoff	1	1834
Horaninoff	4	1843
Owen	3	1859, 1861
Hogg	3	1860
Wilson and Cassin	3	1861
Haeckel	3	1866, 1904
Walton	1	1930
Conard	3	1939
Copeland	4	1938, '47, '56
Barkley	4	1939, '49
Rothmaler	4	1948

Jahn and Jahn	6	1949
Vada	3	1952
Whittaker	4	1957
Whittaker	4	1959
Grant	5	1963
Dillon	1	1963
Whitttaker	5	1969
Jeffrey	7	1971
Dodson	3	1971
Margulis	5	1971
Leedale	4, 19	1974
Margulis, Schwartz	5	1974, 1988, 1998
Edwards	9	1976
Margulis, Whittaker	5	1978
Cavalier-Smith	6	1981, '83, '86, '98, 2004
Cavalier-Smith	8	1993
Jeffrey	5	1982
Takhtajan	4	1983
Möhn	19	1984
Bold et al	4	1987
Kussakin and Drozdov	22	1994, 1998
Blackwell and Powell	7	1995
	8	1999, 2001
Holt and Iudica	22	2007

for eukaryotes only

Cavalier-Smith	7	1978
Stewart, Mattox	2	1980
Cavalier-Smith	9	1981
Lipscomb	7	1985
Lipscomb	7	1989, 1991
Starobogatoff	9	1986
Corliss	6	1994, 1995
Baldauf et al	8	2000
Adl et al	7	2005

Chapter 2 Some Words on Nomenclature

If names are not correct, language will not be in accordance with the truth of things. - Confucius- quotegarden.com.
The beginning of wisdom is calling things by their right names. - Confucius, c. 500 BC –Taxonomicon.

The name Bacteria is used for Eubacteria because there are fundamental differences between them and Metabacteria in ribosomes, cell wall, and lipid type, however, this is impractical, unnecessary, and confusing, as Metabacteria are a very small group, the term "bacteria" used broadly is too inveterate, generic names in Metabacteria often include the –bacterium suffix, and the Bacteriological Code and bacteriology include the group, so it is far preferable to use the name Eubacteria for typical bacteria. Also, "Archea" is easily confused with "Archean", and, like "Eukarya," is a gratuitous synonym, and the group is younger than Eubacteria so it is a misnomer in any case. Especially egregious is using the name Firmictues for low G-C Posibacteria (Mollicutes and Endospora) and not for Actinobacteria and Endospora together, and Choanozoa for protopisthokonts, which is paraphyletic to boot. Entirely gratuitous are "Holozoa" and "Holomycota" as they can easily be called by their usual names and still include the new groups, and what is now called Animalia by some is actually Metazoa.

I'm not a big fan of the requirement that all names of taxons be plural, so I have used -soma (Table 4-2) and -murus (Table 13-2). The adjectival form should also be admitted and was sometimes used in the past, e.g., Regnum Animale and Regnum Vegetabile.

My nomenclatural system is presented in Table 2-1 and Table 2-2. Based on the Farris system (1976) groups can be inserted without changing ranks of the taxons already included, for instance, between and hyper- and mega- one can add superhyper-. For the higher prefixes I have added grand- (probably introduced by McKenna in 1975) in the middle and tera- on top. For the lower prefixes I have added the oft used parv- in the middle, replaced pico- with nano, and placed pico- at the bottom. The 3 highest and 3 lowest prefixes are the same as in the SI system, and the 6 in the middle are the same in number as those of the SI system.

If ranks above tera- are needed they would start with supertera, etc., and if ranks below pico- are needed they would start with subpico-, etc. The abbreviations, which are my own, are given in parentheses. Rank names between basic units should be assigned as half and half, for example, if there are 6 levels the top 3 would be subtertaxic and the bottom 3 suprataxic, and if there is an odd number, the suprataxic would predominate except when there would be an unnecessary increase in redundant or empty ranks. In cladistics, ranks are designated according to branching order, but at some point there is sometimes a ranking that is based on degree of difference and long-standing ranks for particular groups should be conserved for the most part. There should also be parsimony in the number of taxons of any particular rank.

Table 2-1. Suffix Systems

	bacteria	plants	algae	fungi	animals
phyl.	-bacteria	-phyta	-phycota	-mycota	
sbtph.	-bacterina	-phytina	-phycotina	-mycotina	
sprcl.	-arae	-icae		-mycetia	
class	-bacteriae, -ariae	-opsida	-phyceae	-mycetes	-zoa, -acea
sbtcl.	-arinae	-idae	-phycidae	-mycetinae	

sprord	-oidiona	-arae, -florae	-aliona	
order	-oidia	-ales	-alia	-ida
sbtord	-oidina	-ineae	-alina	-ina
sprfam	-ikea	-area	-idiona	-oidea
fam.	-ikae	-acea	-ideae	- idae
sbtfam	-ikinae	-oideae	-idina	-inae
tribe	-ikineae	-eae	-idini	-ini
sbtribe		-inae		

(-bacteria can, as well as a type plural, alternatively be used to designate any rank. spr-(supra) refers to any rank above and sbt-(subter) refers to any rank below the basic rank.)

The suffix –zoa for protozoan groups is inappropriate and inaccurate in phylogenetic taxonomies and should be avoided, at least for new taxons, in phylogenetic nomenclature and likewise for –phyta for non-plants and -mycetes and -mycota for non-funguses, especially for ambiguous groups like Myxobiota, which is both fungal and protozoan, Euglenaria, which has phototrophic, phagotrophic, and osmotrophic forms, and Miozoa, which has both phototrophic and phagotrophic forms, and might be called Dinocomplexa. It is sometimes argued that -zoa means "life", but this is confused semantics and gratuitous. The same goes for -zoic in geological time units; -biotic should be used, but I generally use the former by force of habit.

Table 2-2. Rank Prefix System.

tera-	(tr)
giga-	(gg)
mega-	(mg)
grand-	(grd
hyper-	(hp)
super-	(sp)
sub-	(sb)
infra-	(inf)
parv-	(pv)
micro-	(mc)
nano-	(nn)
pico-	(pc)

Where it concerns terminology, I introduce "pleophyletic", which is far preferable to the awkward and negative "nonmonophyletic." And misuse of language such as "numerical taxonomy" to mean only "phenetics," "evolutionary taxonomy" to mean only "gradism," "fruiting bodies" or "fructifications" for "sporophores", "influorescences" in mosses, "flowers" in gymnosperms, "morphological" for "phenotypic" or "classical", and "carried flora" for "carried bacteria" should be avoided.

Where it concerns citations, initial parentheses are for the original author followed by the citation for a considerably or substantially emended taxon. An "ex" between names means the original author did not publish the name validly (not in accordance with the rules of the Botanical Code of Nomenclature, which requires a Latin diagnosis unlike the zoological and bacteriological codes) with the second name refering to the valid publication. "Orth. mut." or "orth. emend." means the spelling was changed; "emend." means the taxon was changed; "nom. nud." means a name not validly published.

Chapter 3 Taxonomy: Meaning and Methods

Already as early as 1936, Willi Hennig had begun to deviate from conventional systematics and discussed some aspects (Hennig 1936) which later became essential for his method: “relationship” should be defined in terms of phylogenetic, i. e., genealogical, relations, and only newly acquired characters are adequate arguments in favour of closer relationship. Later, when he wrote his fundamental work (Hennig 1950), he insisted that only a concept of genealogical relationship can provide a sound basis for a consistent classification, in contrast to [simple] “similarity”. This strict definition of “relationship” was the first important step towards the so-called Hennigian revolution. - Willi Hennig, the cautious revolutioniser, Michael Schmitt, 2010 - paleodiversity.org.

As defined in the Penguin Dictionary of Botany, taxonomy is the scientific study of the principles and practices of classification, the study and description of variation in the natural world and subsequent compilation of classifications, and includes systematics, which is the scientific study and description of organisms and the relationships between them, and adds that the 2 words are often used synonymously. Strickberger points out that some authors use the terms "taxonomy," "classification," and "systematics" interchangeably, and that some, like Simpson, consider systematics a much broader field--the study of the diversity of organisms and all their comparative and evolutionary relationships. Simpson defined classification as a subtopic of systematics, as the ordering of organisms into groups, and taxonomy as the study of the principles and practices of classification. The Simpson view, which is lamentably widespread, obviously makes artificial, arbitrary, and illogical distinctions as classification necessarily includes comparative biology, the study of diversity, and both the principles and the practices. The 3 terms, are rightfully treated as synonymous by Webster's 9th New Collegiate Dictionary, which also specifies that "classification" was coined in 1790, "taxonomy" c. 1828, and "systematics" in 1888. Obviously, the 3 are essentially the same thing and the differences, if any, trivial. However, "systematics" might be better used as a synonym for "systems science" as it does not refer specifically to classification but to systems, although it is used specifically to mean systems of classification. Taxonomy, then, is the scientific principles and practices of biological classification (although the word is sometimes used in some other sciences, e.g., "soil taxonomy") and therefore includes nomenclature, identification, comparative biology, and the study of biodiversity, and is not necessarily evolutionary but usually is in our time.

Phylogenetics, naturally a subset of taxonomy, is the only method that is truly and completely phylogenetic, being based entirely on monophyly and phylogeny.

It should also be mentioned that, contrary to the bizarre claims of some, who make artificial distinctions between them, *a dendrogram is necessarily a branching diagram, and a cladogram is a type of dendrogram and is necessarily a phylogenetic tree, which are in turn necessarily classifications.*

Gradism (synthetic taxonomy), on the other hand, admits grades, which are paraphyletic, and considers factors external to phylogeny (such as parallel and convergent evolution) so it is arbitrary, contradictory, inconsistent, and subjective, and is a great example of sophistic argument, so it is unscientific. The deliberate inclusion of known paraphyletic groups such as Pteridophyta and Agnatha and known polyphyletic groups such as Protista Protozoa, and Archeozoa, and the deliberate exclusion of taxons from others to which they are known to belong thereby causing them to be truncated, in other words, splitting obviously monophyletic taxons and creating obviously polyphyletic ones, is artificial and hardly a serious attempt at phylogeny. And it is often not based on any system nor even analysis, yet is touted as “phylogenetic”. Gradists do not believe phylogeny should be included in taxonomy in any serious or consistent manner, yet claim their method is phylogenetic and regard paraphyletic groups as monophyletic, but in order to be monophyletic a taxon must have a common ancestor unique to it, in other words, an immediate common ancestor. It is often called evolutionary taxonomy, but this is obviously misleading since phylogenetics is also evolutionary, and to call gradism evolutionary is dubious since it is only partly so. As well, gradism claims that no groups are real, but they are, at least evolutionary ones, since they are branches in a tree, so to speak, and are definable, identifiable, and distinguishable units.

Convenience taxonomy is advisable, especially for unresolved groups, and is stable, simple, and practical, but must be used in parallel with phylogenetics not in combination with it. I present just such a convenience classification in Chapter 8. The purposes of convenience and phylogeny are irreconcilable within the same taxonomy (although there can be some taxons which are both convenient and evolutionary) but are complementary as parallel systems.

There is a disturbing overreliance on genotypic evidence which is considered superior or even foolproof, and groups are regarded as necessarily and automatically phylogenetic based only on this type of evidence. Because of molecular extremism there is a scarcity of evolutionary taxonomies at high levels based on classical data--only 2 previous ones have been done for eukaryotes (Lipscomb, 1985, 1989, 1991) and none before now for prokaryotes. Especially disturbing, and shameful, is the synonymizing of "molecular" with "phylogenetic," which is done by some. But, in fact, genotypic taxonomy is not at all superior to the phenotypic sort, as it is given to many pitfalls: random noise, long-branch attraction, different evolutionary rates, mutational saturation, paralogous genes, insufficient sampling, and the use of different methods (cf. McKenna, 1987; Raff et al, 1987; Wyss, Novacek, and McKenna, 1987; Meyer, Cusanovich, & Kamen, 1998; Philippe & Adoutte, 1998; Doolittle, 1999; Philippe et al, 2000). Random error (random noise) is due to fast-evolving taxons, systematic error can be caused by failure to correctly model rate-across-sites distribution, changing rates at sites over the tree, changing nucleotide, amino acid, and/or codon usage among taxons, and site-specific substitution properties, the most common systematic error being LBA, and conserved insertions or deletions, intron positions, gene fusions or splits, and other complex molecular events each exhibit their own sources of error, such as convergence, recombination, and/or parallel loss (Keeling et al, 2005). Philippe et al (2000) point out that genotypic taxonomy does not provide more resolution than the morphological kind, there is a generalized lack of resolution in genotypic phylogenies for eukaryotes, LBA (long-branch attraction) is hard to detect, rRNA has incorrectly been considered as a reliable chronometer, and many genes used in molecular analyses are highly mutationally saturated. Baker et al (1998) have found that morphological data generally provide equal or greater support than do molecular data, and state that this result, combined with the fact that morphological characteristics generally exhibit higher consistency, indicates that this source of trait information continues to be useful in taxonomic studies despite the increasing volume of available molecular data. Parfrey et al (2006), citing Doyle (1992) and Maddison (1997), state that genetic taxonomies are geneologies, which are the history of genes, and are not necessarily congruent with phylogenies, which are the history of organisms. Lutzoni and Vilgalys (1995), refering to Hillis from 1994, state that despite the homology problem being less prominent with molecular data, there is still room for mistaken inferences about homologous molecular kinships. But the idea that there is a homology problem with phenotypic data is mostly mythical as usually it is the analysis that determines the homology (Nelson and Platnick (1981) equate homology with synapomorphy). The only advantage to molecular taxonomy is that the trait states are always the same--A, C, G, U. Pisani et al (2007) say that their data show molecules are superior to morphology, but it is unclear, as with Philippe and colleagues and Baker and colleagues, if by "morphology" they mean morphology or they misuse it to mean phenotype. While it could well be that molecular data are superior to gross morphological ones, *it certainly has not been demonstrated that other phenotypic data are inferior to molecular data*. Pisani et al also state, "Our results suggest that both molecular and morphological trees are, in general, useful approximations of a common underlying phylogeny and thus, when molecules and morphology clash, molecular phylogenies should not be considered more reliable a priori."

Cladistics was the first truly evolutionary method and was originated in 1901 by British zoologist Peter Mitchell (for birds)(Schuh, 2000; Folinsbee, 2007), who was Secretary of the Zoological Society of London from 1903 to 1935, and subsequently used by British-Australian zoologist and geologist Robert Tillyard (for insects)(Tillyard, 1921), and W. Zimmermann (for plants) in 1943 (Schuh, 2000). And Mitchell in 1901 and Italian zoologist Daniele Rosa in 1909 and 1918 stated its basic elements (Nelson and Platnick, 1981, p. 135; Luzzatto et al, 2000). The method was elaborated by German zoologist Willi Hennig in 1950 in the classic work, Grundzuge einer Theorie der phylogenetischen Systematik ("characteristics of a theory of phylogenetic systematics"), which was translated in '66 as Phylogenetic Systematics by D.D. Davis and Rainer Zangerl. The groundplan divergence method was apparently introduced by Mitchell but formalized by William Wagner (for whom Wagner parsimony is named) in 1952 (Folinsbee, 2007).

Rosa was an evolutionary theorist and proposed hologenesis (his book, in 1918, was titled L'Ologenesi, nuova teoria dell'evoluzione e della distribuzione geografica dei viventi), which was largely incompatible with modern evolutionary theory. It incorporated orthogenesis (also called straight-line or progressive evolution), developed in the late 1800s and fairly popular in Rosa's time, which maintained species arose through internal forces and had a fixed direction, and it had little use for natural selection, and was later refuted. But some aspects--asymmetrical dichotomous splitting, with plesiotypic (early) and apotypic (late) branches, and vicariance biogeography--resonate with cladistics, and his "specific idioplasm" was synonymous with our modern "genome". He promoted the species-as-individual concept, which foreshadowed a modern debate, but, of course, species can no more be individuals than genera can be species. He wrote the entry "Evolution" in the Enciclopedia Italiana in 1932 and was Haeckel's Italian translator. Rosa was aware that the language barrier could be a problem for the dissemination of scientific ideas so he invented an international language, Nov Latin, in 1890, which, like Latino sine flexione by Giuseppe Peano, invented in 1903, was a simplified Latin (Interlingua has a similar basis). (Luzzatto et al, 2000)

The word "clade" was introduced by Haeckel in 1866 in the sense of a category of classification, and Lucien Cuénot in 1940 used it to mean a unit of classification (Williams and Ebach, 2008, p. 3). Mayr and Camin and Sokal introduced "cladogram" in 1965 (Williams and Ebach, 2008, p. 3; Dupuis, 1984). Ho et al (2013) say "cladogenesis" was coined in 1957 by Julian Huxley, who was a proponent of cladistics, succeded Mitchell as Secretary of the Zoological Society of London, and a major player in the modern evolutionary synthesis (the combination of Darwinian natural selection and Mendelian genetics), but that he extended Bernhard Rensch's 1954 proposal, Dupuis (loc. cit.) states Huxley took "clade" from Rensch's "cladogenesis", and Webster's (loc. cit.) states "cladogenesis" was coined in 1953, probably refering to Rensch from 1954. Dupuis states "cladistic" was derived from "clade" sensu Huxley by Cain and Harrison (who were pheneticists) in 1960, and Webster's says it was coined c.1960, and Williams and Ebach say it was coined by Rensch (who was possibly a gradist) in 1958 and was used by Cain and Harrison in 1960, and both Rensch and Cain and Harrison derived it from Huxley (from1958). Publications by Huxley on the matter are apparently only from '57 and '59. Webster's also states "cladistics" was coined in 1966, probably refering to Mayr and/or Camin and Sokal from 1965. "Cladist" was introduced by Mayr in 1965 (Dupuis, loc. cit.). The word "cladism" was invented by Mayr in 1982 (Dupuis, loc. cit.).

The first evolutionary tree is credited to Haeckel in 1866, but geneological branching diagrams were done by Duchesne in 1766 (for strawberries), Lamarck in 1809, Carpenter in 1841 (for vertebrates), Agassiz in 1844, and Darwin in 1859 (Nelson and Platnick, loc. cit.). And the concept of the tree-like scheme to represent relationships among taxons was introduced by German-Russian naturalist P.S. Pallas in 1766 (Luzzatto et al, 2000)(Pallasite [sometimes called Pallas Iron, the name given to it by Ernst Chladni in 1794], a type of stony-iron meteorite, was named for him in 1772). An evolutionary tree was also done by August Schleicher (for languages) in 1853, and Haeckel was a friend of his. Cladistics has also been used in linguistics, e.g., Kateřina Rexova et al in 2003 for Indo-European languages and Kateřina Rexova et al in 2006 for the Bantu languages.

Polarity is the direction of evolution of a characteristic (trait or feature, usually called a character), either primitive (ancestral), the plesiomorphy, or derived (mostly advanced but may include losses or reversals). Symplesiomorphy is a shared ancestral state, synapomorphy is a shared derived state, and autopomorphy applies to characteristics that occur in only 1 taxon in the matrix. Only the advanced states are counted as these define evolutionary groups (this is relative as the characteristic is advanced outside the group, in a more inclusive group, but ancestral in it, the more inclusive group). Steps taken by a characteristic are counted and the tree with the fewest steps is presumed to be the correct one according to the parsimony principle, but in the event of several most parsimonious trees a consensus method, usually strict or majority, is calculated. Divergent characteristics come from the same ancestor so reflect kinship and are homologous, while convergent, i.e., coincidental, characteristics, do not come from the same ancestor so do not reflect kinship and are homoplasious. Terminals are the taxons at the top or side of the cladogram, depending on its shape, which are the same as the taxons in the data matrix.

Cladistics has developed several innovations in 6 major areas (Forey, 1992; Kitching, 1998):

Table 3-1.

parsimony (optimization) criteria

Camin-Sokal, 1965
Wagner (Kluge and Farris, 1969; Farris, 1970)
Fitsch, 1971
Dollo (LeQuesne, 1969; Farris, 1977)
generalized, Sankoff and Rousseau, 1975

search strategies

NNI (nearest-neighbour interchange), Robinson, 1971
branch and bound, Hendy and Penny, 1982
SPR (subtree pruning and regrafting), Swofford and Olsen, 1990
TBR (tree bisection and reconnection), Swofford and Olsen, 1990

consensus methods

Adams, 1972
Nelson, 1979
majority, Margush and McMorris, 1980
strict, Sokal and Rohlf, 1980
combinable components(=semi-strict=loose), Bremer, 1990

(Bryant (2003) identifies 4 classes:

based on clade intersection: Adams and Neumann, 1983 (and the Neumann extensions: Durschnitt, cardinality, and s-consensus)

based on splits and clades (strict; majority; median (a complex mathematical formulation of majority consensus), Barthélemy and McMorris, 1986; combinable components; majority extended (also called greedy (Degnan et al, 2009); Nelson-Page (combining Nelson and Page, 1990); and asymmetrical median: Phillips and Warnow, 1996

based on subtrees (local, Kannan, Warnow, and Yooseph, 1998; prune and regraft (same as GAS (greatest agreement subtree), Gordon, 1979; Q*; and R*)

based on recoding: Buneman, 1971; MRP (maximum representation with parsimony), Ragan, 1992; average, Lapointe and Cucumel, 1997)

(there is also reduced consensus, Wilkinson, 1994, 1995 (Wilkinson, 1994, 1995))

group support (confidence) measures

jackknife, Lanyon, 1985
bootstrap, Felsenstein, 1985
permutation, Archie, 1989
Bremer support (decay index), Bremer, 1994
symmetrical resampling, Goloboff et al, 2007

homology measures

CI (consistency index), Kluge and Farris, 1969
RI (retention index), Farris, 1989
RCI (rescaled consistency index), Farris, 1989
HER (homoplasy excess ratio), Archie, 1989
DDI (data decisiveness index), Goloboff, 1991

a posteriori character weighting

dynamic, Farris, 1969
successive, Farris, 1969
implied, Goloboff, 1993

algorithm, software packages

PHYLIP, Felsenstein, 1980
PHYSIS, Mikevich and Farris, 1982
PAUP, Swofford, year not specified
Mesquite, Maddison and Maddison, year not specified
MacClade, Maddison and Maddison, 1986
Hennig86, Farris, 1988
TNT, Goloboff, Farris, and Nixon, 1998
Winclada, Nixon, 1999

Of the 20 consensus methods majority and strict are the most popular, strict the more popular, devised by Sokal and Rohlf in 1981, but it is majority consensus, devised by Margush and McMorris (1981), I favour as it usually produces better resolution and is simple. Strict consensus is also simple but tends to yield polytomies and can exclude monophyletic groups. The consensus method was first devised for different data sets instead of for multiple MPTs by Adams in 1972.

As a consensus may not be as parsimonious as the MPTs (or is always less parsimonious) (Carpenter, 1988; Lipscomb, Basics of Cladistic Analysis-gwu.edu), a representational method can be used, whereby the most representative MPT (MRT), that is, the one with the closest fit with the majority consensus, is selected. However, there may not always be a single MRT, in which case only the majority consensus would have to be used. Sometimes a representative tree is shown but not necessarily the most representative.

The jackknife is a statistical, resampling method used for variance and bias estimation invented by Maurice Quenouille in 1949 and extended by John Tukey in 1958, who proposed the name as, like a Boy Scout's jackknife, it is a rough and ready tool that can solve a variety of problems (Jackknife (statistics)-Wikipedia).

Bootstrapping is another statistical, resampling method. It assigns measures of accuracy to sample estimates that allow estimation of the sampling distribution of almost any statistic using only very simple methods (Bootstrap (statistics)-Wikipedia). It was introduced in 1979 by Bradley Efron and was inspired by the jackknife. A Bayesian extension was developed in 1981, the ABC procedure in 1992, and the BCA procedure in 1996. It is so-called from the phrase "picking oneself up by the bootstraps," meaning it relies on internal data and operations.

Bremer support is not a resampling method, and it concerns determining whether a group of interest occurs in other trees that are almost equally short. With every tree search, the heuristic must make multiple decisions about which characteristics are true homologies and which are homoplasies. Generally, the grouping that leads to the most homologies is used. The method asks whether there are other ways to analyze the homoplasious characteristics that lead to trees that are only a few steps longer. Usually, a Bremer score of 3 is moderate and a score of 5 is strong. (Lab 10-Bootstrap, Jackknife, and Bremer Support- ib.berkely.edu).

Symmetrical resampling, which is a combination of bootstrapping and jacknifing, which are the result of interaction between characteristics that favour the group and those that contradict it, was devised by Pablo Goloboff et al (2003). When the characteristics have different prior weights or some different state transformation costs, the frequencies under either bootstrapping or jackknifing can be distorted, producing either under- or overestimates of the actual group support. The method avoids the problem as the probability p of increasing the weight of a characteristic equals the probability of decreasing it.

A type of implied weighting is successive weighting (or successive approximations weighting), although sometimes the 2 techniques are considered distinct, invented by James Farris (1969) and is based on the degree to which a characteristic conforms to a heirarchy, which is related to homology, so that homoplasious characteristics are devalued. The CI is used as the weighting factor. The data matrix is reanalyzed for the MPT, and the new tree is compared to the previous one. The process ends when the form of the tree no longer differs between iterations, that is, until weights stabilize between searches. Farris suggested that each trait could be considered independently with respect to a weight implied by frequency of change. However, the final tree depended strongly on the starting weights and the finishing criteria. He stated that there are 4 main categories of functions associated to the trait weight assignment--linear, convex, concave and bounded, and concave and unbounded--varying according to the concavity value k.

Implied weighting is non-iterative, does not require independent estimations of weights, and is based on searching trees with maximum total fit, with trait fits defined as a concave function of homoplasy. When comparing trees, differences in steps occurring in traits which show more homoplasy on the trees are less influential. The reliability of the traits is estimated during the reanalysis as a logical implication of the trees being compared. The fittest trees imply that the traits are maximally reliable and, given trait conflict, have fewer steps for the traits which fit the tree better. If other trees save steps in some traits, it will be at the expense of gaining them in traits with less homoplasy. In other words, the trees which maximize the concave function of homoplasy resolve trait conflict in favour of the traits that have more homology and imply that the average weight for the traits is as high as possible. The method was invented by Pablo Goloboff (1993).

The first algorithmic, software application in cladistics was by Camin and Sokal in '65, but it was unwieldy for large data sets and never effectively programmed (Folinsbee, 2007).

Of the phylogeny programs I am familiar with, the best is TNT. As it is very fast it can do many more rearrangments than most, and in my analyses it has a better CI and is more parsimonious than PAUP, has a presumably superior confidence method, gives the confidence values for all the clades, does synapomorphy plots or lists, and accepts multiple states and polymorphisms. Unfortunately, it has a matrix format that is hard to do and read since the columns aren't numbered, but this is the case for most of these programs. PHYSIS had numbered columns but is no longer available and so does Mesquite but it didn't work properly when I tried it and is not made for large-scale analyses. Hennig86 does not do resampling and PHYLIP does not do CI (although this can be fairly easily calculated) nor RI nor synapomorphy lists or plots, and in some of its programs multiple states can't be coded., and in its programs where polymorphisms can be coded they can't be specified. Unfortunately, I wasn't able to make TNT work.

Phenetics (from "phenotype") is based on overall similarity, and characteristics are considered of equal value. The modern version is numerical and was introduced by Sokal and Michener in 1958 as a response to the subjective procedures of gradism and uses cluster analysis and the pairwise distance method and the generated tree is called a phenogram. It does not distinguish between ancestral and derived. Its Bible is Numerical Taxonomy by Sokal and Sneath from '63, with the 2nd edition by Sneath and Sokal from '73. The molecular version, although contrary to the defintion or at least the etymology, has the same principles and methodology. Cluster analysis was originated in anthropology by Driver and Kroeber in 1932 and introduced to psychology by Zubin in 1938 and Tryon in1939 (Bailey, 1994), and famously used by Raymond Cattell and others for personality typology eventually culminating in the Big 4 (usually seen as the Big 5). Computational complexity made clustering difficult before the advent of computers. Numerical phenetics did, however, pave the way for numerical analysis in cladistics and molecular taxonomy. Of course, "numerical taxonomy" to mean only phenetics is obviously wrong since phylogenetics is also numerical taxonomy.

Certain genotypic taxonomies (e.g., Wolf et al, 2001; Dutihl et al, 2004) have been done based on gene content as opposed to gene sequencing, and these are sometimes considered phenetic. It is claimed by some (see Dutihl et al, 2004) that gene content is a phenotypic trait or intermediate between phenotype and genotype, but it is clearly not phenotypic as it is genotypic by name and defintion and genomes are based on gene sequencing. Although the name "phenetics" comes from "phenotypic", and phenetics is usually or was originally based on the phenotype, molecular, i.e., genotypic, phenetics is obviously based on the genotype.

Cladistics was once equivalent to phylogenetics, but in recent decades some new methods have rendered the latter as inclusive of, but not limited to, the former. These new methods, used exclusively or almost exclusively in molecular taxonomy, are a technique that uses mostly probabilism (ML (maximum likelihood) and BI (Bayesian inference)) and ME (minimum evolution). ME uses a parsimony principle, is derived from molecular phenetics, and uses an optimality criterion and additive trees.

Maximum likelihood is a probability method that, instead of scoring trees with number of steps as is done in cladistics, scores them with probability values. Its equation, which is based on Bayes' theorem, invented by the Rev. Thomas Bayes, English cleric and mathematician, in the 1760s (published 2 years after his decease), goes: posterior odds ratio = likelihood ratio x prior odds ratio.

The first attempt to use ML for taxonomy was by A.W.F Edwards and L.L. Cavalli-Sforza in 1963. Farris and Felsenstein, both in '73, published ML algorithms for phylogeny but they had limited applications because of computational problems. The first successful application of ML to taxonomy, which was for nucleotide sequences, was by Jerzy Neyman in 1974. The first computationally efficient ML algorithm for phylogeny was introduced by Felsenstein in '81. (Folinsbee, 2007).

Bayesian inference is a method that grew out of ML. It attempts to calculate a different portion, the posterior portion (the final), and assigns a value to the prior (initial) odds ratio. A major problem for it was that the priors, probability of data, and probability of hypothesis have to be specified. This was solved by the Markov chain-Monte Carlo (MCMC) simulation, which scans the tree space (the set of all possible trees) and parameter space. The 1st working methods in BI were independently developed by Li, Mau, and Rannalla and Yang in '96, all using MCMC, so it is the new kid on the block. (Folinsbee, 2007).

An advantage of probabilism is that, given the accuracy of a particular tree, a variety of evolutionary models can be tested within a statistical framework. A disadvantage is that a particular evolutionary model has to be decided on prior to the analysis.

In psychology, there is Bayesian learning theory, an important branch of learning theory. And the idea of analyzing rational degrees of belief in terms of rational betting behavior led to the 20th century development of a new kind of decision theory, Bayesian decision theory, which is now the dominant theoretical model for both the descriptive and normative analysis of decisions.

Other applications are in the science of epistemology, for which it did not emerge as a philosophical program until the first formal axiomatizations of probability theory in the first half of the 20th century, and in which it is one of the most important developments in 20th century science and one of the most promising avenues for further progress in epistemology in the 21st century (Bayesian epistemology-SEP (Stanford Enc. of Phil.)).

One important application of Bayesian epistemology has been to the analysis of scientific practice in Bayesian confirmation theory in the philosophy of science, which is certainly better than falsificationism, which is itself falsified, and is hypocritically refered to as (critical) "rationalism" (it is actually empiricism, which is the opposite of rationalism; secular humanism is also falsely called "rationalism"). The idea that a scientific hypothesis must be falsifiable or refutable is nonsensical because if it is able to be falsified or refuted it means that it is the wrong hypothesis and must be discarded. If it can't be falsified or refuted it means it is the right hypothesis. The idea, then, is testability—to prove an hypothesis right or wrong – not falsifiability nor refutability.

Falsificationism was proposed by philosopher Karl Popper in the 1930s, after dissatisfaction with Freudian theory, who also said that no hypothesis can be proven, which is egregiously wrong, unscientific, and anti-intellectual, as clearly the ideas that the Sun is the center of the solar system, that the Earth is round, and that there is evolution have been proven and there is no doubt about them, and they are, in fact, necessary truths, that is, they can't be otherwise. And it is not possible that we can't know anything, in other words, can't be certain of anything, for in order for uncertainty to exist there must also be certainty and vice versa, and in order for fallacy to exist there must also be truth and vice versa, just like there can't be up without down and vice versa nor hot without cold and vice versa, etc., etc. This is a simple and fundamental ontological principle that Popper was apparently unaware of, in his infinite "wisdom," or chose to ignore, again, in his infinite "wisdom." Also, falsificationism (like absolute skepticism, which is a very extreme and irrational view to say the least, for which it is an extension or version), is self-negating: if nothing can be proven then the idea itself can't be proven yet is considered to be definitely true! As well, it says that a theory can't be proven because there might be some future test or information that would falsify it, but this works both ways, because if this is true it must be said also that a theory can't be falsified because there might be some future test or information that would verify it. So falsificationisn is entirely subjective because it is completely one-sided. Many other scientists who may not subscribe entirely to falsificationism are absolutely certain there's no absolute certainty! Naturally, the same case of self-negation. And usually they don't practice what they preach as they treat their probabilities, which are often based on emotion rather than facts, such as Archeoplastida, where gradism and molecular extremism converge, and unikonts as basal, as certainly true. The slightly less extreme and slightly less irrational view, radical skepticism, which says there isn't much we can know or nothing of consequence, is not much better as there are many important things we know, some of them mentioned above. Dogmatic skepticism, whether absolute or radical, which is an integral part of the empiricist ideology, is entirely subjective as it is disbelief for its own sake, as opposed to the evidential approach where skepticism is warranted or not depending on the individual case.

And Popper used sophistic argumentation to arrive at the conclusion, saying that given the proposition "theory T (the antecedent) implies observation O (the consequent)", then the argument "if O is true, then T is true" is an invalid argument, but "if O is false, then T is false" is a valid argument (and a modus tollens, because it is denying the consequent). This, however, is formal logic which concerns itself only with the validity of an argument, not with the content or the truth of a propostion. So we can validly say "theory T implies O" and "if T is true O is true!"; this is a modus ponens and is equally valid. We can also validly say, "observation O implies theory T" and "if O is true then T is true!" -- this is also a modus ponens and means a theory can logically be proven true. We can say, as well, "theory T does not imply observation O" and "if O is false then T is false," which is also a modus tollens and therefore a valid argument and means that T does imply O and is a deductive argument. We can also say "a theory can be proven false implies it can't be proven true" and "if the consequent is false then the antecedent is false", which means a theory can be proven true and can't be proven false!

Popper rejected probabilities but not for the method or rules, which is, of course, contradictory. And as Deborah Mayo points out, "Popperians say things like: it is warranted to infer (prefer or believe) H (because H has passed a severe test), but there is no justification for H (because “justifying” H would mean H was true or highly probable)."(errorstatistics.com-No-Pain Philosophy).

And as the Penguin Dictionary of Philosophy says, under the heading of the philosophy of science, it has been pointed out that the standards he sets may not be met even by respectable science (as opposed to psychoanalysis and Marxism). American physicist, historian, and philosopher of science Thomas Kuhn's influential book, "The Structure of Scientific Revolutions," from 1962, argued that falsificationist methodologies would make science impossible.

So Popper was flagrantly wrong, and his "methodological falsificationism" is devoid of all logic and reason, yet it is widely supported among cladists and among other scientists.

And *Hennig never quoted, cited, nor referenced Popper and never indicated any interest in Popper's views and never cited them as being compatible with his own* (Nelson and Platnick, 1981, p. ix). But Hennig did subscribe to the fantasy of no absolute certainty in science (Hennig, 1966, p. 83).

It should be noted that not all matters in science are subject to experiment or further observation, and they are fundamental and intellectual matters of nature but are accessible only or primarily through reason and are the purlieu of philosophy and math, which are integral parts of science. And secularism and atheism are rational (but not rationalist), but simply being secular or atheist does not make one rational. Also, the empirical method is rational (but not rationalist) only if followed consistently, which is not often the case, as we see, for example, in gradism, which considers only some phylogenetic evidence, and molecular extremism, which does the same, as it considers only molecular evidence.

The following table summarizes the characteristics of the various methods.

Table 3-2.

	pars.	polarity	mnphyly.	data set	optmlty. criterion	stat.	tree score	tree type
cladistics	+	+	+	+	+	+	+	A
clique analysis	-	?	+	+	+	+	+	A
probabilism	-	?	+	+	+	+	+	A
ME	+	?	?	+	+	+	+	A
molecular phenetics	-	-	-	+	-	+	-	U
standard phenetics	-	-	-	+	-	+	-	U
traditional phenetics	-	-	-	-	-	-	-	-
gradism	-	+	-	-	-	-	-	-
convenience	-	-	-	-	-	-	-	-
polythetic	-	-	-	-	-	-	-	-
monothetic	-	-	-	-	-	-	-	-
special purpose	-	-	-	-	-	-	-	-

	chrct.-based	pairwise distance	overall similarity	chrcts. of equal value
cladistics	+	-	-	-
clique analysis	+	-	-	-
probabilism	+	-	?	-
ME	-	+	?	-
molecular phenetics	-	+	+	+
standard phenetics	+	+	+	+
traditional phenetics	+	-	+	+
gradism	+	-	+	-
convenience	+	-	-	-
polythetic	+	-	-	-
monothetic	+	-	-	-
special purpose	-	-	-	-

A=additive, U=ultrametric

The following is a classification of classification:

Table 3-3.

phylogenetics (based on evolution, numerical)
- cladistics (starts in 1901)
- probabilism

ML (starts in '63)
BI (starts in '96)
clique analysis (character compatability) (starts in '65)
ME (starts in 1971)
phenetics
numerical phenetics (based on overall similarity and equal-value traits
molecular phenetics (claims to be evolutionary)
standard phenetics (does not claim to be evolutionary)(begins in '57)
traditional phenetics (starts with Adanson in 1763)
convenience (does not claim to be evolutionary but can include some evolutionary groups or factors)
gradism (claims to be evolutionary but accepts grades)(starts in1800s, especially with Haeckel)
polythetic (use of many traits)(in modern times starts with Cesalpino, Bauhin, and Ray in 1500s and 1600s)
monothetic (single trait, e.g. habit (Theophrastus), sexuality (Linneus))
special purpose-utilitarian (for a specific purpose such as medicinal or culinary uses)

Chapter 4 Prokaryotes: TNT Analyses

If you look at the ecological circuitry of this planet, the ways in which materials like carbon or sulfur or phosphorous or nitrogen get cycled in ways that makes them available for our biology, the organisms that do the heavy lifting are bacteria. - Andrew Knoll, Harvard scientist, who penned Life on a Young Planet, 2004 – brainyquote.

Methods and Materials

For the 1st analysis the data set is divided into 4 sections and 242 characteristics: morphology (19), chemistry (83), physiology (43), and molecular biology (97) in 263 columns. There are 9 complex (branching) ones (cell shape, protein types, peptide bridge for cross-linkage A (deactivated), amino acids at position 3 (deactivated), cytochromes, carotenoids, quinone classes, bacteriochlorophylls, and chlorophylls). And 7 simple ones have been deactivated, as well, (23-27, 84-86, 139, 146, 170, 235, 250) as they were superfluous since the matrix was reformatted and simplified. The TNT (Tree analysis using New Technology) program, Wagner parsimony, TBR branch swapping were used and no a priori weighting nor topological constraints were used. All 4 algorithms were used: ratchet, sectorial search, tree-drifting, and tree-fusing (Goloboff, 1999; Nixon, 1999). Random seed was set at 1, no topological constraints were used, and the search level was set at 100. A symmetrical resampling was done based on 1000 replicates. Implied weighting was also done.

There were a few errors in the data set which are the following: the 1st characteristic 200 was excluded as there were 2 character 200s and bchl g (128) was not placed with chlorophylls. In the matrix, 133 was improperly coded and tends to place Chloroflexi with Thiobacteria, as was 188, so that it comes up as a synapomorphy for Monodermata when it is really for Neosoma, instead, and 112 and 190 are empty characteristics so should be excluded, too. These, however, have little or no effect on the outcome except for 128, which is mentioned earlier, and 133. Also, Thiobacteria is an artificial grouping of Proteobacteria and Chlorobia. And Heliobacteria was coded wrongly as being Gram negative. These errors were corrected for the 2nd analysis. Thermosiphia (with the single genus *Thermosipho*) has the largest number of primitive traits which are: heterotrophy, hyperthermophilia, fermentation, anaerobiosis, nonmotility, with a single membrane, thin wall, unicellularity, small SRP, and absence of LSP layer, outer membrane, spores, cytochromes, quinones, catalase, and carotenoids, so was used as the root.

The data set for the 2nd analysis (277 entries) is essentially the same as in the 1st but with several characteristics added or deactivated. As in the 1st analysis, TNT (Tree analysis using New Technology), Wagner parsimony, and TBR, and all 4 algorithms were used. Togabacteria were chosen as the root. There can be no outgroup for prokaryotes, and in cases like these or when there is no suitable outgroup, the root is referred to as the "outgroup" for convenience. But the hypotheical ancestor can be used as the outgroup as was done for the PAUP analyses.

Also as in the 1st analysis polarities were determined based on the root, absence/presence, and also according to how the characteristics are already considered, mostly based on simplicity/complexity. For example, the spherical cell shape is simplest and indicates primitiveness. For unsaturated fatty acids monoenoic is primitive. Dimers are advanced over monomers and heteromers over homomers. Bacteriochlorophylls are primitive to chlorophylls as they are anoxygenic and were arranged according to their chemical structures, *a, b*, and *g* having COCH3 in ring 1 and/or geranyl-geranyl in ring 6, and being primitive (as *c, d*, and *e* have a more complex ring 1 structure), and with chlorophyll a and b derived from them as they all have CH3 in ring 2 and 4, CO2CH3 in ring 5, and H in ring 7 (see Stanier et al, 1986). Quinone classes and carotenoid synthesis pathways were arranged on the basis of data in Ragan and Chapman (1978). Cytochromes, however, should have been arranged with the a-type as primitive; this was corrected for the PAUP analyses. Togabacteria have only 10 derived traits, the fewest, while Sulfobacteria have 118, the most. Metabacteria as primitive is an artefact of LBA (long-branch attraction)(Philippe et al, 2000).

Thiobacteria were split into 2 terminals, Chlorobia and Proteobacteria, and Acidobacteria, Saprospirae, Aquificae were added. Togabacteria were assumed monophyletic. Prochlorobacteria (Chloroxybacteria), which contain 3 genera (*Prochloron, Prochlorothrix*, and *Prochlococcus*) but may not be monophyletic, were not included as a terminal as they belong in, not just with, Cyanobacteria on account of both classical and molecular data. For example, *Synechococcus* in Chroococcales (which is a grade) has a cytology comparable to *Prochlorococcus*, the latter has phycobilins, which otherwise occur only in Cyanobacteria among prokaryotes, and *Prochloron* and *Synechococcus* are both symbiotes of ascidians. The data set is in Table 1 (in Appendices) and the data matrix in Table 2 (in Appendices).

The terminals include all the important bacterial clades. About a dozen small, rather obscure eubacterial groups, some only recently discovered, were not included for lack of information, but most or all probably belong to larger, traditional clades—for example, Elusimicrobia with Acidobacteria, Caldiserica and Synergistes with Spirochetes, Lentisphera and Verrumicrobia in the Proteinomura clade, Caldithrix, Gemmatimonada, and Fibrobacteria with Chlorobia or Saprospirae, and Chrysiogenes in Proteobacteria (Rappé and Giovannoni, 2003).

Several new genera of metabacteria proposed as independent groups were not included also for lack of information. Korarcheota are basal in Metabacteria in 16S analysis but the sister group of Euryarcheota+Nanarcheota in concatenated ribosomal protein analysis (Pester et al, 2011). *Nanoarchaeum equitans* (Huber et al, 2002; Das et al, 2006) was discovered in 2002 in a hydrothermal vent off the coast of Iceland on the Kolbeinsey Ridge by Karl Stetter. Strains of this microbe were also found on the Sub-polar Mid-Oceanic Ridge and in the Obsidian Pool in Yellowstone. It is a very small (.4 microns, perhaps the smallest organism known), hyperthermophilic, obligate parasite of *Ignicoccus* in Sulfobacteria. In 16S analysis it branches with Crenarcheota and in concatenated ribosomal protein analysis it is instead the sister group of Euryarcheota (Pester et al, 2011). Korarcheota is basal in Metabacteria in 16S analysis but the sister group of Euryarcheota+Nanarcheota in concatenated ribosomal protein analysis (Pester et al, 2011). But both most probably belong to Sulfobacteria (Elkins et al, 2008; Brochier et al, 2005), with Korarcheota related specifically to Thermococcales (Brochier et al, 2005), which is placed with methanogens in genotypic phylogenies but is a sulfobacterial order according to its phenotype. Lokiarcheota Spang et al, 2015 (1 genus) was identified from a candidate genome in a metagenomic analysis of a mid-oceanic sediment sample taken near a hydrothermal vent at a site known as Loki's Castle on Gakkel Ridge in the Arctic Ocean, which disclosed a monophyletic grouping with the eukaryotes and revealed several genes with cell membrane-related functions the presence of which support the theory of the metabacterial host (or eocyte-like scenarios) for the emergence of eukaryotes. Other new groups excluded for lack of information are Aigarcheota Nunoura et al 2011, Aenigmarchaeota Rinke et al 2013, Bathyarchaeota Meng et al 2014, Diapherotrites Rinke et al 2013, Geoarchaeota Kozubal et al 2013, and Parvarchaeota Rinke et al 2013.

Thaumarcheota (Thaumabacteria) (Muller et al, 2010; Schleper and Nicol, 2010; Pester et al, 2011), aerobic ammonia-oxidizers containing *Nitrososphera, Nitrosopumilus,* and *Crenarcheum*, are not new but have been completely overlooked and omitted from textbooks. Jed Fuhrman's team and Ed DeLong reported their discovery in 1992 from ocean surface waters. Initially classified as a mesophilic sister group to hyperthermophilic Crenarchaeota, phylogenetic analyses based on more SSU and SLU rDNA sequences and comparative genomics (Brochier-Armanet et al, 2008; Pester et al, 2011; Gupta and Shami, 2011) have recently suggested they might form a separate and deep-branching taxon within Metabacteria. Information from the first available genomes of the group indicate that its metabolism is fundamentally different from that of their eubacterial counterparts, involving a highly copper-dependent system for ammonia oxidation and electron transport, as well as a novel carbon fixation pathway that has recently been discovered in hyperthermophilic metabacteria. So its phenotype indicates it belongs with Sulfobacteria. Extrapolating from the wide substrate of copper-containing membrane-bound monoxygenases, to which the taxon belongs, the use of substrates other than ammonia for generating energy by some members of Thaumarchaeota seems likely.

Ammonia oxidation, the first and rate-limiting step in nitrification, the 2nd step being the oxidation of the nitrites to nitrates, is the only biological process converting reduced to oxidized inorganic nitrogen. Nitrification, a chemolithautotrophic process, which was discovered by the renowned Ukrainian-Russian microbiologist, microbial ecologist, and soil scientist Sergei Winogradsky in the 1880s (for which he won the Leeuwenhoek Prize; he also invented the enrichment culture technique, also called Winogradsky's columns, still used today), plays a central role in the nitrogen cycle. For over 100 years, this process was thought to be mediated only by autotrophic beta-proteobacteria and gamma-proteobacteria occasionally supported by heterotrophic nitrifiers in soil environments. The widespread distribution of putative metabacterial ammonia monoxygenase (amo) organisms and their numerical dominance over their eubacterial counterparts in most marine and terrestrial environments suggests that thaumabacteria play a major role in global nitrification, but our understanding of their evolution and metabolism is still in its infancy.

Also recently proposed has been the TACK superphylum that comprises Thaumarcheota, Crenarchaeota, and Korarcheota (Guy and Ettema, 2011).

Table 4-1. Trait Totals for Bacteria.

	advanced	primitive	advanced traits %
Sulfobacteria	118	88	57
Halobacteria	108	125	46
Methanobacteria	103	116	47
Thermoplasmata	61	108	36
Archeoglobi	34	111	23
Proteobacteria	101	241	30
Endospora	74	220	25
Actinobacteria	73	220	25
Cyanobacteria	57	204	22
Saprospirae	36		
Chlorobia	33	181	15
Spirochetes	29		
Chloroflexi	27	211	11
Rickettsiae	25		
Thermi	23	208	10
Deinobacteria	23	210	10
Heliobacteria	21	221	9
Mollicutes	20	195	9
Planctobacteria	19	195	9

Chlamydiae	16	201	7
Togabacteria	12	210	5

<u>Results</u>

The majority consensus tree for the 1st bacterial analysis is in Table 4-2 and Figure 1 (the latter in Appendices). There were 8 equally most parsimonious trees, 315 steps, 176 mln. rearrangements, a CI of .67, and an RI of .76. The trees differ only in the positions of *Fervidobacterium* and *Thermotoga*; Halobacteria and Methanobacteria; and Rickettsiae, Cyanobacteria, and Chloroxybacteria. There were 2 MRTs (most representative trees).

In the confidence tree there are 6 bacterial kingdoms (the 3 togabacterial, Metabacteria, and the 2 deuterobacterial) and in Aerobia, Neosoma (with Metabacteria derived) is primitive and sister group to Deuterobacteria, and in the latter Mollifirmicutes are primitive and sister group to Gracilicutes. Deinobacteria+Thiobacteria and Heliobacteria+Chloroflexi are primitive in Gracilicutes. The resulting single tree in the implied weighting was nearly identical to the confidence tree.

Table 4-2. 1st Bacterial Analysis (inluding Eukaryota)-50 % Majority Consensus (all groups were in all 8 trees so the strict consensus was the same as the majority consensus; confidence values are in parentheses; kingdoms are underlined).

<u>*Thermosipho*</u>
Contophora tax. nov. (100)
    <u>*Thermotoga*</u>
    <u>*Fervidobacteria*</u>
    Aerobia tax. nov. (93)
        <u>Gracilicutes</u> (26)
            Protogracilicutes tax. nov. (1)
                Thermus
                Metagracilicutes tax. nov. (7)
                    Spirochetes
                    Neogracilicutes (0)
                        Rickettsiae+Chloroxybacteria+Cyanobacteria (28)
                        Chlamydiae+Planctobacteria (27)
            Anagracilicutes tax. nov. (0)
                Chloroflexi+Heliobacteria
                Thiobacteria+Deinobacteria
        Monodermata (0)
            Mollifirmicutes nom. nov. (69)
                Mollicutes
                Firmicutes (20)
                    Endospora
                    Actinobacteria
            Neosoma nom. nov. (100)
                <u>*Archeoglobus*</u>
                Metasoma tax. nov. (0)
                    <u>*Thermoplasma*</u>
                    Cenosoma tax. nov.

Halobacteria
Methanobacteria
Parkaryota (0)
Sulfobacteria
Eukaryota

For the 2nd TNT analysis there were 347 steps, 64 MPTs, and the CI was .58 and the RI .67. The majority consensus is shown in Figure 2 (in Appendices) and Table 4-3. There were 4 MRTs. As with the 1st analysis, Monoderma, Mollifirmicutes, Metabacteria, and Gracilicutes are monophyletic. Different are the position of Heliobacteria and the internal arrangment for Gracilicutes, Mollifirmicutes, and Metabacteria. Heliobacteria was sister group to the rest of Deuterobacteria in half the MPTs and in the other half, sister group to Gracilicutes. In half the MPTs Mollicutes was the sister group to Firmicutes, as in the last analysis, but in the other half Actinobacteria was sister group to Mollicutes+Endospora. Archeoglobi was sister group to Methanobacteria+Crenobacteria in half the optimal trees, while in the other half it was sister group to Methanobacteria. Metachromobacteria, Neochromobacteria, and Cenochromobacteria are defined by homoplasies. The synapomorphies (Table 4-4) serve also as descriptions of the new names and groups.

Table 4-3. 2nd Bacterial Anlysis, Eukaryota Removed-Majority Consensus (set at 51%; in parentheses are the number of MPTs the group appears in; the kingdoms are underlined).

sbk. Togabacteria Cavalier-Smith 1992 orthog. emend., stat. nov.
sbk. Deuterobacteria tax. nov.(64)
infk. Heliobacteria stat. nov.
infk. Gracilicutes Gibbons and Murray 1978 (syn. Negabacteriobionta Jeffrey 1982, Agrambacteria Möhn 1984, Negibacteria Cavalier-Smith 1987, Didermata Gupta 1998) stat. nov. (64)
pvk. Mureinomurus tax. nov. (40)
mgph. Thermi
mgph. Eugracilicutes tax. nov. (33)
grdph. Spirochetes (syn. Spirochetae Ehrenberg 1855)
grdph. Acidobacteria Cavalier-Smith 2002 stat. nov.
grdph. Rickettsiae stat. nov.
grdph. Aquificae Reysenbach 2001 (syn. Hydrogenobacteria) stat. nov.
grdph. Chromobacteria tax. nov. (60)
hpph. Chloroflexi stat. nov.
hpph. Metachromobacteria tax. nov. (60)
spph. Neochromobacteria tax. nov. (64)
phyl. Deinobacteria Cavalier-Smith 1986
phyl. Saprospirae
phyl. Proteobacteria Möhn 1984, Stackebrandt et al 1986 emend.
spph. Cenochromobacteria tax. nov. (64)
phyl. Chlorobia
phyl. Cyanobacteria Stanier 1973
pvk. Proteinomurus nom. nov., stat. nov. (52)
ord. Chlamydiae stat. nov.
ord. Planctobacteria Cavalier-Smith 1998 stat. nov.

infk. Monodermata Gupta 1998 (syn. Unibacteria Cavalier-Smith 1998) (64)
hpph. Mollifirmicutes (64)
phyl. Mollicutes Gibbons and Murray 1978 (syn. Tenericutes auc. cit.)
phyl. Actinobacteria Margulis 1974
phyl. Endospora
hpph. Metabacteria Hori and Osawa 1979, Hori, Itoh, and Osawa 1982 emend. (syn. Archeobacteria Woese and Fox 1977, Mendosicutes Gibbons and Murray 1978) stat. nov. (64)
spph. Halobacteria Grant et al 2002 stat. nov.
spph. Anabacteria new taxon (64)
phyl. Archeoglobi
phyl. Methanobacteria Boone 2002
phyl. Caldaria Möhn 1984 (64)
cl. Thermoplasmata Reysenbach 2002
cl. Sulfobacteria Cavalier-Smith 1986 (syn. Crenarcheota Woese, Kandler, and Wheelis 1990; Eocyta Lake 1984)

In the resampling tree Monodermata are between Heliobacteria and Gracilicutes. The results out of 1000 replicates are as follows:

Deuterobacteria	100
Metabacteria	100
Caldaria	72
Posibacteria	67
Anabacteria	52
Cenochromobacteria	31
Firmicutes	19
Monodermata	18
Methanocaldaria	13
Proteinomurus	7
Gracilicutes	5
Proteobacteria-Saprospirae	2

So there is strong support for the 1st 3, moderate support for Posibacteria and Anabacteria, weak support for 7 others, and no support for the remaining clades.

Table 4-4. Synapomorphies.

Aerobia: large STK, loss of long chain diabolic acids, aerobism, loss of hyperthermophily, catalase
Deuterobacteria: large STK, aerobism, catalase, loss of long chain diabolic acids, loss of hyperthermophily
Gracilicutes: large citrate synthase, flagellar ring, outer membrane, LPS, invagination, large succinate thiokinase, NADH resistance
Eugracilicutes: thin cell wall, cell wall with DAP acid
Chromobacteria: photosynthesis, autotrophy, carotenoids, sporulation, hydrocarbon degradation
Proteinomurus: proteinaceous wall
Monodermata: non-formylated methionine, glutaminyl synthase tRNA glutamine transamidation, tRNA

mischarging, proteasoman alpha-amylase primary structure, proteasoman serine protease 3D structure, tyrosine kinases, serine/threonine kinases, type 1 fatty acid synthase, Ku with HEH domain, calmodulin homologs, chitin

Mollifirmicutes: DOXY pathway; foliate derivative as RNA methionine methyl donour; actinomycin, novobiocin, and penicillin sensitivity, DNAP uracil, loss of DNAP exonuclease function

Firmicutes: teichoic acid, naphthaquinones

Neosoma: wall with glycoprotein hexagonal array, loss of murein, pilin-like flagellar protein, HSP 90, thermosomes, ether lipids, archeol, mevalonate LFP, N-linked glycosylation, EM pathway with PFK, EM pathway reversal, ribosome SSU with bill, ribosome LSU with lobe, ribosome LSU with bulge, EF-1 aminocyl, tRNA-to-ribosome catalysis with EF-1, EF-2 with diphthamide, peptidyl tRNA translocation, with EF-2, EF-2 compatability, ribosome subunit, multiple RNAP enzymes, 8 or more RNAP subunits, RNAP subunit A, RNAP subunit B, mRNA with tail cap and tail

Metabacteria: wall with glycoprotein hexagonal array, pilin-like flagellar protein, HSP 90, thermosomes, ether lipids, archeol, mevalonate LFP, N-linked glycosylation, EM pathway with PFK, EM pathway reversal, ribosome LSU with lobe, ribosome LSU with bulge, EF-1 aminocyl, tRNA-to-ribosome catalysis with EF-a, EF-2 with diphthamide, peptidyl tRNA translocation, with EF-2, EF-2 compatability, ribosome subunit, 8 or more RNAP subunits, RNAP subunit A, RNAP subunit B, mRNA with tail cap and tail, nonformylated methionine, multicomponent RNAPs, introns, B DNAPs, PCNA sliding clamp, high no. of r-proteins, RNAP, DNAP, and protein synthesis antibio resistance, DNAP VI, DNA 10b MCM, N-linked glycosylation, neosoman 5S 3D structure, co- translational protein secretion, vacuolar proton-pumping ATPase, oligosaccharyl transferases, SECIS- binding protein, ubiquitin- directed proteolysis, fibrillarin, replication factor C, RNA-binding proteins, ribosomal subunit compatability, EF-2 with diphthamide, IF-2 and 5A, absence of EF compatability, loss of murein, HSP 10, and SEC A

Metasoma: potentially coaxial helices, core histones

Anabacteria: ribosome with bill, gap, lobe, and platform split

Parkaryota: LSU bulge, SSU lobes.

Chapter 5 Prokaryotes: PAUP Analyses

Bacteria are the model organisms for everything that we know in higher organisms. There are 10 times more bacterial cells in you or on you than human cells. - Bonny Bassler, American molecular biologist at Princeton-brainyquote.com.

Methods and Materials

In PAUP, multistate taxons are in theory enclosed in parentheses or curly braces (in TNT it is brackets), but in practice the matrix must be aligned so symbols are used to represent this or that combination of states, but this did not work so I coded the highest state for each taxon. Togabacteria were split into 3 taxons as were Proteobacteria and Methanobacteria and Saprospirae were split into 2 to test the monophyly of the 4 groups, as Togabacteria are paraphyletic in the 1st analysis, Proteobacteria are a heterogeneous group, Methanobacteria are polyphyletic in molecular taxonomies, and there is a dichotomy in Saprospirae not recognized in molecular taxonomies that might indicate pleophyly (see Chapter 10).

For the analysis with PAUP for bacteria+Eukaryota there were 31 taxons and 281 characteristics and Fitch parsimony, TBR, random addition, were used. In all my PAUP analyses the trait-state optimization was with ACCTRAN (accelerated transformation), and a Dell Dimension 3100 was employed. The one with bacteria alone also had 281 characteristics, with Fitch parsimony, TBR, and random addition again used, and no a priori weighting nor topological constraints were used in either analysis nor with the eubacterial analyses.

I also did a cladistic analysis for Eubacteria with Proteobacteria split into 12 groups besides Rickettsiae, using Phylip's Penny program (Felsenstein, 2005), which does branch and bound and uses Wagner parsimony, and set at 100,000 cycles instead of the default 1000. Branch and bound is an exact search technique, but it broke off by itself before the search could be terminated. There were 29 groups and 78 characteristics. Phylip does not do CI nor RI nor synapomorphy lists nor plots and I wasn't able to work the bootstrap and jackknife programs. I did the same analysis but run on PAUP with Fitch parsimony and TBR. The data set and matrix are in Tables 5 and 6 in Appendices.

Results

The PAUP analysis for all prokaryotes plus Eukaryota generated 10 trees with a length of 654 after 4.3 mln. rearrangements, with a CI. of .41 and an RI of .61. There was high confidence for Togabacteria, Saprospirae, Firmicutes, Proteobacteria, Metabacteria, Metaerobia, Neosoma, and Methanobacteria, and moderate confidence for Deino-Thermi and Aquificae-Chlorobia. In contrast to the 2 previous analyses the Didermata clade was not recovered. The classification is presented in Table 5-1 (along with the bootstrap-jackknife tree) and Figure 3 (the latter in Appendices).

Table 5-1. Prokaryotes plus Eukaryota with PAUP and Fitsch.

a) 50% Majority Consensus for the 10 optimal trees using PAUP and Fitch (in parentheses are the MPT frequencies in %)(there are 6 bacterial kingdoms (underlined)).

Hypothetical Ancestor
Togabacteria (100)

Thermosipho-Thermotoga (60)
Fervidobacterium
Metaerobia (100)
Protometaerobia (80)
Thermaria tax. nov. (80)
Deinothermi nom. nov. (100)
Thermi
Deinobacteria
Paradeinobacteria (80)
Subspirochetes (80)
Rickettsiae
Proteinomurus (80)
Chlamydiae
Planctobacteria
Spirochetes
Chlorobacteria tax. nov. (100)
Chloroflexi-Acidobacteria tax. nov. (100)
Aquificae-Chlorobia tax. nov. (100)
Neaerobia (80)
Subheliobacteria tax. nov. (100)
Subcyanobacteria tax. nov. (80)
Saproprotei tax. nov. (100)
Saprospirae (100)
Flavobacteria
Bacteroidetes
Proteobacteria (100)
Campylo-Ubiquinonae (100)
Myxospora
Neosoma (100)
Metabacteria (100)
Halobacteria
Anabacteria nom. nov. (100)
Methanaria nom. nov. (100)
Archeoglobi
Methanobacteria (100)
Methanobacteriae
Methanomicrobia
Methanosarcinae
Caldaria (100)
Thermoplasmata
Sulfobacteria
Eukaryota
Terrabacteria Battistuzzi and Hedges, 2008, emend. (80)
Cyanobacteria
Mollifirmicutes (100)
Mollicutes
Firmicutes (100)

- Endospora
- Actinobacteria

<u>Heliobacteria</u>

b) bootstrap-jackknife tree (50% majority, 100 replicates)(there were 1.5 mln. rearangements for the bootstrap and 1.2 mln. for the jackknife)

- Hypothetical Ancestor
- Togabacteria (84/82)
  - *Thermosipho*
  - *Thermotoga*
  - *Fervidobacterium*
- Metaerobia (87/76)
  - Deino-Thermi (65/80)
  - Rickettsiae
  - Chlamydiae
  - Planctobacteria
  - Spirochetes
  - Chloroflexi
  - Aquificae-Chlorobia (58/59)
  - Saprospirae (97/97)
  - Acidobacteria
  - Proteobacteria (70/66)
    - Campylobacteria
    - Myxospora
    - Ubiquinonae
  - Cyanobacteria
  - Heliobacteria
  - Mollicutes
  - Firmicutes (85/87)
  - Neosoma (100/100)
    - Metabacteria (69/79)
      - Halobacteria
      - Anabacteria (-/59)
        - Methanaria (53/63)
          - Archeoglobi
          - Methanobacteria (100/96)
            - Methanobacteriae
            - Methanococci (50/50)
              - Methanomicrobia
              - Methanosarcinae
        - Caldaria (-/51)
          - Thermoplasmata
          - Sulfobacteria
    - Eukaryota

The PAUP analysis for only bacteria generated a single tree with a length of 522 after 1.8 mln. rearrangements, with a CI. of .50 and an RI of .71. Fitch parsimony was used. There was high confidence for Togabacteria, Deuterobacteria, Proteobacteria, Firmicutes, Metabacteria, and Methanobacteria. Deino-Thermi, Aquificae-Chlorobia, Posibacteria, and Caldaria had moderate confidence. Heliobacteria branched near Posibacteria. Again the Didermata clade was not recovered. The classification is presented in Table 5-2 and Figure 4 (the latter in Appendices).

Table 5-2. Bacteria using PAUP and Fitch.

a) 50% Majority Consensus for only 1 optimal tree using PAUP and Fitch (there are 8 kingdoms (underlined)).

- Hypothetical Ancestor
- Togabacteria
  - *Thermosipho-Thermotoga*
  - *Fervidobacterium*
- Deuterobacteria
  - Hypertoga tax. nov.
    - Subchlorobacteria tax. nov.
      - Subspirochetes tax. nov.
        - Subplanctobacteria tax. nov.
          - Subchlamydiae tax. nov.
            - Deino-Thermi
              - Thermi
              - Deinobacteria
            - Paradeinothermi tax. nov.
              - Rickettsiae
              - Pararickettsiae tax. nov.
                - Saproprotei tax. nov.
                  - Saprospirae
                    - Flavobacteria
                    - Bacteroidetes
                  - Proteobacteria
                    - Campylo-Ubiquinonae tax. nov.
                      - Campylobacteria
                      - Ubiquinonae nom. nov.
                    - Myxospora
                - Cyanoposibacteria
                  - Cyanobacteria
                  - Mollifirmicutes
                    - Mollicutes
                    - Firmicutes
                      - Endospora
                      - Actinobacteria
          - Chlamydiae
        - Planctobacteria
      - Spirochetes

<u>Chlorobacteria</u> tax. nov.
Chloroflexi-Acidobacteria tax. nov.
Aquificae-Chlorobia tax. nov.
Heliometabacteria
<u>Heliobacteria</u>
<u>Metabacteria</u>
Halobacteria
Anabacteria nom. nov.
Methanaria nom. nov.
Archeoglobi
Methanobacteria
Methanobacteriae
Methanomicrobia
Methanosarcinae
Caldaria
Thermoplasmata
Sulfobacteria

b) bootstrap-jackknife tree (50% majority, 100 replicates)(there were 4.4 mln. rearrangements for the bootstrap and 1.8 mln. for the jackknife)

Hypothetical Ancestor
Togabacteria (83/87)
Thermosipho
Thermotoga
Fervidobacterium
Deuterobacteria (89/81)
Deino-Thermi (65/68)
Rickettsiae
Chlamydiae
Planctobacteria
Spirochetes
Chloroflexi
Aquificae-Chlorobia (67/58)
Saprospirae (99/97)
Acidobacteria
Proteobacteria (75/75)
Campylobacteria
Myxospora
Ubiquinonae
Cyanobacteria
Heliobacteria
Mollifirmicutes (61/58)
Mollicutes
Firmicutes (96/97)
Metabacteria (100/100)

Halobacteria
Methanaria (55/54)
    Archeoglobi
    Methanobacteria (97/93)
        Methanobacteriae
        Methanomicrobia
        Methanosarcinae
Caldaria (59/65)
    Thermoplasmata
    Sulfobacteria

The analysis with Penny generated 27 MPTs, 163 steps, and the CI was .48. The majority consensus was the same as the strict consensus and is in Table 5-6.

For the PAUP analysis for Eubacteria there were 6 mln. rearrangements, 194 steps, and only 1 MPT, with a CI of .43 and an RI of .58. Myxospora were coded with sphingolipids (which occur in only 1 genus) and Saprospirae with spores (which occur in only 1 genus) and characteristic 24 was excluded. I also ran the same matrix but with Myxospora coded without sphingolipids and Saprospirae without spores and characteristic 24 included, which resulted in the same topology and there were also 6 mln. rearrangements but 199 steps and a CI of .32 and an RI of .31. As with the Penny taxonomy, there was no Proteobacteria nor Photobacteria clade, but there was a Chlorobacteria clade and a Clatobacteria 1 (Pseudomonada-Nitrobacteria) and a Clatobacteria 2 (Siderobacteria, Thiobacteria, Hydrogenobacteria-Desulfobacteria, Methylomonada).

Table 5-3. Eubacterial Classification, 29 x78, with Phylip's Penny program, 50% Majority Consensus.

spk. Togabacteria
spk. Paratoga
    Mollicutes
    SG 2
        Heliofirmicutes
            Heliobacteria
            Firmicutes
                Endospora
                Actinobacteria
        Gracilicutes
            Planctobacteria-Campylobacteria
            SG 5
                Chlamydiae-Rickettsiae
                SG 6
                    SG 7
                        Thermi-Spirochetes
                        SG 8
                            Myxospora
                            SG 9
                                Siderobacteria
                                Methylomonada
                                Thiobacteria

Enterobacteria
Nitrobacteria
SG 10
Deinobacteria-Caulobacteria
Pseudomonada
SG 11
Cyanobacteria
Porphyrobacteria
Saprospirae
SG 12
Chloroflexi
SG 13
SG 14
Chlorobia
SG 15
Hydrogenobacteria-Desulfobacteria
Acidobacteria

Table 5-4. Classification for Eubacteria, 30x78, using PAUP and Fitch.

a) the single MPT

Hypothetical Ancestor
Togabacteria
Paratoga
Submollicutes
Gracilicutes
Subchlamydiae
Subrickettsiae
Subcampylobacteria
Subplanctobacteria
SG 8
SG 9
SG 10
SG 11
SG 12

Th-Dn.
Caulob.
Enterobacteria
SG 13
Chlorocyanobacteria

Chlorobacteria

Chlrsm

Cx

Cb

Ac

Cy

Porphyrabacteria
Psdmda-Nitrobacteria
SG 15
Spirochetes
Saprospirae-Myxospora
Clatobacteria 2
Siderobacteria
SG 17
Thiobacteria
Hydrogenobacteria-Desulfobacteria
Methylomonada
Planctobacteria
Campylobacteria
Rickettsiae
Chlamydiae
Heliofirmicutes
Heliobacteria
Firmicutes
Mollicutes

b) the confidence value tree (50% majority, 100 replicates)(the 1st figure is the bootstrap, 2nd is the jackknife)(the number of bootstrap rearrangements was about 1 mln. And the number of jackknife rearrangements was also about 1 mln).

Hypothetical Ancestor
Togabacteria
Paratoga (59/67)
Submollicutes (81/89)
Gracilicutes (70/78)
Thermi-Deinobacteria (61/-)
Caulobacteria
Enterobacteria
Chloroflexi
Chlorobia
Acidobacteria
Cyanobacteria
Porphyrabacteria (80/78)
Pseudomonada

Nitrobacteria
Spirochetes
Saprospirae-Myxospora (54/53)
Siderobacteria
Thiobacteria
Hydrogenobacteria-Desulfobacteria (-/53)
Methylomonada
Planctobacteria
Campylobacteria
Chlamydiae
Rickettsiae
Heliofirmicutes (59/57)
Heliobacteria
Firmicutes (89/87)
Mollicutes

Discussion

Classical evidence in phylogeny for bacteria (=prokaryotes) is universally considered without merit, so no cladistic analysis has ever been done based on this kind of evidence for this group at any level until now. There are no reasonable grounds for such a belief, even considering LGT (lateral gene transfer).

The first 2 phylogenies basically are in accordance with that of McMaster's Radhey Gupta (1998a,b, 2000), who uses conserved signature sequences of proteins. They agree on 4 fundamental points: the monophyly of Gram negatives and Monodermata and the position of Metabacteria. They disagree principally in being branching instead of linear, in having Metabacteria as monophyletic, and in Togabacteria being separate from Mollifirmicutes. Hori and Osawa (1987) also found Mollifirmicutes and Negibacteria but in a Eubacteria clade.

Gupta (1998a, pp. 16-17) points out that most eubacterial taxonomies based on rRNA do not give any confidence measures and the few that do have most of the critical nodes at values in the range of 25-50%, which means they are not reliable (and the lower the confidence estimates the higher the homoplasy), and that Woese himself acknowledges the lack of resolution of branching order. (The CI and RI aren't usually given either, but this is because they do not apply to probablistic methods.) The Gupta result in which Metabacteria arose from various subgroups of Gram positives (some methanogens and some thermoacidophiles from Clostridia and halophiles from Actinobacteria) is not supported by the phenotypic evidence.

Some molecular data indicate *Thermotoga* might be part of Firmicutes (Gupta, 1998b), but it is basal in at least 4 molecular taxonomies, and usually Togabacteria are monophyletic, including also several other genera. *Thermosipho* has the largest number of primitive traits which are: heterotrophy, hyperthermophilia, fermentation, anaerobiosis, nonmotility, with a single membrane, thin wall, unicellularity, small SRP (signal recognition particle), and absence of LPS (lipopolysaccharide) layer, outer membrane, spores, cytochromes, quinones, catalase, and carotenoids, so is probably the ancestral group as indicated also by molecular evidence and was used as the root. The Proteinomura (Pirellulae) clade concurs with genotypic data, as well, but because of the adenylate system Chlamydiae might go with Rickettsiae.

The primitive position of Heliobacteria in the 2nd analysis is especially congruent with the results of Gupta's analysis of phototrophic bacteria (Gupta, 2003), but in contrast to this they are the crown group in the 3rd analysis. Endospores were coded as primitive in the 1st analysis but in the others as advanced yet Heliobacteria still did not branch with Mollifirmicutes except in the Eubacteria-alone analyses. And also contrary to most molecular results, Thermi and Deinobacteria do not form a clade in the first 2 classifications, but in accord with them they do in the prokaryotes+Eukaryota analysis using PAUP, in the Bacteria-alone analysis using PAUP, and the Eubacteria analysis using PAUP, probably due to the addition of ATP coupling factor A. Rickettsiae do not turn up close to Proteobacteria in most of the analyses nor do they in Wolf et al (2001). Surprises were also the relatively low position of Acidobacteria in the 2nd analysis, which shows up close to Proteobacteria in molecular analyses (as it does in the bootstrap-jackknife tree) and has phenotypic similarities with them, and the low confidence for Proteinomurus. But there is no surprise in the low placement of Chloroflexi, Aquificae, and Spirochetes--the primitive nature of Chloroflexi among Gram negative phototrophic groups corroborates most molecular analyses.

Gracilicutes show up in the first 2 analyses and in the eubacterial analyses but with low confidence and do not show up in the 2 PAUP analyses for all bacteria. Gupta (2000) states that it seems highly unlikely a complex characteristic like the double-membrane, which created a new compartment (periplasm) and in the process had a major effect on cellular physiology, evolved more than once, however, this is compatible also, as in some of the classifications, including the 2 PAUP analyses for all bacteria here, with a larger clade for which one of the special similarities (synapomorphies) is the double membrane but which is lost 1 or more times and with 1 loss of the wall.

Gupta (2000) explains the double membrane of Gracilicutes as evolution by normal mechanisms in response to the strong selection pressure exerted by antibiotics produced by certain groups of Gram-positives. However, new information suggests something very different. Analyzing the flows of protein families, the Lake Lab (2009) has obtained evidence that Gram-negatives were formed as the result of a symbiosis between an ancient actinobacterium and an ancient clostridium. As Lake points out, the resulting taxon has been extraordinarily successful and has profoundly altered the evolution of life by providing endosymbiotes necessary for the emergence of eukaryotes and generating Earth's oxygen atmosphere, that their double-membrane architecture and the observed genome flows into them suggest a common evolutionary mechanism for their origin. But Gupta (2011) states the Lake proposal is based on a number of false assumptions and the data presented in its support are also of questionable nature, so he claims there is no reliable evidence to back up the endosymbiotic origin of double-membrane bacteria.

The primitive position of Mollicutes and Mollifirmicutes in the support trees for the 2 TNT classifications and the 1st eubacterial classification is in accord with some molecular phylogenies, e.g., Gupta (1998), Wolf et al (2001), Ciccarelli et al (2006), and Lienau et al (2006)(see Table 10-3), and also the view of some others, e.g., Jeffrey (1982). However, this is contradicted by the high percentage of advanced traits in Firmicutes, as well as the parsimony trees and most molecular taxonomies. Also noteworthy is the sharp contrast between the support values for Firmicutes, only 20 in the 1st TNT analysis and only 19 in the 2nd, and high in the PAUP analyses, and for Mollifirmicutes, moderate in the TNT analyses and low or unrecovered in the PAUP analyses.

Halobacteria have the most primitive ribosomal structure in Metabacteria (they have only the bill)(Lake et al, 1984) and this shows up in the 3rd and 4th analyses. *Halobacterium* was found in the same (basal) position by Olendzenski et al in 2001 and Korbel et al. in 2002 (Gogarten et al, 2002) as well as in Wolf et al (2001). Euryarcheota s.l. are weakly supported in Pisani et al (2007) having a bootstrap estimate of 0 including Thermoplasmata, 52 including *Pyrococcus* and *Thermococcus* (which are in subgroup 2 of Sulfobacteria in Bergey's Manual [Holt et al, 1994]), 40 including Methanobacteria 2, but a core Euryarcheota is strongly supported at 97 including Archeoglobi and 99 including only Methanobacteria 1 and Halobacteria. In Brown et al (2001) it is supported at a modest

64 in the MP bootstrap and below 50 in the NJ and ML bootstrap but 100, 100, and 83 including *Pyrococcus*-Methanobacteria+*Archeoglobus*. It is weakly supported also in the NJ phylogeny of Daubin et al (2002) and also weakly supported excluding *Thermoplasma*, but *Halobacterium-Archeoglobus*+Methanobacteria is at 74 and *Halobacterium-Archeoglobus* is at 93, while in the ML phylogeny it is strongly supported at 93, is at 71 excluding *Pyrococcus*, and including *Thermoplasma* +*Halobacterium-Archeoglobus* is at 76, and *Halobacterium-Archeoglobus* is at 99. Its bootstrap support in Battistuzzi et al (2004) is 92, 36, and 98 for ME, ML, and Bayesian, respectively. In Gao and Gupta (2007), which has a large sampling (29 genera) and is based on a concatenation of 31 proteins, it is at 100/90/95 for NJ, ML, and MP, respectively, including *Thermococcus* and *Pyrococcus*, with strong support also for the *Archeoglobus*+Halobacteria-Methanobacteria clade at 100/92/71 and for Halobacteria-Methanobacteria at 100/98/100. In view of the generally low confidence for Anabacteria in the analyses here and the putative phenotypic synapomorphies for Euryarcheota s.s. (halophilia (for Halobacteria and Methanobacteria) and B'B" RNA polymerase (split RNAP)(for Archeoglobi, Halobacteria, and Methanobacteria) which was coded as primitive in the 1st analysis but as advanced in the others), the latter can't be said to be convincingly contradicted.

Metabacteria as directly related to Posibacteria is supported also by Skophammer et al (2007) and Valas and Bourne (2011). Lorraine Olendzenski et al (2000) also say G+ genes transfered laterally to Metabacteria. Cheryl Andam, David Williams, and J. Peter Gogarten say the same thing and specifically, "Acetoclastic methanogens in *Methanosarcina*, which today generates about 60% of the biogenic methane, became possible when the genes that encode enzymes to convert acetate to acetyl-S coenzyme A were acquired through HGT from cellulolytic Clostridia (Fournier and Gogarten, 2008)." But they also state, "Horizontal gene acquistion is not a random process--more transfers occur between more closely related organisms than between divergent ones. This preference of exchange partners consequently reinforces the identity of higher taxonomic groups." (Andam et al, 2010, p. 590-91 and 596).

But, if Eukaryota is factored in, the 1st analysis results show a paraphyletic Metabacteria. However, an analysis excluding Eukaryota would result in a monophyletic Metabacteria clade (as in the 2nd and 4th analyses and with high confidence), and even with Eukaryota included they can show up as monophyletic (as in the 3rd analysis, and with high confidence in the jackknife), and they are monophyletic in the confidence tree for the 1st analysis but with nearly no confidence, an even sharper contrast than for Firmicutes.

Metabacteria or Caldaria are sister group to Eukaryota in the molecular taxonomies of Gogarten et al (1989) and Iwabe et al (1989), Brown et al (2001), Battistuzzi et al (2004), Ciccarelli et al (2006), and Pisani et al (2007). The molecular classifications of Hori and Osawa (1987), Iwabe et al (1989), Dutihl et al (2004), Qi et al (2004), and Lienau et al (2000) also support Metabacteria and Eukaryota as sister groups. In Battistuzzi et al (2004) eukaryotes branch with Sulfobacteria. In Pisani et al (2007) the strongest signal for eukaryotes is with Cyanobacteria, the 2nd strongest with Alpha-Protei, and the 3rd strongest with Thermoplasmata. Also, newer molecular evidence indicates there are only 2 domains, Eubacteria and Metabacteria, and that the metabacterial host for eukaryotes is to be found in a paraphyletic Metabacteria (Pace, 2006; Willliams et al, 2013; Willliams et al, 2014; Spang et al, 2015). As there are key characteristics between Sulfobacteria and eukaryotes, a direct relationship between between them is maintained especially by Lake (1983,1986) and Lake et al (1984). The vertical descent theory is also supported by Cavalier-Smith (1987, 2002, 2006). Mindell (1992) says that, if we factor in endosymbioses, eubacteria are paraphyletic and eukaryotes are polyphyletic.

However, Parkaryota are unsupported in resampling in the 1st analysis, and an endosymbiotic origin for eukaryotes is generally accepted, but there is no consensus on details. Katz (2002) outlines 7 major theories: Zillig's already-mentioned fusion; the contemporaneous by Lang et in 1999 and Roger in the same year (acquistion of mitochondria occured at same time as origin of eukaryotes); the hydrogen by Martin and Muller (1998)(combining observations on the genetic makeup of eukaryotic nuclei, the biochemistry of hydrogenosomes, and the nature of extant syntrophic relationships (symbiosis based on metabolic interactions), proposes symbiosis between alpha-proteobacterium and a methanogen); syntrophy by Lopez-Garcia and Moreira (2004)(symbiosis between alpha-proteobacterium and a methanogen; mitochondria acquired as independent symbiosis); SET (serial endosymbiosis theory) by Margulis in 1996 and Margulis et al in 2000 (Thermoplasmata-Spirochetes symbiosis providing flagella, symbiosis with an alpha-proteobacterium providing mitochondria, and the symbiosis for chloroplasts), Woese's genetic annealing (1998)(LGTs mostly occured in a progenote era in a community of progenotes which eventually gave rise to the 3 domains), and Doolittle's continous tempo (continuous transfer of genes into nuclei of microbial eukaryotes from organelles and ingested prey; multiple donor lineages) in 1998 and 1999. She concludes that there is no hypothesis supported by current data, except for the Doolittle model, and that it is unclear whether this is because of inadequate sampling, obscuring effects of more recent LGTs, or the inappropriateness of the models, except for genetic annealing, which she says is due to the 3rd. There is also *Thermoplasma*-(S-dependent) Alphaprotei syntrophy by Searcy and Hixon in 1991, going back to Searcy and colleagues in 1978. In Searcy and Hixon and Martin and Muller, the origin of eukaryotes concurs with the origin of mitochondria.

A fusion event, that is, a fusion between a metabacterium and a eubacterium, was first proposed by Wolfram Zillig and colleagues (1989) and is also put forward by Gupta and Singh (1994)(but it would be with an actinobacterium, instead of a Gram negative as that is where most of the similarities lie), and Maria Rivera and James Lake (2004) as the ring of life model and a Crenarcheota-Alphaprotei symbiosis. It provides an explanation for the nuclear double membrane and the discrepancy between molecular evidence categories, but the nucleus is part of the cytomembrane system, which evolved through infoldings of the plasma membrane; this is explained by Gupta (1998).

A major difference between the Lopez-Garcia and Moreira model and other chimeric models is that a selective force for the origin of the nucleus is advanced: metabolic compartmentalization. In this, the nucleus originated not to isolate the genetic material from the cytoplasm, as is generally believed, but to allow the coexistence of 2 interdependent metabolic pathways in the protoeukaryotic cell; the cytoskeleton and other eukaryotic properties are products of symbiotic innovation; and the primary symbiosis leading to eukaryotes took place in microbial communities thriving in the widespread anaerobic environments that characterized the Archeozoic.

Another model has been proposed by Hartman and Federov (2002). This involves a separate ancestral cell type, which they call a chronocyte, that engulfed both eubacteria and metabacteria to form eukaryotes, making today's eukaryotes a combination of all 3 cell types. The chronocyte had a cytoskeleton that enabled it to engulf prokaryotic cells and a complex internal membrane system where lipids and proteins were synthesized. It also had a complex internal signaling system involving calcium ions, calmodulin, inositol phosphates, ubiquitin, cyclin, and GTP-binding proteins. The nucleus was formed when a number of eubacteria and metabacteria were engulfed by a chronocyte.

When Jere Lipps, a paleontologist at the University of California at Berkeley, adjusted a tree showing the emergence of eukaryotes to remove long-branch attraction artifacts, the tree turned into a bush with a long stem, with all eukaryotes emerging in a geologically short period of time. In the PNAS article, Hartman presents evidence for his chronocyte hypothesis: a unique set of eukaryotic proteins not found in any eubacterium or metabacterium. He found 347 proteins he and coauthor Alexei Federov, of Harvard, call eukaryotic signature proteins (ESPs). Among these proteins are several associated with the cytoskeleton, leading Hartman to the conclusion that the chronocyte had the mechanisms necessary to engulf other cells.

Clues gleaned from this protein set and other research lead Hartman to the conclusion that the chronocyte stored its genetic information in RNA rather than DNA. This further supports his contention that the eukaryotic nucleus, a DNA-based structure, arose from a symbiosis of the chronocyte and a DNA-based metabacterium and that the symbiotic event in which chronocytes captured both eubacteria and metabacteria happened around 2 bya. "And that's precisely when oxygen comes into the atmosphere," Hartman notes. While we thrive in an oxygen-rich atmosphere, the gas was poisonous to many of the organisms on Earth when photosynthesis arose. "It forced cells into horizontal transfer due to the fact that they had to react to this new poison, oxygen."

Another theory, viral eukaryogenesis, proposed by Villarreal and DeFilippis (2000), Bell (2001, 2004, 2006, 2009), Forterre (2001, 2005, 2006, 2008), Takemura (2001), Pennisi (2004), Raoult et al (2004), Villarreal (2007), and Witzgany (2015) suggests that precursors of today's NCLDVs (nucleocytoplasmic large DNA viruses), a complex, enveloped DNA virus, led to the emergence of eukaryotic cells in an endosymbiotic-like event in which the virus became a permanent resident in an emerging eukaryotic cell. Bell posulates that a complex DNA virus established a persistent presence in the cytoplasm of a methanogen and evolved into the eukaryotic nucleus by acquiring a set of essential genes from the host genome, eventually usurping its role; several characteristic features of the eukaryotic nucleus derive from a viral ancestry, including mRNA capping, linear chromosomes, and separation of transcription from translation; and phagocytosis and other membrane fusion-based processes are derived from viral membrane-fusion processes and evolved in concert with the nucleus. The coevolution of phagocytosis and the nucleus rendered much of the host metabacterial genome redundant since the protoeukaryote could obtain raw materials and energy by engulfing bacterial syntrophs/prey. This redundancy allowed loss of the metabacterial chromosome, generating an organism with eukaryotic features. The evolution of phagocytosis allowed the eukaryotes to be the first organisms to be predators. He identifies several similarities between eukaryotes and one or another type of virus: cytoplasmic replication, linear chromosomes, chromosome ends with short tandem repeats, a membrane-bound genome compartment, mRNA transport to the cytoplasm, and mRNA capping. He also proposes the eukaryote is a composite of a DNA virus, a metabacterial cell that evolved into the cytoplasm, and an alpha-proteobacterium that evolved into the mitochondrion.

Villareal (2007) also suggests DNA viruses and retroviruses were involved in the origin of all 3 domains, and are ancestral to the innate and adaptive immune systems, to flowering plants, and to placental mammals.

Viruses have 2 completely different life strategies: acute (pathogenic) and persistent (compatible interactions with, and harmless during most life stages of, the host)(Witzgany, 2015).

There are 3 possible explanations for the origin of viruses: the progressive or escape (viruses as arisen from genetic elements that gained the ability to move between cells), the regressive or reduction (viruses as remnants of cellular organisms), the virus-first or coevolution (viruses as predating [using self-replicating macromolecules as their first hosts] or coevolving with their current cellular hosts) (Wessner, 2010a).

In the 1st of these, mobile genetic elements, pieces of genetic material capable of moving within a genome, gained the ability to exit one cell and enter another, which is very similar to retroviuses and retrotransposons (especially viral-like retrotransposons). The latter is an important, though somewhat unusual, component of most eukaryotic genomes; they make up an astonishing 42% of the human genome and can move within the genome via an RNA intermediate. The genetic structures of retroviruses and viral-like retrotransposons show remarkable similarities.

The 2nd is boosted by the fact that certain bacteria, like *Chlamydia* and *Rickettsia*, are, like viruses, obligate intracellular parasites, and it is generally agreed that they evolved from free-living ancestors. It follows, then, that existing viruses may have evolved from more complex, free-living organisms that lost genetic information over time, as they adopted a parasitic life-style. NCLDVs best illustrate this idea. In addition to their large genome, they exhibit greater complexity than other viruses and depend less on their host for replication. Because of the size and complexity of NCLDVs, some virologists have hypothesized that they may be descendants of more complex ancestors. So, in this model, autonomous organisms initially developed a symbiotic relationship, eventually turning parasitic, and as the parasite became more dependent on the host, it lost previously essential genes, and over time it was unable to replicate independently, becoming a virus.

The giant viruses such as Mimivirus (the first discovered, isolated from *Acanthameba* in a hospital water-cooling tower in 1992 in Bradford, England; the name comes from "mimicking microbe"), Megavirus, and the largest, Pandoravirus (at 2.5 mln. base pairs)(Raoult et al, 2004; Wessner, 2010b; Yong, 2013), add credibility to this theory, since Mimivirus contains a relatively large repertoire of putative genes associated with translation--genes that may be remnants of a previously complete translation system--and is strikingly similar to parasitic bacteria such as *Rickettsia*. They have also led to another intriguing proposal: that viruses are derived from a 4th domain (Raoult et al, 2004; Villareal, 2007). Villareal says it can be argued for or against at this time because of limited information.

The virus-first notion was first proposed by Canadian microbiologist Félix d'Herelle in his 1921 book Le bactériophage, son rôle dans l'immunité, and was revived by Koonin and Martin in 2005 (Claverie, 2006; Félix d'Herelle-Encyclopedia Britannica-britannica.com).

The first theory is the favoured one, but it is possible that all 3 are right (Wessner, 2010a).

Frank Ryan (Fox, 2004), in his 2003 book Darwin's Blind Spot: Evolution Beyond Natural Selection, suggests the retroviral content of the human genome probably reflects "plague culling", i.e., a series of catastrophic epidemics in which only certain human genetic types survived, in mutually beneficial partnership with the viruses, and in this sort of evolution the plagues probably arose when viruses crossed a species barrier, and a new retroviral transfer from ape to man has been reported from Africa. In addition, retroviruses may have contributed to saltations by causing deletions, duplications, and rearrangements of chromosomes. So endogenous human retroviruses are attracting enormous attention. He expands on the notion of cooperation as fundamental to life on earth, including the more familiar phenomenon of exosymbiosis in which different genera cooperate to their mutual advantage—trees with funguses, funguses with ants, humans with intestinal bacteria. The concept is connected to James Lovelock's Gaia hypothesis first put forward in 1967—the idea that the planet is "homeostatically controlled by active feedback processes operated automatically by the biota."

The endosymbiosis hypothesis goes back some way, originating with German pathologist, histologist, and anatomy professor Richard Altmann (1890), who discovered the mitochondrion (naming it "bioblast"; he developed an histological staining method for this organelle, and coined the term "nucleic acid" as a synonym or replacement for Friedrich Miescher's "nuclein" when it was found to have acidic properties), with Russian botanist Konstantin Merezhkovsky (1910) doing important research in it and coining the term "symbiogenesis" and perhaps 1st formulating the theory, French zoologist and marine biologist Paul Portier (1918) presenting the 1st detailed description, Russian botanist Boris M. Kozo-Polyansky's explaining it in terms of Darwinian evolution in his book Symbiogenesis: a New Principle of Evolution in 1924, Wallin (1927) proposing it for mitochondria, more detailed electron microscopic comparisons between cyanobacteria and chloroplasts notably by Hans Ris published in 1961, and Lynn Margulis popularizing, elaborating, and synthesizing it in 3 articles in the late '60s.

The notion may actually have originated with German-French botanist and phytogeographer Andreas Schimper, who had observed in 1883 that the division of chloroplasts in green plants closely resembled that of free-living cyanobacteria, and who had himself tentatively proposed (in a footnote) that green plants had arisen from a symbiotic union of 2 organisms (Symbiogeneis-Wikipedia).

Lynn Margulis (2010) calls the modern evolutionary synthesis, also called NeoDarwinism, which was developed from the 1920s to the 1940s, a "widely touted but undocumented explanation of the origin of evolutionary novelty by "gradual accumulation of random mutations;"" and predicted it would be replaced by the details of symbiogenesis: genetic mergers, especially speciation by genome acquisition, karyotypic fissions (neocentromere formation, related chromosome change), and D.I. Williamson's larval transfer concept for animals, and that Kozo-Polyansky's work would be lauded in the same way Mendel's studies were.

Robin Fox (2004) says endosymbioses led to the origins of at least 26 phyla, and that with each new endosymbiosis, evolution took a leap forward—in the saltations that were ill-explained by Darwin's competition theory. Retroviral components contributed both beneficially and harmfully to mammalian life.

Rhawn Joseph (2010) says, possibly around 4 bya, eukaryotic cells were shaped by repeated invasions by prokaryotes thereby creating the eukaryotic nucleus, compartments, and mitochondria, that viruses provided elements which enabled nucleotides to be copied, duplicated, and then expressed and, in conjunction with prokaryotes, played a key role in eukaryogenesis, and that viruses may have also served as mobile "RNA Worlds" which acted on pre-biotic cells, splicing together and duplicating nucleotides, thereby kick starting DNA-based life capable of replicating itself.

The notion that prokaryotes evolved from eukaryotes (Poole et al, 1998) is completely contradicted by all kinds of evidence--morphological, physiological, chemical, molecular, and fossil. as well as by the endosymbiosis hypothesis. It is difficult to imagine how chloroplasts, given they evolved from bacteria, could have appeared before them, how a complex structure like the eukaryotic ribosome could be primitive, and how the wholesale loss of a complex structure like the cytomembrane system could occur. The idea has little credibility and has gained little acceptance.

LGT (as opposed to vertical gene transfer, i.e., from parental generation to offspring), also called HGT (Horizontal Gene Transfer), is maintained to be highly exaggerated by the Swedish team of Kurland, Canbeck, and Berg (2003) at the U. of Uppsala, who state that rampant global LGT was significant only in the progenote population, and that, although LGT is an important evolutionary force, in modern organisms both its range and frequencies are constrained most often by selective barriers, and as a consequence those LGT events that do occur generally have little influence on genome phylogeny, and classical Darwinian lineages are the dominant mode of evolution. They point out that an estimate of frequencies of events such as gene loss, gene duplication, novel sequence genesis, and LGT in the cohort of fully sequenced genomes done by Snel and colleagues found less than 15% of such phylogenetically troublesome events as ascribable to HGT in eubacterial and metabacterial genomes. An explicit estimate of the influences of biased and variable mutation rates was not presented in the study, so even this modest estimate is likely to be inflated. They also underline that anomalous phylogenetic reconstructions may be generated, as well, by improper clade selection, which often means examining too few clades or relying on inadequate phylogenetic methods. The contribution of LGT has been estimated to vary between 0% and 17% with a mean of 6% in eubacterial genomes.

Recently, the 1000s of coding sequences found in 5 taxons of photosynthetic bacteria were reduced to a set of 188 orthologous lineages by Raymond et al that could not be resolved into a single tree. Kurland et al point out that from this failure alone it was concluded that HGT was responsible for the phylogenetic ambiguity of the selected data set, yet there was a complete absence of any direct indication of HGT. No attempts were made to identify the impact of segregating paralogs, gene loss, extreme mutation rates, nor biased mutation rates on the reconstructions of this subset of orthologs.In their review of the HGT data, Gogarten and colleagues present a selected cohort of 30 instances of putative gene transfer. This list is used to support the notion that the genomes of bacteria that share a common environment may be mosaics created by the rampant exchange of sequence domains from both rRNA and proteins. Kurland and colleagues note that even if each of these had been rigorously identified as examples of HGT, the list is far too short to justify the conclusion that HGT is rampant and that these examples were culled from nearly as many genomes, which means that they represent a minute fraction of the many 1000s of proteins encoded by those genomes, so their impact on genome phylogeny would be correspondingly minute.

Kurland and colleagues point out that another systematic source of inflated HGT estimates has come from the uncritical use of BLAST (Basic Local Alignment Search Tool) to identify the most similar homologs in pairwise comparisons of eubacterial, metabacterial, and eukaryotic sequences. The most closely related pairs in the different domains are identified then as members of orthologous lineages. For example, in this protocol a protein from the genome of a eukaryote found to be most similar to a protein from a eubacterium is identified as a eubacterial gene that has been transferred horizontally, so in this way eukaryotic proteins could be classified as more closely related to bacterial homologs. It was found that eukaryotic operational genes (those involved in intermediary metabolism) often were related more closely to eubacterial homologs, whereas those with informational functions (transcription, translation, and related processes) (which transfer much more often than the former; the complexity hypothesis [based on the fact that informational processes are more complex] has been put forward by Jain, Rivera, and Lake [1999] to explain this) were identified more closely with metabacterial sequences, which gave rise to the fusion theory. In contrast, phylogeny by more demanding methods casts doubt on such simplistic interpretations of the best-match searches for alien sequences. BLAST does not distinguish the different phylogenetic anomalies from genuine HGT events, so they maintain that there are no indications in global phylogenetic reconstructions of these intermediary metabolism enzymes suggesting that they were transfered from eubacteria to eukaryotes.

Jin and colleagues (2007) make the same point--even advanced approaches, such as whole-genome sequencing, are sensitive to differential selection pressures, uneven evolutionary rates, and biased sampling, which can all give rise to false identification of HGT events.

They also point out that views as to the extent of HGT in bacteria vary between 2 extremes--researchers who believe that HGT is so rampant that a prokaryotic evolutionary tree is useless and those who believe HGT is of litle consequence and does not affect phylogeny reconstruction--and that there are supporting arguments and evidence for these 2 views, e.g., Welch et al from 2002 for the former view, and the well supported phylogeny reconstructed by Lerat et al from 2003 from about 100 "core" genes in gamma-proteobacteria for the latter.

If the former is the case, a convenience classification is necessary (which sometimes partly coincides with phylogeny) and would be advisable at any rate and would be basically the taxonomy by Gibbons and Murray (1978), that is, Gracilicutes (Scotobacteria and Photobacteria), Firmicutes (Endospora and Actinobacteria), Tenericutes (Mollicutes), and Mendosicutes.

In any case, what is particularly noticeable is that, across the 6 analyses, all clades are unstable excepting Deuterobacteria or its equivalent, Firmicutes, Gracilicutes (with or without Gram positives nested it), and Metabacteria. But taking into account molecular evidence what is most likely is the domains Eubacteria and Metabacteria and their monophyly, as well as the monophyly of Togabacteria, Paratoga, Firmicutes, and Gracilicutes (with or without Mollifirmicutes nested in it) in that order, and the paraphyly of Gram positives.

Chapter 6 Eukaryotes: TNT Analysis

Les protistologues s'accordent, aujourd'hui, à considérer les Flagelles autotrophes, comme les plus primitifs des Protozoaires à noyau vrai, des Eucaryotes (ensemble qui embrasse aussi les Végétaux et les Métazoaires), parce qu'ils sont les seuls à pouvoir faire la synthèse totale de leur protoplasme à partir du milieu minéral. Les organismes hétérotrophes sont donc subordonnés à leur existence, ainsi que celle des Procaryotes chimiotrophes et autotrophes (Bactéries nitrifiantes et sulfureuses, Cyanophycés)." - Edouard Chatton, 1937, p. 50, in "History of the Terms Prokaryote and Eukaryote' -RedOrbit.Com (first account of the distinction between prokaryotes and eukaryotes).

Materials and Methods

The data set contained 27 taxons and 260 characteristics, 20 branched (complex), in 283 columns excluding 14 deactivated characteristics and was divided into 4 sections, chloroplasts (29), morphology (139), chemistry (34), and physiology (58), excluding the addenda section. The computer calculations were done by Dr. James Carpenter in 2007 using the TNT (Tree analysis using New Technology) computer program (Goloboff, 1999; Nixon, 1999). All 4 new technologies were used (ratchet, tree drifting, fusing, sectorial search) and TBR (tree bisection and reconnection) branch swapping was employed, random seed was set at 1, constraints were off, and there were 208 mln. rearrangements. *Cyanidioschyzon* was used as the root.

Some groups included should be specified as to how they are circumscribed and are as follows:

Glaucobiota: *Glaucocystis, Cyanophora, Gleochete.* (*Glaucosphera*, formerly included, has GB (Golgi body) association with the nucleus, like *Dixoniella* and *Rhodella*, the 3 forming a clade in genotypic classifications, and unlike true glaucophyceans, which have the GB associated with the basal body, and also unlike true glaucophyceans has no cyanelles and no flagella).

Plasmodiophorae: includes *Phagomyxa,* etc.

Spongomonada: includes only *Spongomonas* and *Rhipidiodendron*; excludes *Pseudodendromonas* and *Cyathobodo,* which go to Bicosecales in Heterokonta.

Plantae: green algae as well as higher plants.

Cryptomonada: katablepharids as well as the usual forms.

Heterokonta: hydromyxians, opalinates-proteromonads, mycelial algae (pseudofungi), labyrinthulans actinophryds-*Ciliophrys*, and bicosecaleans (including pseudodendromonads), as well as the photosynthetic forms.

Cercomonada: Cercomonadidae and Heteromitidae.

Animalia: Choanozoa (choanoflagellates), *Spheroforma, Capsospora, Ministeria,* Myxozoa (in celenterates), Mesozoa, as well as the usual Metazoa.

Fungi: corallochytrids, ichtyosporeans, chytrids, microsporidians, haplosporidians, which are also diplokaryotic, and the probably related ascetosporans, as well as the usual funguses.

Dinoflagellata: probably includes *Colponema*, Ebrida, and Ellobiopsida.

Pelomyxida: Pelomyxidae (*Pelomyxa, Mastigina*), Mastigamebae (*Mastigameba, Mastigella*), and Entamebidae (*Entameba, Endolimax*).

Eulobosa: the testate groups Trichosida, Cochliopodida, Arcellinida, and Phryganellida, as well as the usual Gymnamebia (Gymnolobosa, nom nov.).

Polymastigota: Metamonada and Parabasalia.

Heterolobosa: Acrasiales (cellular slime molds) and Schizopyrenida (Vahlkampfidae, Percolomonadidae).
Euglenaria: Euglenoida, Diplonema, Kinetoplastida, Pseudociliata, and Hemimastigota.

Information for the data set and matrix came largely from Heath (1980), Parker (1982), Raikov (1982), Lipscomb (1989, 1991), Margulis et al (1990), and Lee et al (1985, 2000).

Results

There were 448 steps, and 13 most parsimonious trees (mpts) resulted. The CI was .58 and the RI .45. There were c. 208 mln. rearrangements. The majority consensus is in Table 6-1 and Figure 5 (the latter in Appendices) and the MRT is in Table 6-2. The confidence tree is in Table 6-3. Implied weighting was also done, with c. 208 mln. rearrangements, and produced only 1 MPT (Table 6-4, Figure 6, the latter in Appendices). The synapomorphies are in Table 6-5.

Table 6-1. Majority Consensus Tree (set at >50%) for the First Eukaryotic Analysis (in parentheses are the number of MPTs the taxon appears in; there are 14 kingdoms, which are underlined).

infemp. *Cyanidioschyzon* stat. nov.
infemp. Metakaryota tax. nov. (13)
  pvemp. *Cyanidium+Galdieria* tax. nov. (13)
  pvemp. Neokaryota (13) tax. nov.
    mcemp. Glaucobiota nom. nov. (syn. Glaucophyceae Bohlin 1901, Glaucophyta Skuja 1954, Glaucocystophyta Kies & Kremer 1986)
    mcemp. Cellulosa (13) nom. nov.
      nnemp. Rhodobiota Jeffrey 1971 stat. nov.
      nnemp. Anakaryota tax. nov. (13)
        pcemp. Plasmodiophorae G.M. Smith 1955 (syn. Phytomyxea Cavalier-Smith 1993) stat. nov.
        pcemp. Metanakaryota tax. nov. (13)
          sptrk. Spongomonada stat. nov. (syn. Spongomonadida D.J. Hibberd 1983)
          sptrk. Neanakaryota tax. nov. (13)
            trk. Sublobosa tax. nov. (13)
              ggk. Subdinaria tax. nov. (11)
                mgk. Subretaria tax. nov. (8)
                  grdk. Histonia-Chromista nom. nov., stat. nov. (13)
                    hpk. Histonia Haeckel 1904, emend., stat. nov.
                      spk. Plantae Treviranus 1822 (syn. Vegetabilia Linneus 1753, Isokontae Blackman and Tansley 1902, Chlorophyta Pascher 1914, Euchlorophyta Whittaker 1959, Chlorobiota Jeffrey 1971, Phyta Bold et al, 1987)
                      spk. Opisthokonta (Copeland 1956) Cavalier-Smith 1992 (13)
                        kgdm. Fungi Linneus 1753 (syn. Mycobiota Jeffrey 1971, Eumycota Möhn 1984, Mychota Starobogatoff 1986, Myceteae Bold et al

1987)

kgdm. Animalia Linneus 1753 (syn. Zoodeae Walton 1930, Zoobiota Jeffrey 1971)

hpk. Chromista Cavalier-Smith, 1981 (12)

sbk. Chlorarachnia Cavalier-Smith, 1993 (syn. Chlorarachniophyceae Hibberd and Norris, 1984, Chlorarachniophyta auct.)

sbk. Euchromista Cavalier-Smith, 1993 (12)

infk. Haptomonada Cavalier-Smith, 1989 (syn. Haptophyta Christiansen 1962, Prymnesiophyta Casper 1972 ex Hibberd 1976)

infk. Neochromista tax. nov. (13)

phyl. Cryptomonada Ehrenberg 1838 (syn. Cryptophyta Pascher 1914, Cryptophyceae Thuret in Le Jolis 1863 ex (Pascher) Fritsch in West 1927)

phyl. Heterokonta Copeland 1956, emend. Cav-Sm. 1986

grdk. Cercomyxa tax. nov. (13)

phyl. Cercomonada Poche 1913 (as order)

phyl. Myxobiota Edwards 1976 emend., orthog. emend., Mycetozoa de Bary 1859, Myxomycetes Sachs 1874, Myxomycota, Zopf 1885)

mgk. Retaria Cavalier-Smith, 1999 (13)

phyl. Foraminifera d'Orbigny 1826 (syn. Granuloreticulosa De Saedeleer 1934)

phyl. Actinopoda Calkins 1902

ggk. Dinaria tax. nov. (13)

spph. Apicomplexa Levine 1970 (Sporozoa Leuckart 1879)

spph. Excavata Cavalier-Smith, 2002 (syn. Discicristata (Cavalier-Smith, 1998) Cavalier-Smith, 2002) (8)

phyl. Protoexcavata tax. nov. (8)

Jakobida Cavalier-Smith 1993

Euglenaria nom. nov.

phyl. Neoexcavata tax. nov. (13)

Heterolobosa Page and Blanton 1985

Polymastigota Blochman 1895

spph. Dinociliata nom. nov., stat. nov. (11)

phyl. Dinoflagellata Butschli 1885 (syn. Pyrrhophyta Pasher 1914, Dinophyceae Fritsch 1935, Pyrrhophycophyta Papenfuss 1946, Dinomastigota, Dinophyta, Dinozoa Cav.-Sm. 1981)

phyl. Ciliophora Doflein 1901 (syn. Ciliata Perty 1852, Infusoria Doflein 1901, Heterokaryota Hickson 1903)

trk. Lobosa Carpenter, 1861 stat. nov. (13)

phyl. Pelomyxida Schulze, 1877 (as fam.) nom. nov., stat. nov. (syn. Pelobiotida Page 1976, Archamebae Cavalier-Smith 1983)

phyl. Eulobosa nom. nov.

Table 6-2. MRT for the 1st Eukaryotic Analysis (in parentheses are the number of MPTs the taxon appears in; there are the same 14 kingdoms as above).

- Cyanidioschyza
- Metakaryota (13)
  - Cyanidiobiota (13)
  - Neokaryota (13)
    - Glaucobiota
    - Cellulosa (13)
      - Rhodobiota
      - Anakaryota (13)
        - Plasmodiophorae
        - Metanakaryota (13)
          - Spongomonada
          - Neanakaryota (13)
            - Sublobosa (13)
              - Subdinaria (11)
                - Subretaria (8)
                  - Histonia-Chromista (13)
                    - Histonia (13)
                      - Plantae
                      - Fungi+Animalia (13)
                    - Chromista (12)
                      - Chlorarachnia
                      - Euchromista (12)
                        - Haptomonada
                        - Neochromista (13)
                          - Cryptomonada
                          - Heterokonta
                  - Cercomyxa (Cercomonada+Myxobiota) (13)
                - Foraminifera+Axopoda (13)
              - Dinaria (13)
                - Protodinaria (6)
                  - Apicomplexa
                  - Excavata (8)
                    - Euglenaria+Jakobida (8)
                    - Heterolobosa+Polymastigota (13)
                - Dinociliata (11)
                  - Dinoflagellata
                  - Ciliata
            - Lobosa (13)
              - Pelomyxida
              - Eulobosa (13)

Five groups are based on homoplasies, Metanakaryota, Sublobosa, Subdinaria, Subretaria, and Protodinaria.

Table 6-3. Symmetrical resampling (for 1000 replicates).

Cyanidioschyzon
Metakaryota (100)
    Galdieria+Cyanidium (92)
    Neokaryota (78)
        Glaucobiota
        Cellulosa (59)
            Rhodobiota
            Anakaryota (58)
                Plasmodiophorae
                Metanakaryota (4)
                    Haptomonada
                    Chlorarachnia
                    Spongomonada
                    Apicomplexa
                    Euglenaria
                    Plantae
                    Animalia+Fungi (75)
                    Heterokonta+Cryptomonada (43)
                    Myxobiota+Cercomonada (42)
                    Eulobosa+Pelomyxida (43)
                    Ciliata+Dinoflagellata (19)
                    Axopoda+Foraminifera (79)
                    Metexcavata (5)
                        Jakobida
                        Polymastigota+Heterolobosa (28)

Metakaryota	100
Cyanidium-Galdieria	92
Retaria	79
Neokaryota	78
Opisthokonta	75
Cellulosa	59
Anakaryota	58
Neochromista	43
Eulobosa-Pelomyxida	43
Cercomyxa	42
Neoexcavata	28
Dinociliata	19
Metexcavata	5

So there is strong support for the top 5, moderate support for Cellulosa and Anakaryota, and weak support for the other clades. Unsupported groups are excluded which means there is no confidence for Chromista, Histonia, nor Alveolata.

Table 6-4. Eukaryota-TNT Analysis-Implied Weighting Tree (with symmetrical resampling frequencies for those above 50; 14 kingdoms)

- Cyanidioschyza
- Metakaryota
  - Cyanidium+Galdieria
  - Neokaryota
    - Glaucobiota
    - Cellulosa
      - Rhodobiota
      - Anakaryota
        - Plasmodiophorae
        - Metanakaryota
          - Pelomyxida+Eulobosa
          - Neanakaryota
            - Spongomonada
            - Cenanakaryota
              - Dinaria
                - Dinoflagellata+Ciliata
                - Excavata
                  - Euglenaria-Apicomplexa
                  - Jakobida+(Heterolobosa+Polymastigota)
              - Epidinaria
                - Retaria
                  - Foraminifera+Axopoda
                - Epiretaria
                  - Cercomyxa (Cercomonada+Myxobiota)
                  - Histonia-Chromista
                    - Histonia
                      - Plantae
                      - Fungi+Animalia
                    - Chromista
                      - Chlorarachnia
                      - Euchromista
                        - Haptomonada
                        - Neochromista
                          - Cryptomonada
                          - Heterokonta

Table 6-5. Synapomorphies for the Eukaryotic Clades.

Metakaryota - larger genome, larger chloroplast DNA size, sporulation
Cyanidiobiota - intermediate genome, intermediate chloroplast DNA size
Neokaryota - large genome, large chloroplast DNA size, multiple mitochondria, open mitosis, flagella, MLS, cruciate rootlets
Cellulosa - cellulose, pyrenoids, chloroplast DNA arrangement as scattered nodules, multiple

chloroplasts
Anakaryota - gametic meiosis, tubular cristas
Neanakaryota - stigmas
Histonia - gibberelins, ergolines, and the MSD myosin subfamily
Chromista - chloroplast in ER (endoplasmic reticulum), nucleomorphs
Euchromista - PR (periplastidial reticulum), chloroplast in SER (smooth ER), silica in walls, transitional helix (including helical band in haptophyceans)
Neochrista - epsilon-carotene, mastigonemes formed in nucleaer envelope and ER, tripartite mastigonemes
Cercomyxa - ppks, striated fiber emanates from anteriorly directed flagellum extending as cone and terminating in MTOC, cercomyxan semicircle complex
Retaria - pheodarian-type pseudopods
Dinaria - articulins, mastigoneme rows 1 and 0, kinetochore location on nuclear envelope, permanently condensed chromosomes
Excavata - discoid cristas, linked mts underlie entire cell membrane, feeding groove (ventral groove used for suspension feeding), composite fiber, I fiber, B fiber, C fiber
Neoexcavata - polymastigote rootlets, singlet rootlet between right rootlet and BB runs along floor of groove
Dinociliata - dinociliate thecal vesicles
Opisthokonta - TZ constriction and striation, opisthokontic histones, ophiobolins, tryptophan pathway with nicotinic acid

The Lipscomb results are presented below.

Table 6-5. Diana Lipscomb's Classification (1989, 1991)-8 kingdoms.

Rhodophyceae
Supergroup 1 (Contophora)
  Plantae
  Supergroup 2 (Anisokonta)
    Cryptomonada
    Supergroup 3
      Supergroup 4 (Amebae+Oomycota+Chromobiota+Heliozoa)
      Supergroup 5
        Polymastigota
        Animalia+(Choanozoa+Fungi)
        Opalinida+Dinoflagellata-Ciliata+Euglenaria

Hennig86 Wagner parsimony analysis with branch swapping was used on 86 genera and 137 phenotypic features resulting in 7 optimal trees of 475 steps with a CI of .52; Nelson consensus was applied; Supergroup 4 includes heterokonts, haptomonads, rhizopods, forams, heliozoans, *Cercomonas*, *Cyanophora,* and *Colponema*; Hennig86 does not do confidence measures.

Chapter 7 Eukaryotes: PAUP Analysis

The original nucleate organism must have been unicellular rather than multicellular, and autotrophic rather than dependent. - H.F. Copeland, 1938, p. 396.

Methods and Materials

In the 2nd analysis, Paup 4-Beta 10 (Swofford, 2003) was employed. Fitch parsimony, TBR, random addition, were in effect, and no a priori weighting nor topological constraints were used. The hypothetical ancestor was used as the outgroup based on the ancestral condition, which is marine, unicellular, coccoid, autotrophic but capable of heterotrophy, thermophilic, nonsporulating, and nonsexual, with a 2-membrane chloroplast envelope, chlorophyll a only, a proteinaceous wall, lamellar cristas, glycogen starch, closed mitosis, phycobilins, and GB-ER association, and still *Cyanidioschyzon* turns up as the stem group.

Rhodobiota was split into 4, Actinopoda into 5, 6 taxons were added, and the hypothetical ancestor was the root, making a total of 41. 18 characteristics were added, making a total of 320. Genome size and chloroplast DNA size were recoded as small and large only. The data set, retaining the TNT format, which starts at 0 instead of 1, and the data matrix are presented in Appendices. Paup has no synapomorphy plot or list.

Results

There were 3 MPTs, 4.3 mln rearrangements, 1050 steps, and the CI was .40 and the RI .44. The 50% majority consensus is presented in Table 7-1 and Figure 7 (the latter in Appendices). With Wagner the results and number of rearrangements were the same but there were 4 optimal trees, 1179 steps, and the CI was .36 and the RI was .43. The confidence tree is in Table 7-2, and the synapomorphies are in Table 7-3.

Table 7-1. 50% Majority Consensus for 2nd Eukaryotic Analysis (all clades are in all 3 trees except for Subfilosa which was in 2; there are 16 kingdoms which are underlined).

Hypothetical Ancestor
Core Eukaryotes new taxon
- *Cyanidioschyzon*
- Metakaryota
  - *Cyanidium+Galdieria*
  - Neokaryota
    - Glaucobiota
    - Cellulosa
      - Rhodobiota
        - Stylonemata nom. nov.
        - Metarhodobiota nom. nov.
          - Rhodellae nom. nov.
          - Neorhodobiota tax. nov.
            - Porphyridiales
            - Eurhodobiota

- Anakaryota
  - <u>Plasmodiophorae</u>
  - Neanakaryota
    - Subcerci tax. nov.
      - Subparachromista tax. nov.
        - Subdinochromista tax. nov.
          - <u>Spongomonada</u>
          - Histonia-Retaria tax. nov.
            - Histonia
              - <u>Plantae</u>
              - Opisthokonta
                - <u>Animalia</u>
                - <u>Fungi</u>
            - <u>Retaria</u>
              - Foraminifera
              - Actinopoda
                - Heliocelest. tax. nov
                  - Desmo-Cntrhl.
                  - Celestina
                - Pheod. Haeckel 1879
        - <u>Dino-Chromista</u> tax. nov.
          - Chlorarachnia
          - Euchromista tax. nov.
            - Cryptomonada
            - Metachromista tax. nov.
              - Haptomonada
              - Hetk.-Dnflag. tax.nov.
      - <u>Parachromista</u>
        - Myxobiota
        - Pelomyxida
    - <u>Cercobiota</u> nom. nov.
      - Cercomonada
      - Metacercobiota nom. nov.
        - Apusomonadida Karpov and Mylnikov 1989
        - Thaumatomonadida Shirkina 1987
    - <u>Subfilosa</u> tax. nov.
      - Eusubfilosa tax. nov.
        - Ciliophora-Euglenaria new taxon
      - Metasubfilosa tax. nov.
        - Jakomastigota tax. nov.
          - Jakobida-Polymastigota
          - Heterolobosa
        - Apicomplexa
    - <u>Filosa</u> Leidy 1879 orthog. emend.
      - Gymnofilosa nom. nov. (syn. Aconchulinida De Saedeleer 1934)
        - Nuclearida Cavalier-Smith 1993 (syn.

Cristidiscoidia Cavalier-Smith 1993)
Vampyrellida Starobogatov ex Krylov et al 1980 (syn. Cristivesiculatia Page 1987)
Testafilosa De Saedeleer 1934 orthog. emend.
Gromida Claparède and Lachmann 1858
Euglyphida Copeland 1956

<u>Taxolobosa</u>
Taxopoda Fol, 1883
Eulobosa

The following are the confidence estimates under Fitch parsimony:

	bootstrap (3.9 mln. rearrangements)	jackknife (2.4 mln. rearrangements)
Heliocelestina	100	100
Rhodobiota	92	82
Desmocentra	91	83
Jakomastigota	86	87
Cyanidiobiota	74	75
Metacercobiota	73	69
Actinopoda	70	68
Taxopoda-Eulobosa	65	60
Retaria	63	60
Opisthokonta	56	62
Neokaryota	57	52
Metakaryota	59	-
Testafilosa	-	52

With Wagner the estimates for Opisthokonta were below 50.

Table 7-2. Bootstrap-Jackknife Tree (50% majority, 100 replicates) for 2nd Eukaryotic Analysis.

Hypothetical Ancestor
Cyanidioschyzon
Metakaryota 59/-
Cyanidium+*Galdieria* 74/75
Neokaryota 57/52
Glaucobiota
Rhodobiota 92/82
Stylonematales
Rhodellae
Porphyridiales
Eurhodobiota
Plasmodiophorae

Spongomonada
Plantae
Chlorarachnia
Cryptomonada
Haptomonada
Heterokonta
Cercomonada
Myxobiota
Metacercobiota 73/69
    Apusomonadida
    Thaumatomonadida
Opisthokonta 56/62
    Animalia
    Fungi
Retaria 63/60
    Foraminifera
    Actinopoda 70/68
        Heliocelestina 100 /100
            Desmothoracida Hertwig and Lesser 1874-Centrohelida Hartmann 1913 91/83
            Celestina
        Pheodaria
Taxopoda-Eulobosa 65/60
Dinoflagellata
Ciliophora
Apicomplexa
Pelomyxida
Nuclearida
Vampyrellida
Testafilosa -/52
    Gromida
    Euglyphida
Jakomastigota 86/87
    Jakobida
    Polymastigota
Heterolobosa
Euglenaria

With *Cyanidioschyzon* as the root and using Fitch the only major differences were a Plasmodiophorae-Parachromista clade, with low confidence, DinoChromista and Subfilosa switched places, and there was dichotomy throughout. There were 8 mln. rearrangements and 7 optimal trees. With Wagner there was the same topology, 10 mln. rearrangements, and 9 optimal trees. In both these runs there was no Metakaryota clade and the Taxopoda-Eulobosa clade was in Neokaryota, the only major difference in confidence values was higher confidence for Neokaryota (74/77), and the RIs were the same, but the CIs were higher by 1 point.

Discussion

Cyanidioschyzon merolae has the fewest advanced traits with 12 and highest in percentage of primitive traits at 91 % (excluding inapplicable and missing data). It is supported as basal by Seckbach (1994) and Nagashima et al (1993), having the most primitive chloroplast, only 1 mitochondrion, no vacuoles, no trienoic acids, and the smallest eukaryotic genome at 8 Mbp. Prerhodophyceae (after Seckbach (1987)), includes all 3 genera. *Cyanidium* is considered as a bridge alga between Cyanobacteria and Rhodophyceae by Klein (1970), Frederick (1976), and Seckbach (loc. cit.). It also lacks vacuoles and trienoic acids, and has only 1 chloroplast and 1 mitochondrion. *Galdieria* has only 1 chloroplast but numerous mitochondria, a vacuole, and trienoic acids. For the root, Lipscomb (loc. cit.) used red algae but these have fewer primitive features than some other groups like Parabasalia, Pelomyxidae, and Glaucophyceae, the first 2 sometimes considered as sister group to all other eukaryotes. *Cyanidium+Galdieria* form the sister group to red algae in Saunders and Hommersand (2004) based on the GB (Golgi Body, aka dictyosome) association with the ER and presence of peripheral thylakoids, and a chloroplast dividing ring occurs in *Cyanidioschyzon* and *Cyanidium*, the 1st and 3rd characteristics not included in the 1st analysis (but included in the 2nd) and probably primitive. Peripheral thylakoids occur also in *Glaucosphera* and higher algae, and *Cyanidium*, *Galdieria*, and Rhodophyceae have endospores, which might be apomorphic, but the monophyly of Prerhodophyceae and Rhodophyceae sensu lato is further contradicted by the analysis.

The other groups that would be possible candidates for the root are Parabasalia or Polymastigota and *Pelomyxa*, Pelomyxidae or Pelomyxida, but their seemingly primitive traits are reductions or losses due to their parasitic or otherwise symbiotic lifestyle and have affinities to other taxons. There are 4 apomorphies linking *Pelomyxa* to *Mastigina*, *Mastigella*, or *Mastigameba*: nuclei surrounded by apposed bacteria within vacuolar membranes linked to the nuclear membrane by vesicular membranes, densly vesiculated cytoplasm, bacteria apposed to the plasmalemma (Griffin, 1988), and fountain flow (Walker et al, 2001). If Pelomyxidae was separated as a terminal taxon, as it would need to be since Mastigamebidae has open mitosis, or if these traits were included as autopomorphies, then there would be 5 extra apomorphies making a total of 33. Even if we were to consider some chracteristics as plesiomorphic instead of apomorphic, and there would be 6 of these (single basal body, mt cone, radiating mts, intranuclear spindle, lobose pseudopods, and eruptive motion), then it would still have 23 derived traits, almost double *Cyanidioschyzon*, making 88% primitive traits, lower than *Cyanidioschyzon*. As well, most molecular phylogenies support Pelomyxida as derived rather than basal as does some of the morphology (pseudoflagella suggesting reduction, ED (electron dense) material, or cryptons suggesting mitochondrial derivatives). And *Mastigameba* and *Entameba* uniquely share neo-inositol polyphosphates, the 2 together also well supported by molecular evidence (Bapteste et al, 2002). Dinoflagellates, often touted as the basal eukaryotes, have a lack of histones or of typical histones, but this is a reversal and are now recognized as part of Alveolata, in any case. Their advanced position, along with that of Polymastigota and Euglenaria also occurs in Lipscomb's arrangement.

Below is a list of all the taxons with their totals of derived characteristics. The ones with the highest number of primitive traits are Plantae with 222 (and 38% advanced traits) and Heterokonta with 217 (and 41% advanced traits, the highest). The taxon with the fewest number of primitive traits is *Cyanidioschyzon* with 127.

Table 7-3. Number of Derived Traits.

Heterokonta	151
Plantae	135
Euglenaria	99

Dinoflagellata	94
Haptomonada	74
Cryptomonada	71
Fungi	71
Rhodobiota	68
Actinopoda	65
Ciliophora	63
Myxobiota	61
Foraminifera	54
Animalia	51
Polymastigota	50
Eulobosa	47
Thaumatomonada	40
Chlorarachnia	39
Heterolobosa	35
Glaucobiota	34
Jakobida	32
Apusomonada	30
Pelomyxida	30
Cercomonada	30
Spongomonada	28
Plasmodiophorae	27
Euglyphida	26
Gromida	22
Cyanidium	21
Nuclearida	20
Vampyrellida	20
Galieria	17
Cyanidioschyzon	12

Red algae as basal is supported by molecular evidence (Hori and Osawa, 1987; Hori et al, 1990; Luttke, 1991; Nozaki et al, 2007) and Archeoplastida as polyphyletic is also supported by molecular evidence (Hori and Osawa, 1987; Hori et al, 1990; Luttke, 1991; Olsen, 1994; Bhattacharya, 1995; Nozaki et al, 2007; Yoon et al, 2008; Kim and Graham, 2008; Tekle et al, 2008; Parfrey et al, 2010; etc.) as well as previous classical evidence (Lipscomb, 1985, 1989, 1991). Red algae as the most primitive eukaryotes is agreed to also by Starobogatov (who, as well, separated *Cyanidium* as phylum Cyanidiophyta, and placed Glaucophyceae close to Cryptophyceae and Centrohelida, a radiolarian group), Vada, Taylor, Cavalier-Smith in 1978, Edwards, Leedale, Takhtajan, Jeffrey, Pascher (1931), Chadefaud (1960), Copeland in 1947 and 1956, Margulis (1974), and Möhn (1984). In treatments of algae the red kind is on top in 11 of 14 authors from 1935-80 included in the appendix in Darley (1982). Van Den Hoek and colleagues (1995) have glaucophytes as the earliest eukaryotes, with red algae next.

Table 7-4. Summary of 14 molecular analyses regarding typical red algae and prerhodophyceans (the probabilities and bootstrap values, respectively, except where otherwise stated, are in parentheses, and for Rhodobiota s.l. are given beside the citation; all are part of larger analyses).

Ciniglia, et al (2004) Rhodobiota s.l. >95% Baysian posterior probablity, ME bootstrap 100, ML bootstrap 99, Cyanidiales (all 3 genera) >95%/100/99

Stiller and Harrell (2005) *Cyanidioschyzon*+Rhodobiota s.s. 100/100/100 (Bayesian probability, ML bootstrap, NJ bootstrap, respectively, using JTT)

Moreira et al (2006) (using both LSU and SSU rDNA and the GTR model of evolution).

Cyanidioschyzon-Thorea (.89/74)

Rodriguez-Ezpeleta et al (2006) (ML tree using 143 genes and the WAG model)

Rhodobiota s.l. (1/100)

Galdieria

Rhodobiota s.s. (1/100) *Gracilaria+Porphyra*

Burki et al (2007) (using the RtREV and WAG models). Rhodobiota s. l. (1/100)

Porphyra

Cyanidioschyzon-Galdieria (1/90)

Hackett et al (2007) (ML classification based on 16 proteins and using the WAG model).

Rhodobiota s. l. (1/100)

Porphyra

Galdieria-Cyanidioschyzon (1/94)

Nozaki et al (2007)(1st figure=probability, 2nd figure=MP bootstrap, 3rd=ML bootstrap with WAG, 4th=ML bootstrap) Rhodobiota s. l. 1/100/100(100)

Cyanidioschyzon+Galdieria .99/-/71(-)

Porphyra

Kim and Graham (2008) (ML using WAG model).

EEF 2 alignment

a) Rhodobiota s.l. (.52/61/92) (the bootstrap values are RAxML and FastME in that order)

Cyanidioschyzon

Rhodobiota s.s. (.81/69/71)

Porphyra

Botryocladia-Bonnemaisonia (1/100/100)

concatenated protein data (EEF-2, actin, cytosolic HSP 70, cytosolic HSP 90, alpha- and beta-tubulin)

b) Rhodobiota s.l. (1/100)

Cyanidioschyzon

Porphyra

Yoon et al (2008) (ML analysis; model not specified). Rhodobiota s. l. (90 or more/66)

Galdieria

Rhodobiota (90 or more/52)

Porphyra

Cyanidium+Cyanidioschyzon (90 or more/100)

Burki et al (2009)(Bayesian inference under CAT model). Rhodobiota s. l. (1/100)

Porphyra-Gracilaria (1/100)

Galdieria

Okamoto et al (2009) (ML based on HSP 90 under RaxML and RTREV model). Rhodobiota s. l. (1/84)

Cyanidioschyzon

Rhodobiota s.s. (.84/89)

Hemiselmis (nucleomorph)+*Guillardia* (nucleomorph)(1/over 85)

Griffithsia

Minge et al (2009)(78 concatenated genes; ML under rtREV model; BI under CAT model)
Porphyra+Cyanidioschyzon (1/100)
Brown et al (2012)(159 proteins, CAT-Poisson model). Rhodobiota s. l. (1/100)
Chondrus+Porphyra (1/100)
Cyanidioschyzon
Burki et al (2012)(CAT model). Rhodobiota s. l. 1/100
Galdieria+Cyanidioschyzon (1/100)
Rhodobiota s.s. (1/100)

Judging from these Rhodobiota sensu lato are well supported, however, the sampling is limited in most. Red algae most likely evolved their plastids from Cyanobacteria which has chlorophyll a only and phycobilins, while plant plastids probably evolved from Chloroxybacteria which has chlorophyll a and b and no phycobilins. There were at least 4 secondary plastid symbiogeneses: chlorophyll c in chromistans (from a red alga), chlorophyll b in chlorarachnians (from a green alga), chlorophyll b in euglenoids (also from a green alga), and chlorophyll c in dinoflagellates (from a heterokont), hence the 4 fucoxanthins found also in these, and the symbiote is still recognizable.

Red algae as basal eukaryotes is supported in Lipscomb (1989, 1991) and by molecular evidence (Hori and Osawa, 1987; Hori et al, 1990; Luttke, 1991; Nozaki et al, 2007) and regarded as such also in the taxonomies of Jeffrey (1971), Edwards (1976), Taylor (1978), Cavalier-Smith (1978), Parker (1982), Möhn (1984), and Starobogatov (1986)(who, as well, separated *Cyanidium* as phylum Cyanidiophyta).

The generally-considered pleophyletic Porphyridiales in Rhodobiota contain Stylonemataceae (Goniotrichaceae)(several genera), Rhodellaceae (3 genera)), and Porphyridiaceae (3 genera), designated as Porphyridiales 2, 1, and 3, respectively. The first and second of these might be considered as separate from Rhodobiota and would be kingdoms. Stylonemataceae lack sexuality and pit plugs and have the GB association with the ER, and, along with Rhodellaceae, which lacks sexuality, multicellularity, and pit plugs, might not share any apomorphies with the rest of Eurhodobiota. Both families are putatively the most primitive. *Porphyridium* the GB associated with the mitochondrion, a common feature in higher red algae, and sulfated polysaccharides similar to carrageenan (Gabrielson et al, 1990), which also occurs in several higher orders (Bangiales, Gelidiales, Palmariales, and Gigartinales). I tested the monophyly of Rhodobiota s.s. by adding the 3 families and 2 extra features and the clade is strongly supported at 92 and 82 in bootstrap and jackknife estimates, respectively, and is in all MPTs.

Some or all of the genera in the 3 families are sometimes considered as being reduced from multicellular red algae (Gabrielson et al, 1990; van den Hoek et al, 1995). In the molecular classifications of Saunders and Bailey (1997) Stylonemataceae are basal or on top with low confidence and in Yoon et al (2006) they are basal with high confidence or sister group to Compsogonales with low confidence. Rhodellaceae appear as the sister group to Florideophyceae in Saunders and Bailey (1997) with high confidence and as sister group to Florideophyceae+Bangiales in Yoon et al (2006) with low or moderate confidence. In Yoon et al (2006) Porphyridiaceae are either basal with high confidence or sister group to Rhodellaceae + (Florideophyceae+Bangiales) with low confidence.

Pueschel remarked in 1990, "Being an ancient lineage, the red algae have undergone a broad range of modifications in cellular organization. Even the spectrum of morphological possibilities, from unicellular forms . . . to complex . . . parenchymatous thalli, fails to convey the degree of cellular diversity." (Saunders and Hommersand, 2004). As noted in Saunders and Hommersand, the morphological and fine structural diversity of red algae is as striking as their genetic variation revealed in molecular studies. There are at least 3 patterns of Golgi association, 3 methods by which cells achieve multinuclearity in development, several methods of establishing intercellular connections through cellular fusions and pit-plug formation, 5 distinct patterns of mitosis, 3 patterns of cytokinesis, and 3 patterns by which the various reproductive structures are formed.

Van Den Hoek et al (1995) remark that a comparison of evolutionary distances based on 5S rRNA (Hori and Osawa, 1987) between genera within various groups show that in red algae they are higher overall than those between genera of other algae, higher plants, and animals, which attests to rhodobiotes' great antiquity and that this agrees with the fossil record. Red algae have 22-56, green algae <.02-45, and brown algae 1-5. Mammals have 0.

The fossil record corroborates red algae as the ancestral eukaryotes (Table 7-5) since the oldest undisputed eukaryotic fossil identified as a member of a modern group is *Bangiomorpha* Butterfield 2000 (Rhodobiota s.s.) from 1.2 bya in the Middle Proterobiotic found in the Hunting Formation on Somerset Island in Nunavut, Canada (Javaux, 2007). There are no fossils of *Cyanidioschyzon*, nor of the other 2 Prerhodophyceae genera, but if it is basal it would probably be about 2 bln. yrs. old, since the oldest eukaryote is an acritarch (*Valeria* Horsfield, 1829) at 1.8 bya (Javaux, 2007). Dinoflagellate chemical markers have been reported from several Proterobiotic and even Archeobiotic units as old as the 2.5-2.8-bln.-yr.-old Mount Bruce Supergroup, Pilbara Craton, Australia (Porter, 2006), but, given its age, the Archeobiotic occurrence is attributed to an independent (non-dinoflagellate) origin and the Proterobiotic occurrences have been interpreted as either possible contaminants or dinosteroid precursors that do not by themselves indicate dinoflagellates were present. Interestingly, the Paleobiotic record of dinosteroid abundance correlates well with that of acritarch diversity, suggesting that many acritarchs may represent dinoflagellate cysts. Many modern dinoflagellate cysts lack diagnostic characteristics, and would probably be grouped with the acritarchs if found as fossils. Several research papers have suggested certain Proterobiotic acritarchs might be dinoflagellate cysts, which showed that some Ediacaran acanthomorphic acritarchs have chemical and fine structural traits consistent with a dinoflagellate affinity, although this is contested by others. Porter adds that because the taxonomic distribution of these traits is not well documented, it is impossible to know whether their occurrence in both fossil and modern groups is due to homology or convergence, and, if due to homology, whether their occurrence reflects an apomorphic or plesiomorphic state. The earliest phagotrophs known are arcellinids (lobose testate amebas) dated at 742 mya (in the Neoproterobiotic), found in the Chuar Formation, Grand Canyon (Porter et al, 2003).

Table 7-5. Fossil Record for the Major Eukaryotic Groups

	apprx. mya	geological time	genus	reference
Pheodarea	140	Cretaceous		Porter
Dinoflagellata	220	Triassic		Moldowan
	420	Silurian	putative	Loeblich, Moldowan
	580	Late Proterozoic	biomarkers putative	Moldowan
Haptomonada	300	Carboniferous		Loeblich, Van Den Hoek
Ciliophora	300	Carboniferous		Porter
	500	Ordovician		Loeblich
Euglenoida	500	Ordovician		Porter
Polycistina	550	Late Proterozoic		Porter

Forams	580	Late Proterozoic		Loeblich, Porter
	650	Late Proterozoic	putative	Porter
Fungi	600	Late Proterozoic	chytrids	UCMP
Animalia	700	Late Proterozoic		Porter
Eulobosa	750	Late Proterozoic		Porter
Euglyphids	740-770	Late Proterozoic	*Melicerion* putative; putatively assigned to the hypothetical Cercozoa	Porter
Plantae	600	Late Proterozoic		Van Den Hoek
	750	Late Proterozoic	*Proterocladus*	Javaux
	1500	Middle Proterozoic	disputed	Emiliani
Heterokonta	1000	Middle Proterozoic	*Paleovaucheria*	Javaux
Rhodophyceae	1200	Middle Proterozoic	*Bangomorpha*	Javaux

Myxobiota date to the Eocene but have a poor fossil record (Porter, 2007). Green algae are sometimes said to be the oldest eukaryotes at 1.5 bln. yrs. (Emiliani, 1995), and Loeblich placed the origin at 900 or 1000 mya (at the beginning of the Late Proterozoic), but these are disputed and UCMP states fossils superficially resembling green algae date back to the Precambrian. MicrobeWorld.Org says they go back 500-600 mln. yrs. Prasinophycean phycomas are readily fossilizable and their distinctive morphologies have permitted their recognition in rocks as old as Precambrian age (Margulis et al, 1990). The calcified Dasycladales fossils are the 2nd oldest of the green algae, dating back to the Precambrian (Margulis et al, 1990). The oldest eukaryote is an acritarch (*Valeria*) at 1.8 bya (Javaux, 2007). Dinoflagellate biomarkers have been reported from several Proterozoic and even Archeozoic units as old as the 2.78-2.45 bln. yr. old Mount Bruce Supergroup, Pilbara Craton, Australia (Porter, 2006), but, given its age, the Archeozoic occurrence is attributed to an independent (non-dinoflagellate) origin and the Proterozoic occurrences have been interpreted as either possible contaminants or dinosteroid precursors that do not by themselves indicate dinoflagellates were present. Interestingly, the Paleozoic record of dinosteroid abundance correlates well with that of acritarch diversity, suggesting that many acritarchs may represent dinoflagellate cysts. Indeed, many modern dinoflagellate cysts lack diagnostic characteristics, and would probably be grouped with the acritarchs if found as fossils. Several articles have suggested certain Proterozoic acritarchs might be dinoflagellate cysts, which showed that some Ediacaran acanthomorphic acritarchs have chemical and fine structural traits consistent with a dinoflagellate affinity, although this is contested by others. Porter adds that because the taxonomic distribution of these traits is not well documented, it is impossible to know whether their occurrence in both fossil and modern groups is due to homology or convergence, and, if due to homology, whether their occurrence reflects an apomorphic or plesiomorphic state. There are no fossils of *Cyanidioschyzon* but if it is basal it would be about 2 bln. yrs. old. The fossil record has an abundance of autotrophs in the Precambrian and a paucity of heterotrophs in that same eon, which indicates autotrophs are the most primitive eukaryotes, contrary to the prevailing and probably erroneous opinion.

Porter says heterotrophs are necessarily the earliest eukaryotes but does not substantiate this extraordinary claim. Presumably she is refering to molecular phylogenies and therefore sees them as foolproof, which they are definitely not. She offers 2 possible explanations for the absence of heterotrophs in the Precambrian. One is that heterotrophic diversity may have been low due to limited primary productivity in Mesoproterozoic oceans. Evidence for this is primarily hypothetical, and empirical evidence is even more uncertain. The other is, according to her, more likely and is taphonomic (fossilization) bias due to algae having organic walls which make them more preservable, while these are supposedly rare in heterotrophs, and mineralized eukaryotes from the Precambrian are rare. But organic walls are not rare in funguses and Fungi is a very large group, so the argument is weak. Of course, the most probable explanation is that autotrophs are the earliest eukaryotes, and prerhodophyceans are the most primitive eukaryotes.

My phylogenies are mostly in accord with the fossil record as only 3 taxons, Heterokonta, Eulobosa, and Haptomonada, are not in agreement with it, the first 3 appearing earlier in the fossil record than indicated in the phylogenies, out of 10 (Retaria counted as 1). In the typical molecular taxonomies 6 of 10 are in disagreement with the fossil record: Rhodophyceae, Heterokonta, Plantae, Eulobosa, Animalia, and Fungi, the 1st 3 later than they should be and the 2nd 3 earlier than they should be.

Table 7-6. Taxons Indicated by Geological Time (indicated where known) and Trophic Mode; the start years are given for each period in mya).

		autotrophic	osmotrophic	phagotrophic
Cretaceous	66			
	140			Pheodaria
Jurassic	144			
Triassic	210			
				Filosa
	220	Dinoflagellata		
Permian	250			
Carboniferous	286			
	300	Haptomonada		Ciliophora
Devonian	360			
				Apicomplexa
Ordovician	440			
	500	Euglenoida		
				Polymastigota
				Jakobida
				Acrasia
				Schizopyrenida
Cambrian	505			
Late Proterobiotic	543			
	550			Polycistina
	580			Forams
	600		Fungi	
	700			Animalia
	750	Plantae		Eulobosa
				Pelomyxida
				Cercobiota
				Myxobiota
Middle Proterobiotic	900			
	1000	Heterokonta		
		Cryptomonada		
		Chlorarachnia		

				Spongomonada
			Plasmodiophorae	
	1200	Rhodobiota		
		Glaucobiota		
Early Proterobiotic	1800	acritarchs		
		Cyanidiobiota		
		Cyanidioschyzon		
			Metabacteria	
			Posibacteria	
	3600	Cyanobacteria		
			Togabacteria	

Moreover, the molecular evidence for Archeoplastida is very weak (see later). An evolutionary link between red algae and funguses had been proposed by Sachs (1874), Chadefaud (e.g., 1957), Cain (1972), and Demoulin (1974), among others, who pointed out similarities between the 2 groups: trichogynes, spermatia, perforated septa, and trehalose storage, and favoured fungal polyphyly. This has been contradicted by deBary (1881), Atkinson (1915), Linder (1940), and Savile (1968), among others, who favoured fungal monophyly, as they are cases of convergence rather than actual kinship, and no phylogenetic analyses (including here, which includes the 4 traits), either phenotypic or genotypic, have found such a link, and the monophyly of Fungi is well established.

Glaucophyceans are considerd a separate lineage by Kies and Kremer, Patterson, Graham and Wilcox, van den Hoek et al, and Lee (Phylum Glaucophyta-comenius.susque.edu) and are a separate lineage in Burki et al (2008), Tekle et al (2008), and Parfrey et al (2010). They are positioned among chromobiotes in Lipscomb (including only *Cyanophora*)(1989, 1991) and close to Cryptophyceae and Centrohelida by Starobogatov (1986). In genotypic classifcations the group is variously placed, usually with unconvincing support (Table 7-7).

Table 7-7. Position of Glaucophyceae in Genotypic Classifications.

Bhattacharya et al (1995) Glaucophyceae+Cryptomonada (59 (bootstrap value using distance analysis/68 (bootstrap value using weighted maximum-parsimony (MP) analysis))
Moreira et al (2006) Glaucophyceae+Cryptomonada (.79/62)(using both LSU and SSU rDNA)
Kim et al (2006) Glaucophyceae+ Cryptomonada (1/74)
Nikolaeav et al (2004) .67
    Cryptomonada
    (Glaucophyceae+Centrohelida) .68
Minge et al (2009) Glaucophyceae + (Cryptomonada+Haptomonada) (<.50/<50)
Yoon et al (2008) Glaucophyceae + (Cryptomonada+Plantae) weak support
Rodriguez-Ezpeleta et al (2006) Glaucophyceae+Plantae (1/54)
Burki et al (2007) Glaucophyceae+Plantae (1/70)
Hackett et al (2007) Glaucophyceae+Plantae (<1/<60)

Burki et al (2009) Glaucophyceae+Plantae (.85/86)
Hampl et al (2009) (Haptomonada+Rhodobiota)+Plantae)+
Glaucophyceae (1/100)
Kim and Graham (2008) *Cyanophora*+Polymastigota (<.50/<50)
Brown et al (2012) sister group to about half of eukaryotes
(1/52)
Burki et al (2012) with picobiliphyceans sister group to most eukaryotes
(1/61)

Retaria, combining actinopods and forams, which receive a high confidence value of nearly 80 in TNT, are partly corroborated in several molecular studies. Details on these and molecular classifications for Rhizaria are given in Table 7-8. The lower confidence values for Metakaryota, Retaria, Opisthokonta, and Cyanidiobiota in the PAUP analysis are probably due to the smaller number of replicates and this is because a higher number was not well manageable in PAUP.

Rhizaria are also not recovered, not surprisingly, as it is ill-defined and only moderately supported in molecular phylogenies (Palfrey et al, 2007), statistical support for it is inconsistent in multigene genealogies with larger taxon sampling (Yoon et al, 2008), and it is ambiguously supported in Goloboff et al (2009).

Rhizaria are an artificial group, but there is possibly a core or Cercobiota taxon, as Thaumatomonada and Spongomonada have 2 parallel kinetosomes interconnected by fibrillar bridges (Karpov, 2010); Cercomonada, Thaumatomonada, and Actinopoda have kinetocysts; and Cercomonada and Thaumatomonada have similar rootlets. This clade, however, is not found in the analysis, but a smaller Cercobiota clade is, but with low confidence.

Table 7-8. Molecular Classifications for Rhizaria and Retaria (figures are only for values above .50 and 50).

Cavalier-Smith and Chao (2003)(116 18S rRNAs, GTR model; bootstrap values over 50 and for only novel clades are in parentheses; nearly identical results were obtained by Bass and Cavalier-Smith (2004) and Bass et al (2005)).

Apusozoa (95)
    Apusomonada
    Ancyromonada
Centrohelida
Retaria-Cercozoa (96)
    Retaria
        Foraminifera-Polycistina (99)
        Acantharia
    Cercozoa
        Endomyxa
            Phytomyxa
            Haplosporidia (Ascetospora)-*Gromia*
        "Filosa" (95)
            Chlorarachnia
            Spongomonada
            Monadaofilosa

Cercomonada-Cryomonada
Thaumatomonada
Euglyphida

Nikolaev et al (2004) (supertree based on 111 SSU rRNA gene sequences and 68 actin gene sequences and BI; figures in parentheses: SSU rRNA and actin probabilities, respectively, and x indicates the node is absent in one of the trees due to the differences in available taxon sampling between the two genes; confidence values were omitted).

Rhizaria (1/1)
Acantharia-Taxopoda-Polycistina (1/x)
Middle Clade (1/.98)
Gromia
Foraminifera-Haplosporida (1/x)
Remaining Rhizarians (.66/x)
Plasmodiophorida
Crown Clade 1/- (Cercomonada, Desmothracida, Euglyphida, Pheodarea)

Moreira et al (2006) (using Bayesian inference, LSU rDNA, and GTR model; only values over .50 or 50 are given).

Rhizaria (1/-)
Retaria (.97/-)
Rotaliella (Foraminifera)
Sticholonche-like clone KW16-Polycistine-like cloneHA2 (.49/89)
Cercomonada-Thaumatomonada (.89/100)

Parfrey et al (2010)(ML, 16 genes (SSU rDNA+15 proteins), GRT, WAG, rtREV, and CAT models)

Rhizaria (1/100)
"Filosa " (93) (Chlorarachnia, Desmothoracida, Pheodaria, Thaumatomonada-Spongomonada, Ebrida, Euglyphida, Cercomonada)
Endomyxa-Retaria
Endomyxa (Plasmodiophorida, Haplosporida, *Gromia,* Vampyrellidae)
Retaria (79) (Acantharia, Taxopoda, Polycistina, Foraminifera)

Brown et al (2012) (using 159 proteins and the CAT-Poisson model)

Rhizaria (1/100)
Retaria-Plasmodiophorida (.99/59)
Retaria (1/52)
Acantharia+Foraminifera (1/100)
Gromia
Plasmodiophorida
Cercomonada+*Guttilinopsis* (.99/99)

Sierra et al (2013) (ML under the LG model; 36 genes)

Rhizaria (1/100)
  Clade 1 (-/60)
    Clade 2 (.99/78)
      Retaria s.s. (1/100)
        Foraminifera-Acantharia (.99/98)
        Polycistina
      Gromia-Filoreta
    Cercomonada-Pheodarea
  Plasmodiophorida

Perkinsus considered as a dinoflagellate based on both classical and molecular information and its putative affinities with Apicomplexa (i.e., possession of an apical complex and molecular evidence) have led many to recognize a Miozoa group. The clade is recovered with a probablity of 1 and strong bootstrap support in 11 of 11 recent studies, which are impressive figures, as follows:

Moreira et al (2006)	1/83
Rodriguez-Ezpeleta et al (2006)	1/100
Burki et al (2007)	1/100
Hackett et al (2007)	1/100
Yoon et al (2008)	.95-1/75
Burki et al (2009)	1/100
Hampl et al (2009)	1/100
Minge et al (2009)	1/100
Okamoto et al (2009)	1/over 90
Brown et al (2012)	1/100
Burki et al (2012)	1/100

Dinociliata are found also in both Lipscomb phylogenies (apicomplexans were not included). Molecular support for dinociliates appears to be rare, but the clade is found in Lenaers et al (1989). I added the apical complex in the second analysis, but still Alveolata are not recovered. Alveolates are possibly a grade as the cortical alveoli occur also in Glaucophyceae and forams, as well as Raphidophyceae and *Telonema* (Shalchian-Tabrizi et al), and the micropores occur only in Apicomplexa, and these are the only putative synapomorphies.

Genotypic classifications for the internal arrangement of Excavata do not compare with those here as most often (7 of 12) Euglenaria branches with Heterolobosa and with strong support. Otherwise, Euglenaria branches twice with Polymastigota, and twice with Jakobida. Only 1 (Hackett et al, 2007) corroborates Jakomastigota. The list is as follows (figures for Excavata are beside the citations):

Table 7-9. Genotypic Classifications for Excavata.

Rodriguez-Ezpeleta et al (2006) 1/100
  Jakobida + (Euglenaria-Heterolobosa) 1/100
Hampl et al (2007) 88/-/79

Polymastigota +1/89 (Jakobida (Heterolobosa-Euglenaria)(1/89)
Yoon et al (2008) weak support
Malawimonas+strong support (Jakobida +(Euglenaria-Heterolobosa) (strong support)
Burki et al (2009) 1/100
(Euglenaria-Heterolobosa)(1/100)+Jakobida
Burki et al (2012) 1/100
Jakobida+(Heterolobosa-Euglenaria) (1/88)
Parfrey et al (2010) 83
Polymastigota+43 (*Malawimonas* + (Jakobida (Heterolobosa -Euglenaria)(over 95)
Brown et al (2012) (1/100)
Jakobida (Euglenaria-Heterolobosa)(1/100)
Moreira et al (2006) .87/-
Jakobida (Euglenaria- Polymastigota)(.64/-)
Okamoto et al (2009) not given
Heterolobosa +(Jakobida-Euglenaria)(.92/-)
Hackett et al (2007) 1/97
(Polymastigota-Jakobida)<1/65)+ Euglenaria)
Minge et al (2009) -/63/55
Malawimonas (Jakobida) (1/93/74)
Kim and Graham (2008) .96/51/-
(Euglenaria (Heterolobosa- Jakobida)(.99/75/93))(1/96/97)+ *Malawimonas*

In recent molecular phylogenies, alveolates, rhizarians, or excavates are usually on top and opisthokonts or amebozoans usually at the bottom, the former similar to Lipscomb's phylogenies and mine, and red algae usually come right after unikonts.

Taxopoda-Eulobosa are an artefact as Taxopoda logically belongs in Actinopoda s.s. because of common possession of the hexagonal axonemal microtubular pattern, and the high confidence for them is probably overestimated, and for Opisthokonta it is probably underestimated. The Histonia-Retaria clade is also an artefact.

Supergroups are often touted as certain or probable but most are only moderately, ambiguously, or weakly supported. Philippe et al (2000) state that only opisthokonts, archeoplastids, and alveolates are unambiguously recognized by molecular taxonomists, but it is only the first of these that is strongly supported. Here is a summary from the review/survey of genotypic analyses by Parfrey et al (2006) (the 1st column is the number of studies the taxon is supported in, the 2nd is the number of studies including the taxon, and the 3rd is the percentage; figures include both nuclear and plastid genes):

Opisthokonta	43	51	84.3
Rhizaria	19	29	65.5
Amebozoa	20	42	47.6
Archeoplastida	26	61	42.6
Chromalveolata	16	60	26.6
Excavata	9	39	23

Alveolates, chromistans, unikonts, bikonts, and retarians were not included, but apparently, alveolates are usually recovered and chromistans not. Chromalveolates, like archeoplastidans, are recovered with plastid genes but not nuclear genes.

Baldauf et al (2000) present an analysis of various proteins, single, pairwise, 3-way, and all-4, in 15 categories. The 4 proteins are EF-1alpha, actin, alpha-tubulin, and beta-tubulin. Of the 18 groups included, bootstrap values showed moderate (50-74) to strong (75-100) support, except for Chromalveolata (Heterokonta+Ciliata-Apicomplexa), Archeoplastida (plants, red algae, and glauophycans), Plantae+Rhodobiota, Miozoa (ciliophores and apicomplexans), and Excavata. Conosa was not included, however, Amebozoa (Eulobosa-Myxobiota) was. Bikonta and Chromista were not included either. The following is a summary of the 8 supergroups included (figures include both nuclear and plastid genes).

	strongly supported	moderately supported	total	%
Opisthokonta	11	2	15	80
Amebozoa	3	0	4	75
Unikonta	6	5	15	57
Miozoa	4	2	15	33
Discicristata	4	0	15	27
Chromalveolata	1	2	8	25
Plantae+Rhodobiota	0	1	8	6
Archeoplastida	0	0	8	0

The percentage I calculated from the score--2 points for strongly supported, 1 for moderately supported, and 0 for weakly supported--out of the total possible score. The following are the score percentages including also other proteins and rRNA (SSU, LSU, and combined) for resampling support presented in the article from other analyses:

Amebozoa	87.5
Opisthokonta	62
Miozoa	50
Unikonta	33
Chromalveolata	27.5
Discicristata	27
Plantae+Rhodophyceae	21
Archeoplastida s.l.	12

Combining Baldauf et al with Parfrey et al we get the following averages:

Opisthokonta	73
Amebozoa	68
Archeoplastida	27
Chromalveolata	25
Excavata	25

Parfrey et al say that their analysis demonstrates that supergroup taxonomies are unstable and that support for them varies tremendously, indicating the current classification for eukaryotes is likely premature.

Opisthokonta have the strongest support from genotypic characteristics of any eukaryotic supergroup and the same goes for phenotypic characteristics. In spite of the name, opisthokonty (as in posterior insertion) is not unique to animals and funguses as it occurs also in some dinoflagellates and some ciliates, but the term is ambiguous since it is unclear if "posterior flagellum" refers to direction or insertion. Amebozoa have strong support but from only 4 studies and have no phenotypic synapomorphies.

Parfrey et al (2010) point out that Archeoplastida support comes primarily from phylogenomic analyses and these may be picking up misleading EGT (endosymbiotic gene transfer) signal of genes independently transfered from the plastid to the host nucleus in the 3 archeoplastid clades.

Stiller and Harrell (2005) emphasize that the "clade" can be explained by "short-branch exclusion" and "subtle and easily overlooked biases can dominate the overall results of molecular phylogenetic analyses of ancient eukaryotic relationships. Sources of potential phylogenetic artifact should be investigated routinely, not just when obvious 'long-branch attraction' is encountered."

Tekle et al (2009) point out that red and green algae have different rubisco protein complexes and light-harvesting compounds and state that support for the monophyly of Archeoplastida host genomes is generally high in analyses of large data sets (phylogenomics) with *limited taxon sampling,* and reanalysis of Rodriguez-Ezpeleta and colleagues' data, including additional taxons of interest and with removal of fast-evolving sites, which may produce spurious relationships, did not support the monophyly of the group, and that their (Tekle et al's) *taxon-rich* analyses of the 4 most-sampled markers (SSU, actin, alpha-tubulin, and beta-tubulin) similarly failed to recover it. They also state that these results suggest incongruence due to the conflicting signals of the host and the plastid genomes, but they also show that a cause of the artifact is limited taxon sampling.

Archeoplastida are not supported in my analysis as expected as they have no synapomorphies, are weakly supported genotypically, and are contradicted by Lipscomb's classical analysis, by 35 molecular analyses, and by Goloboff et al's combined analysis.

Histonia and Histonia-Chromista are also recovered in the molecular analyses of Sogin et al (1989) and Bhattacharya et al (1990) and the Histonia clade is recovered in Knoll (1992) and Wainright et al (1993). A plant-animal or plant-fungal grouping is recovered by Gouy and Li (1989), Wolf et al (2004), Philip et al (2005), and Yang et al (2005).

The following are clades not found in the analyses here nor in Lipscomb but which have (unique) synapomorphies:

Bikonta (Glaucophyceae+(Plantae (this excludes, of course, red algae)+Ochrobiota)): DHFR-TS gene fusion, green-yellow photoaction spectrum, pyrenoid type D, stacked thylakoids, vesicular vacuoles, MLS (multi-layered structure), pantonematic flagella, and cortical (pellicular) alveoli (this last feature occurs in Glaucophyceae and Actinopoda as well as Alveolata andd appaerntly also in Raphidophyceae and *Telonema*);

Unikonta (Conosa+Opisthokonta): unikonty, radiating perpendicular mts, posterior-anterior flagellar transformation, cartwheel in BB (basal body), PNB (paranuclear body), CSP (carotenoid synthesis pathway) L, 3 myosin features (myosin TH2, class II myosins, SH3 domain tails);

Ochrobiota (Chromista, Retaria, and Dinobiota): chlorophyll c, beta-1-3 storage, xanthophyll type C and D, B – D type stigma structure, cytoplasmic stigma, flagellar photoreceptor, flagellar swelling, paraxial rod, protein import mechanism by GB vesicles, CSP T, cytostome;

Chromaria (Chromista+Retaria): axopods; 2 mastigoneme rows on 1 flagellum, none on the other; 2 mastigoneme rows on 1 flagellum and the other reduced; ED (electron dense) plaque in the TZ (transition zone); ED bodies with rectangular arrays;

Ramicristata (Conosa, Testafilosa, Gromida, Cryptomonada): branching cristas.

Anisokonta (Chromista, Opisthokonta, amebas, cercomonads, Myxofilosa, Retaria, Dinaria,

Excavata): tubular cristas, AAA lysine synthesis pathway, MYTH4FERM myosins, and TH 1 myosins.

Bikonta (but including also red algae) are corroborated by molecular evidence (Baldauf et al, 2000; Nozaki et al, 2007; Hampl et al, 2009) but, in spite of the name, cannot be defined by dikont flagella as this is a primitive trait and occurs also in Unikonta. Unikonta are found in Burki et al (2007), Hampl et al (2007), Burki et al (2009), and Brown et al (2012). Both Unikonta and Bikonta are supported in the molecular phylogeny of Nikolaev et al (2004), Hampl et al (2009), Kim and Graham (2008), and Minge et al (2009). Details are provided as follows:

Table 7-10.

Unikonta

Burki et al (2007)	1/100
Burki et al (2009)	1/100
Burki et al (2012)	1/100
Brown et al (2012)	1/100
Hampl et al (2009)	1/96
Kim et al (2006)	-/.94
Kim and Graham (2008)	<.50/<50
Minge et al (2009)	<.50/<50
Nikolaev et al (2004)	.54 (RNA)/ - (actin), no confidence values given

Bikonta

Nozaki et al (2007)	1/96/97(97)
Minge et al (2009)	1/83/84
Hampl et al (2009)	values not specified
Nikolaev et al (2004)	values not given

The Burki et al (2009) study found the proposed chromalveolate clade, extended as the SAR group (stramenopiles (heterokonts), alveolates, and rhizarians), which includes Halvaria (heterokonts+alveolates), Rhizaria (Cercomonada+Retaria), and Hacrobia. In Parfrey et al (2010), SAR is also supported but without Cryptomonada and Haptomonada. Kim and Graham (2008) maintain that their analysis strongly refutes it. Halvaria are supported also by Janouskovec et al (2010) based on a red algal symbiosis, i.e., the common occurence of red algal plastids in apicomplexans, dinoflagellates, and heterokonts. A red algal symbiote also occurs in cryptomonads. And Ramicristata corresponds to the molecular Lobosa/Amebobiota, but without cryptomonads. Anisokonta were recovered in the second Lipscomb analysis (1989, 1991).

The Hacrobia grouping (Haptomonada-Cryptomonada) is recovered by Shalchian-Tabrizi (2006) (as (*Telonema*+Cryptomonada)+Haptomonada), Hackett et al (2007), Burki et al (2009), and Okamoto et al (2009) (including *Telonema* and centrohelids), but is contradicted by Burki et al (2012). *Telonema*, which is anisokont, tubulocristate, and with a unique subcortical lamina (a multilayered cytoskeleton), its defining characteristic, and possibly with cortical alveoli, has tripartite mastigonemes (Klaveness et al, 2005), which is a synapomorphy for Neochromista.

There are over 200 unstudied genera of eukaryotes about which so little is known they are so far unclassifiable and which are placed in 4 categories: free-living heterotrophic flagellates, parasites, ameboids, and algae, most being in the first 3 (Patterson, 1999). We can't know at this time how many, if any, of these or groupings of them, might eventually be accorded regnal rank, but the number would not be dramatic as many certainly belong in known taxons (Baldauf, 2008). Two new lineages of microalgae, picobiliphytes (mentioned in Table 7-7) and rappemonads, are known only from environmental sequence data, so very little of their phenotype is known. They may be related to glaucophyceans and haptophyceans, respectively (Not et al, 2007; Kim et al, 2011).

Pawlowski (2013) identifies 8 eukaryotic micro-kingdoms, which are Apusomonadidae-Ancyromonadidae, Collodictyonidae-Rigidifilida, Katablepharidae-Cryptophyta, Glaucophyta-Picobiliphyta, Centrohelida, rappemonads, Haptophyta, and Telonemia, but all are with established groups in at least some genotypic results and/or according to the phenotype. Collodictyonidae-Rigidifilida have affinities with excavates, Apusomonada branch with Cercomonada in my classification, and Ancyromonada have affinities with cercomonads. Contrary to Pawlowski, Ancyromonadidae do not form a clade with Apusomonada in molecular classifcations (Atkins et al, 2000). *Breviata*, another group considered of uncertain position, may have affinities with Amebozoa and branches with them in genotypic classifications.

As for the number of known kingdoms, if we go by confidence values, including moderately supported, there are 28 in phenotypic results (Table 7-2), and 19 in genotypic results, and mostly polytomously arranged in both types. The lower number in the latter is because of Rhodobiota s.l., Amebozoa, Rhizaria, and Miozoa, and Nuclearidae in Fungi. A phylogeny where there is agreement between the genotype and phenotype without regard necessarily to confidence measures would have 18 kingdoms as follows: *Cyanidioschyzon*, *Cyanidium*, *Galdieria*, Rhodobiota, Glaucobiota, Plasmodiophorae, Plantae, ((Animalia+Fungi)+Lobosa+Myxobiota), Chlorarachnia, Chromalveolata, Retaria+Cercobiota s.l., Nuclearida, Vampyrellida, Testafilosa, and Excavata. An arrangement based on agreement between my classifications and molecular ones, again without regard necessarily to confidence measures, has 2 dozen groups of regnal rank.

As can be seen there is a lack of resolution for eukaryotes at high levels in both phenotypic and genotypic phylogenies, which is attributed to a rapid radiation by Philippe et al (2000): extant eukaryotic phyla emerged from a massive multifurcation in a short time, which implies that no actual primitive eukaryotes are known. They also state that missing data and inefficient tree reconstruction methods are valid alternative explanations.

Chapter 8 A Convenience Classification

Natural bodies are divided into 3 kingdoms of nature: viz. the mineral, vegetable, and animal kingdoms. Minerals grow, Plants grow and live, Animals grow, live, and have feeling. — Carolus Linnaeus, 'Observations on the Three Kingdoms of Nature', Nos. 14-15. Systema Naturae (1735). As quoted (translated) in Étienne Gilson, From Aristotle to Darwin and Back Again: a Journey in Final Causality (2009), p. 42-43 - todayinsci.com.

If we are to design a convenience classification as was the convention prior to recently, but touted as "phylogenetic", and which we should, for the sake of practicality but also because phylogeny is largely uncertain and the fossil record incomplete, but separate and distinct from phylogenetic classification, as it is as important and as valid as a phylogenetic system, then a 4-kingdom arrangement (Figure 8-2), without a protistan kingdom but with a protistan level, is more useful and informative, unlike most modern alternative systems, which had a protistan kingdom of some sort. This particular taxon, maintained by gradists as monophyletic and phylogenetic, is especially artificial even by convenience standards.

The 4 kingdoms are Bacteria (prokaryotic, mostly osmotrophic, and with typically a murein wall), Phyta (usually autotrophic, producing, and with a typically cellulose wall), Mycota (osmotrophic, decomposing, mostly with a chitinous wall), and Zoa (phagotrophic, consuming, with no wall). The 2 superkingdoms are Prokaryota and Eukaryota. It is biaxial so horizontal lines separate cristal types and vertical lines separate trophic modes (a double line separates nuclear type; a diagonal line can be drawn to separate histonian (mostly lamellar cristas, mostly multicellular, and mostly terrestrial) and protistan (mostly tubular cristas, mostly unicellular, and mostly acquatic) levels. Terrestrial and usually multicellular forms are placed higher and acquatic, usually unicellular ones lower. Similar forms across kingdoms are placed on the same line or lines. The bases are, then, trophic mode/functional community, wall composition, presence or absence of a wall, cristal shape, chlorophyll type, ecology, and nuclear type. Exceptions are Acantharia with lamellar cristas; Hemimastigota, Mesozoa, water snakes, *Aphelidium* (in Fungi), and trypanosomes (in Kinetoplastida) with tubular cristas (trypanosomes also have discoid cristas in their life cycle and *Aphelidium* also have lamellar cristas in theirs); and Mycetozoa and Sporozoa with a walled stage. Similar taxonomies were Takhtajan's, but ochrophyceans were lumped with green algae, and blue bacteria were separated from eubacteria, and Leedale's pteropod scheme, and the internal arrangement for the component kingdoms is different from theirs.

In Heterogracilicutes, Neganaerobia are parasitic (except for Desulfobacteria) and unpigmented (except for Enterobacteria), while Negaerobia are free-living (except for a few in Pseudomonada and Flavobacteria) and pigmented (except for Planctobacteria and Caulobacteria). In Autogracilicutes, the exceptions are Chloroflexi and Rhodobacteria, which are heterotrophic. Clatabacteria are the chemolithotrophic groups.

Another arrangement (Table 8-1 and Figure 8-2) would take phylogeny more into account, so the lamellar groups would be at the bottom, and the tubular on top, with Apicomplexa replacing Sporozoa and Myxozoa going to Animalia, and Haplosporidia and Ascetospora going to Fungi, which is basically the pattern followed in Part 2. Also presented is a synoptic key (Table 8-2).

Table 8-1. Convenience Classification for Empire Biota.

spk. Prokaryota
  kgdm. Bacteria

sbk. Eubacteria
spph. Gracilicutes
phyl. Heterogracilicutes tax. nov.
cl. Neganaerobia tax. nov.
cl. Negaerobia tax. nov.
phyl. Autogracilicutes tax. nov.
cl. Clatabacteria tax. nov.
cl. Photobacteria
spph. Mollicutes (Tenericutes)
spph. Firmicutes
cl. Togabacteria
cl. Endospora
cl. Actinobacteria
sbk. Metabacteria (Mendosicutes)
spk. Eukaryota
kgdm. Phyta
sbk. Rhodophyta (Prerhodophycota, Rhodophycota, Glaucophycota)
sbk. Chlorophyta
spph. Chlorophycota (Prasinophyceae, Euchlorophyceae, Charophyceae)
spph. Cormophyta
phyl. Bryophyta
phyl. Pteridophyta
phyl. Spermophyta
sbk. Euglenophyta (Euglenales)
sbk. Chlorarachnia
sbk. Chromophyta
phyl. Ochrophyta
phyl. Dinophyta (phototrophic/phagotrophic orders)
kgdm. Mycota
sbk. Fungi
hpph. Chytridiomycota
hpph. Amastigomycota
spph. Zygomycota
spph. Dikaryomycota
phyl. Ascomycota
phyl. Basidiomycota
sbk. Aphagea
sbk. Plasmodiophorae
sbk. Heteromycota
kgdm. Zoa
sbk. Animalia
grdph. Choanozoa
grdph. Metazoa
hpph. Diploblastica (Parazoa, Placozoa, Mesozoa, Actinozoa)
hpph. Triploblastica
spph. Protostomia (Helminthes, Arthropoda, Mollusca)
spph. Deuterostomia (Lophophorata, Echinodermata, Chordata)

sbk. Discozoa (phagotrophic excavates)
sbk. Tubulozoa
  phyl. Marinozoa (actinopods, forams)
  phyl. Heterozoa (opalinates, bicesids)
  phyl. Dinociliata (Dinozoa (phagotrophic dinoflagellates), Ciliata)
  phyl. Spongomonada
  phyl. Apicomplexa
  phyl. Amebozoa (eulobosans, filosans; conosans (ameboflagellates))

Figure 8-1. Traditional, Convenience Classification for Empire Biota.

Cormophyta Chlorophycophyta Rhodophyta	Amastigomycota Chytridiomycota	Metazoa Choanozoa
Euglenophyta	Aphagea	Acrasia, Schzpyrnd. Jakobida, Polymastig. Euglenozoa
Chromophyta Ochrophycota Dinophycota Chlorarachnia	Plasmodiophorae Pseudofungi, Hydromyxa	Conosa Amebae Sporozoa Spongomonada Opalinata, Bicosecida Dinozoa Ciliata Forams, Actinopoda
Photobacteria Clatobateria	Metabacteria Firmicutes Mollicutes Heterogracilicutes Negaerobia Neganaerobia	

Figure 8-2. Evolution-Oriented Convenience Classification.

Phyta	Mycota	Zoa
		Conosa
		Amebae
		Apicomplexa
	Plasmodiophorae	
		Spongomonada
Chromophyta		
Ochrophycota	Pseudofungi, Hydromyxa	Bicosecida, Opalinata
Dinophycota		Dinozoa, Ciliata
Chlorarachnia		
		Forams, Actinopoda
		Acrasia, Schyzopd.
		Jakobida, Polymastig.
Euglenophyta	Aphagea	Euglenozoa
		Metazoa
		Choanozoa
	Amastigomycota	
	Chytridiomycota	
Embryophyta		
Chlorophycophyta		
Rhodophyta		
	Bacteria	
	Metabacteria	
	Firmicutes	
	Mollicutes	
Photobacteria Clatobateria	Heterogracilicutes	
	Negaerobia	
	Neganaerobia	

Table 8-2. Key to the Kingdoms.

1 Nucleus not membrane bounded; flagella composed of only 1 fiber and with flagellin and rotary motion (proterokontic); ribosomes 70S, dispersed in cytoplasm, and MW 2.8 mln. daltons; 16S RNA; wall mostly with murein; chloroplasts lacking; no ER; mitosis, meiosis, mitochondria, dictyosomes, peroxisomes, and lyzosomes all absent; no nucleolus; chromosomes lacking; DNA without protein coat; plasmids present. Prokaryota

1 Nucleus membrane bounded; flagella with 9+2 arrangement, without flagellin, and no rotary motion (pecilokontic); chloroplasts in phototrophs; ribosomes 80S, in cytosol, and with MW 4.5 mln. daltons; 18S RNA; wall without murein; chloroplasts in phototrophs; with ER; mitosis, meiosis, mitochondria, dictyosomes, peroxisomes, and lyzosomes usually present; with nucleolus; chromosomes present; DNA with protein coat; plasmids rare. Eukaryota

 2 Wall rarely absent; nutrition autotrophic or osmotrophic; generally with sporulation; largely sessile.

  3 Wall mostly cellulosic; nutrition mostly autotrophic; kinetids typically present and usually biflagellate; LSP DAP; no hyphal organization. Phyta

  3 Wall mostly chitinous; nutrition osmotrophic; kinetids typically absent and usually uniflagellate; LSP AAA; usually with hyphal organization. Mycota

 2 Wall mostly absent; nutrition phagotrophic; generally without sporulation; mostly mobile. Zoa

Chapter 9 Conclusions

For now we see through a glass, darkly; but then face to face: now I know in part; but then shall I know even as also I am known. - 1 Corinthians 13: 12 (King James Version).

In summary, several points should be emphasized concerning the analyses presented here:

Prokaryotes

1. They show that phenotypic evidence in bacterial phylogeny is highly important.
2. They corroborate the monophyly of Paratoga, Togabacteria, Gracilicutes (with or without Mollifirmicutes), and Metabacteria.
3. There is the probablity that Gram positives are paraphyletic and primitive.
4. Halobacteria as the basal metabacterial clade is a position found also in some genotypic taxonomies.
5. Metabacteria as primitive is an artefact of LBA.
6. The Monodermata clade appearing in the 2 TNT analyses is evidence for its monophyly, but its weak support and failure to be recovered in the 2 PAUP analyses are evidence for Mollifirmicutes and Metabacteria as separate clades, as indicated in most molecular analyses.
7. The lack of confidence for groups within Gracilicutes, the disagreement of these results in eubacteria with molecular taxonomies, and the discordance for this group in molecular taxonomies, indicate lack of resolution in the internal taxonomy of eubacteria, so is therefore an area of considerable uncertainty.
8. There is the possibilty of another domain, being Gracilicutes, as evolved through endosymbiosis, as suggested by Lake, and of yet another domain, as suggested by giant viruses.
9. The importance and extent of LGT may be exaggerated.
10. There appear to be 2 basic competing phylogenies in molecular classifications for eubacteria, one represented by RNA and Woese and the other by proteins and Battistuzzi et al.

Eukaryotes

1. Red algae s.l. are basal, a position supported by phenotypic and fossil evidence and some genotypic evidence.
2. Red algae s.l. are paraphyletic or polyphyletic, but in molecular analyses they (prerhodophyceans +rhodobiotes s.s.) are usually a strongly supported clade.
3. Archeoplastids are a paraphyletic group, with no support from classical evidence and weak support (in nuclear genes) from molecular data.
4. The phylogenies are mostly (70%) congruent with fossil evidence, while the typical, recent molecular phylogeny is mostly not (only 30%).
5. Metakaryota, Neokaryota, Cyanidiobiota, Opisthokonta, and Retaria have strong support.
6. Cercozoa are a heterogenous assemblage with no support from classical data and only weak support from molecular data; Rhizaria have no support from classical data either and only moderate support from molecular data.
7. In phenotypic analyses Dinociliata are usually recovered but with weak support, and in molecular analyses Miozoa are usually recovered and with strong support.
8. The only supergroups strongly supported by both classical and molecular evidence are Opisthokonta and possibly Retaria, so that relationships for Eukaryota at high levels are largely unresolved.

Three more points to emphasize are that phenotypic data are just as reliable as genotypic data, clades in prokaryotes are generally unstable (but Deuterobacteria or Paratoga, Firmicutes, Gracilicutes [with or without Mollifirmicutes], and Metabacteria are always recovered) and the importance of using convenience classifications but separate from phylogenetic ones.

References

Adl, Sina M. et al. 2005. The new higher level classification of eukaryotes with emphasis on the taxonomy of protistans. J. Euk. Microbiol. 52: 399-451 (rutgers.edu).

Altmann, R. 1890. Die Elementarorganismen und ihre Beziehungen zu den Zellen. Veit & Co, Leipzig (as cited in Dodson, 1971).

Ammonius Hermiae. 400s AD. Porphyri Isagogen (as cited in Ragan, 1997).

Andam, C, Williams, D., Gogarten, J.P. 2010. Natural taxonomy in light of HGT. Biol. Phil. 25: 589-602 (efn.uncor.edu).

Aristotle. 300s BC. History of Animals.

Atkins, Margaret Stella; McArthur, Andrew Grant; Teske, Andreas. 2000. Ancyromonadida: a new phylogenetic lineage among the Protozoa closely related to the common ancestor of Metazoans, Fungi, and Choanoflagellates (Opisthokonta). Journal of Molecular Evolution 51:278-85 (researchgate.net).

Atkinson, G.F. 1915. Phylogeny and Relationships in Ascomycetes. Ann. Miss. Bot. Gard. 2: 315-76 (as cited in Demoulin, 1974).

Bailey, Ken. 1994. Typologies and Taxonomies-Numerical Taxonomy and Cluster Analysis (sagepub.com).

Baker, R.H, Yu, X., DeSalle, R. 1998. Assessing the relative contribution of molecular and morphological characters in simultaneous analysis trees (Abstract). Mol. Phylogenet. Evol. 9: 427-36.

Baldauf, S.L., Roger, A.J., Wenk-Siefert, I., & Doolittle, W.F. 2000. A kingdom-level phylogeny of eukaryotes based on combined protein data. Science 290: 972-977 (researchgate.net).

Baldauf, S.L. 2008. An overview of the phylogeny and diversity of eukaryotes. J. Syst. Evol. 46, 263–273 (plantsystematics.com).

Bapteste E., Brinkman H., Lee J.A., Moore D.V., Sensen C.W., Gordon P., Duruſl L., Gaasterland T., Lopez P., Muller M., Philippe H. 2002. The analysis of 100 genes supports the grouping of 3 highly divergent amebas: *Dictyostelium, Entameba*, and *Mastigameba*. Proc. NAS 99: 1414-1419.

Barkley, F.A. 1939. Keys to the Phyla of Organisms. U. of Montana Press, Massoula, Montana (as cited in Blackwelder, 1964, and WorldCat.Org).

Barkley, F.A. 1949. Un esbozo de clasificacion de los organismos. Rev. Fac. Nac. Agron. 10: 83-103 (as cited in Lipscomb, 1991).

Barns, S.M., Fundyga, R.E., Jeffries, M.W., Pace, N.R. 1995. Remarkable archaeal diversity detected in a Yellowstone National Park hot spring environment. Proc. Natl. Acad. Sci. USA 91: 1609–1613.

Barns, S.M., Delwiche, C.F., Palmer, J.D., Pace, N.R. 1996. Perspectives on archaeal diversity, thermophily, and monophyly from environmental rRNA sequences. Proc. Natl. Acad. Sci. USA 93: 9188–9193.

Bass, D. & Cavalier-Smith, T. 2004. Phylum-specific environmental DNA analysis reveals remarkably high global biodiversity of Cercozoa (Protozoa). Int. J. Syst. Evol. Microbiol. 54, 2393-2404.

Bass, D., Moreira, D., López-García, P., Polet, S., Chao, E. E., von der Heyden, S., Pawlowski, J. & Cavalier-Smith, T. 2005. Polyubiquitin insertions and the phylogeny of Cercozoa and Rhizaria. Protist 156, 149-161 (u-psud.fr).

Battistuzzi, F. U., Feijao, A., Hedges, S. B. 2004. A genomic timescale of prokaryote evolution: insights into the origin of methanogenesis, phototrophy, and the colonization of land. BMC

Evolutionary Biology 4: 44.

Bell, P. J. L. 2001. Viral eukaryogenesis: was the ancestor of the nucleus a complex DNA virus? Journal of Molecular Evolution 53: 251–256 (Abstract-researchgate.net).

Bell, P.J.L. 2004. The Viral Eukaryogenesis Theory. In Origins, Vol. 6, ed. J. Seckback of series Cellular Origin, Life in Extreme Habitats, and Astrobiology, pp 347-367, Kluwer Academic (as cited in Springer.Com).

Bell, Philip J.L. 2006. Sex and the eukaryotic cell cycle is consistent with a viral ancestry for the eukaryotic nucleus. Journal of Theoretical Biology 243: 54–63 (Abstract-researchgate.net).

Bell, P.J.L. 2009. The viral eukaryogenesis hypothesis: a key role for viruses in the emergence of eukaryotes from a prokaryotic world environment. Ann. NY Acad. Sci. 1178: 91-105 (researchgate.net).

Bhattacharya, D., Elwood, H.J., Goff, L.J., Sogin, M.L. 1990. Phylogeny of *Gracilaria lameneiformis* (Rhodophyta) based on sequence analysis of its SSU rRNA coding region. J. Phycol. 26: 181-86.

Bhattacharya, D., Helmchen, T. Bibeau, C., Melkonian, M. 1995. Comparisons of Nuclear-Encoded Small-Subunit Ribosomal RNAs Reveal the Evolutionary Position of Glaucocystophyta. Mol. Biol. Evol. 12: 415-20.

Blackwell, W.H. 2004. Is It Kingdoms or Domains? Confusions and Solutions. The Am. Biol. Teacher 66: 268-75 (nabt.org).

de Bary, A. 1881. Untersuchengen uber die Peronosporeen und Saprolegnien und die Grundlagen ienes naturalischen der Pilze. In Beitr. Morphol. Physiol. Pilze, A. de Bary and M. Woronin, 4: 1-145, Winter, Frankfurt (as cited in Demoulin, 1974).

Blackwelder, R.E. 1964. The Kingdoms of Living Things. Systematic Zoology 13: 74-75.

Bold, H.C., Alexopoulos, C.J., Delevoryas, T. 1987. Morphology of Plants and Fungi. Harper and Row, New York.

Bory de Saint-Vincent, J-B. 1824. Psychodiaire (Règne), pp. 657-63, in Encyclopédie Méthodique (J-V Lamouroux, J-B Bory de Saint-Vincent, E. Deslongchamps, eds.), Vol. 138,. Paris (as cited in Ragan, 1997).

Bory de Saint-Vincent, J-B. 1824. Microscopiques, pp. 515-43, in Encyclopédie Méthodique (J-V Lamouroux, J-B Bory de Saint-Vincent, E. Deslongchamps, eds.), Vol. 138, Paris (as cited in Ragan, 1997).

Brochier, Céline, Gribaldo, S., Zivanovic, Y., Confalonieri, F., Forterre, P. 2005. Nanoarchaea: representatives of a novel archaeal phylum or a fast-evolving euryarchaeal lineage related to Thermococcales? Genome Biol. 6: R42.

Brochier-Armanet, C., Boussau, B., Gribaldo, S., Forterre, P. 2008. Mesophilic Crenarchaeota: proposal for a third archeote phylum, Thaumarchaeota. Nat. Rev. Microbiol. 6: 245–52.

Brown, M.W., Kolisko, M., Silberman, J.D., Roger, A.J. 2012. Aggregative Multicellularity Evolved Independently in the Eukaryotic Supergroup Rhizaria. Curr. Biol. 22, 1123–1127 (cell.com).

Brugerolle, G., Mignot, J.-P. 1984. The cell characters of two Helioflagellates related to the Centrohelidian lineage: *Dimorpha* and *Tetradimorpha.* Origins of Life 13: 305- 314.

Bryant, David. 2003. A Classification of Consensus Methods for Phylogenetics, pp. 163-84, in Bioconsensus (DIMACS Series Vol. 61), M.F. Janowitz et al, eds., AMS (books.google; mathnet.or.kr mathnet.or.kr).

Burki, F., Shalchian-Tabrizi, K., Minge, M., Skjaeveland, A., Nikolaev, S.I., Jacobsen, K.S., Pawlovsky, J. 2007. Phylogenomics reshuffles the eukaryotic super-groups. PLoS ONE 2e: 790 (plosone.org).

Burki, F., Shalchian-Tabrizi, K., Pawlovsky, J. 2007. Phylogenomics reveals a new 'megagroup' including most photosynthetic eukaryotes. Biol. Letters 4: 366-69 (researchgate.net).

Burki F , Inagaki Y, Bråte J, Archibald J M, Keeling PJ, Cavalier-Smith T, Sakaguchi M, Hashimoto T , Horak A, Kumar S, Klaveness D, . Jakobsen KS, Pawlowski J, and Shalchian-Tabrizi, K. 2009. Large-Scale Phylogenomic Analyses Reveal that 2 Enigmatic Protistan Lineages, *Telonemia* and Centroheliozoa, are Related to Photosynthetic Chromalveolates. Genome Biology and Evolution 2009: 231-55 (gbe.oxfordjournals.org).

Burki, F., Okamoto, N., Pombert, J.F., Keeling, P.J. 2012. The evolutionary history of haptophytes and cryptophytes: phylogenomic evidence for separate origins. Proc. Biol. Sci. 279: 2246-54.

Cain, R.F. 1972. Evolution of the fungi. Mycologia 64: 1-14 (as cited in Demoulin, 1974).

Carpenter, J.M. 1988. Choosing Among Multiple Equally Parsimonious Cladograms. Cladistics 4: 291-296 (researchgate.net).

Cavalier-Smith, T. 1978. The evolutionary origin and phylogeny of microtubules, mitotic spindles, and eukaryote flagella. BioSystems 10: 93-114.

Cavalier-Smith, T. 1981. Eukaryote kingdoms: 7 or 9? BioSystems 14: 461-481.

Cavalier-Smith, T. 1983. A 6-kingdom Classification and a Unified Phylogeny, pp. 1027-1034, in Schwemmler, W., Schenk, H. E. A. (eds.), Endocytobiology II, deGruyter, Berlin.

Cavalier-Smith, T. 1986. The kingdoms of organisms. Nature 324: 416-417.

Cavalier-Smith, T. 1992. Bacteria and eukaryotes. Nature 356: 570.

Cavalier-Smith, T. 1993. Kingdom Protozoa and its 18 phyla. Microbiol. Revs. 57: 953-954.

Cavalier-Smith, T. 1998. A revised 6-kingdom system of life. Biol. Rev. 73: 203-266.

Cavalier-Smith, T. 2004. Only 6 kingdoms of life. Proc. Roy. Soc. Lond. B 271: 1251-62.

Cavalier-Smith, T, Chao, E. 1996. 18S RNA sequence of *Heterosigma carterae* (Raphidophyceae) and the phylogeny of heterokont algae (Ochrophyta). Phycologia 35: 500-10.

Cavalier-Smith, T., Chao, E.E. 2003. Phylogeny and classification of phylum Cercozoa (Protozoa). Protist 154: 341–58 (eurekamag.com).

Chadefaud, M. 1957. Les champignons et les algues. Ann. Univ. Paris 27: 5-22 (as cited in Demoulin, 1974).

Chadefaud, Marius & Emberger, Louis. 1960. Traité de botanique systématique, Tome 1 (Les végétaux non-vasculaires). Masson, Paris (as cited in Caratini, 1971).

Chatton, E. 1925. *Pansporella perplexa*: reflexions sur la biologie et la phylogénie des protozoaires. In Annales des Sciences Naturelles, Zoologie: l'anatomie, la physiologie, classification, et 1'histoire naturelle des animaux. 10me séries, volume 8. edited by Bouvier, M.E.L. Masson, Paris (as cited in Whittaker, 1959).

Chatton, E. 1937. Titres et travaux scientifiques. E. Chatton, Sète, France (as cited in Whittaker, 1959).

Ciccarelli, F. D., Doerks, T., Von Mering, C., Creevey, C.J., Snel, B., Bork, P. 2006. Toward Automatic Reconstruction of a Highly Resolved Tree of Life. Science 311: 1283–1287 (binf.bio.uu.nl).

Ciniglia, Claudia; Yoon, Hwan Su; Pollio, Antonino; Pinto, Gabriele; Bhattacharya, Debashish. 2004. Hidden biodiversity of the extremophilic Cyanidiales red algae. Molecular Ecology 13: 1827–1838 (rutgers.edu).

Claverie, Jean-Michel. 2006. Viruses take center stage in cellular evolution. Genome Biology 7: 110 (genome biology).

Conard, H. S. 1939. Plants of Iowa (Grinnell Flora, 5th ed.). Iowa Acad. Sci., Biol. Surv. Publ. 2, 1-92 (as cited in Whittaker, 1957) .

Copeland, E B. 1927. What is a plant? Science 65: 388-390 (as cited in Lipscomb, 1991).

Copeland, HF. 1938. The kingdoms of organisms. Quart. Rev. Biol. 13: 383-420.

Copeland, HF. 1947. Progress report on basic classification. Am. Nat. 8: 340-61.

Copeland, HF. 1956. The Classification of Lower Organisms. Pacific Books, Palo Alto, Cal (as cited in Blackwelder, 1964).
Corliss, J.O. 1984. The Kingdom Protista and its 45 Phyla. BioSystems 17: 87-126.
Corliss, J.O. 1994. An interim Utilitarian ("User-friendly") Hierarchical Classificarion and Characterization of the Protists. Acta Protozoologica 33: 1-51.
Corliss, J.O.1995. The Need for a New Look at the Taxonomy of the Protists. Rev. Soc. Mex. Hist. Nat. 45: 27-36.
Cuvier, G. Règne animal. 1817 (as cited in Ragan, 1997).
Darley, W. M. 1982. Algal Biology: a Physiological Approach. Blackwell, Oxford.
Daubin, V., Gouy, M., Perrière, G. 2002. A Phylogenomic Approach to Bacterial Phylogeny: Evidence of a Core of Genes Sharing a Common History. Genome Res. 12: 1080-90 (genome.cshlp.org).
Das, Sabyasachi; Sandip, Paul; Bag, Sumit; and Dutta, Chitra. 2006. Analysis of *Nanoarchaeum equitans* genome and proteome composition: indications for hyperthermophilic and parasitic adaptation. BMC Genomics 7: 186 (researchgate.net).
Degnan, James; DeGiorgio, Michael; Bryant, David; Rosenberg, Noah. 2009. Properties of Consensus Methods for Inferring Species Trees from Gene Trees. Syst Biol. 58: 35–54 (oxfordjournals.org).
De Luca, P., Taddei, R., Varano, L. 1978. "*Cyanidoschyzon merolae*": a new alga of thermal acidic environments. Webbia 33: 37-44 (as cited in Merola et al, 1981).
Demoulin, V. 1974. The Origin of Ascomycetes and Basidiomycetes-the case for a red algal ancestry. Bot. Rev. 40: 315-45.
Dillon, L. 1963. A re-evaluation of the major groups of organisms based on comparative cytology. Syst. Zool. 12: 71-82.
Dodson, E.O. 1971. The kingdoms of organisms. Syst. Zool. 20: 265-281.
Doolittle, W. F. 1999. Phylogenetic classification and the universal tree. Science 284: 2124-2128.
Dougherty, E. Neologism needed for structures of primitive organisms 1- Types of nuclei. J. Protozool. 4: 14.
Doyle, J.J. 1992. Gene trees and species trees: molecular systematics as one-character taxonomy. Systematic Botany 17: 144-163.
Dutilh, B.E., Huynen, M.A., Bruno, W.J., Snel, B. 2004. The Consistent Phylogenetic Signal in Genome Trees Revealed by Reducing the Impact of Noise. J. Mol. Evol. 58: 527-39 (cmbi.ru.nl).
Edwards, P. 1976. A classification of plants into higher taxa based on cytological and biochemical criteria. Taxon 25: 529-542.
Elkins, J.G. 2008. A korarchaeal genome reveals insights into the evolution of the Archaea. Proc. Natl. Acad. Sci. USA 105: 8102–8107.
Emiliani, Cesare. 1995. The Scientific Companion. Wiley.
Farris, J.S. 1969. A succesive approximations approach to character weighting. Syst. Zool. 18: 374-85.
Farris, J.S. 1976. Phylogenetic classification of fossils with recent species. Syst. Zool. 25: 271-282.
Felsenstein, J. 2005. PHYLIP (Phylogeny Inference Package) version 3.6. Distributed by the author. Department of Genome Sciences, University of Washington, Seattle.
Folinsbee, Kaila et al. 2007. 5 Quantitative Approaches to Phylogenetics. Rev. Mex. Div. 78: 225-52 (kfolinsbee.iastate.edu).
Forey, P. L., Humphries, C. J., Kitching, I. J., Scotland, R. W., Siebert, D. J., Williams, D. M. (eds.). 1992. Cladistics: a practical course in systematics. (Systematics Association Publications, 10). Oxford U. Press.

Forterre, Patrick. 2001. Genomics and early cellular evolution. Comp. Rend. Acad. Sci III 324: 1067-76 (Abstract-nih.gov).
Forterre, P. 2005. The two ages of the RNA World and the transition to the DNA World: a story of viruses and cells. Biochimie 87: 793-803 (Abstract-nih.gov).
Forterre, P. 2006. The origin of viruses and their possible roles in major evolutionary transitions. Virus Res. 117: 5-16 (Abstract)(nih.gov).
Forterre, P. 2008. Origin of Viruses. In: Encyclopedia of Virology (3rd Ed., 5 vols.), Editors: B.W.J. Mahy and M.H.V. van Regenmortel, pp. 472-79, Academic Press (Abstract-sciencedirect.com).
Fournier, G.P. and Gogarten, J.P. 2008. Evolution of Acetoclastic Methanogenesis in *Methanosarcina* via Horizontal Gene Transfer from Cellulolytic Clostridia. J. Bacteriol. 190: 1124–1127.
Fox, Robin. 2004. Symbiogenesis. J. Roy. Soc. Med. 97: 559 (nih.gov).
Frederick, J.F. 1976. *Cyanidium caldarium* as a bridge alga between Cyanophyceae and Rhodophyceae: evidence from immunodiffusion studies. Plant Cell Physiol. 17: 317-22.
Fries, E. 1821. Systema mycologicum, vol 1. Berlinger, Lund.
Gabrielson, P.W., Garbary, D.J., Scagel, R.F. 1985. The nature of the ancestral red alga: inferences from a cladistic analysis (abstract). BioSystems 18: 335-346.
Gabrielson, P.W., Garbary, D.J., Sommerfeld, M.J., Townsend, R.A., Tyler, P.L. 1990. Phylum Rhodophyta, pp. 102-118. In: Margulis, L., Melkonian, M., Corliss J.O., Chapman D.J. (eds.), Handbook of Protoctista. Jones and Bartlett, Boston.
Gaillon, B. 1833. Aperçu d'histoire naturelle, pp. 95-114, in Mémoires et notices, Soc. agric. com. arts, Boulogne-sur-mer.
Galdieri, A. 1899. Su di un'alga che cresce intorno alle fumarole della solfatura. Rend. R. Accad. Sci. Fis. Mat. Napoli 6: 160-64 (as cited in Merola et al, 1981).
Gao, B., Gupta, R.S. 2007. Phylogenomic analysis of proteins that are distinctive of Archaea and its main subgroups and the origin of methanogenesis. BMC Genomics 8: 86 (susqu.edu).
Geitler, L., Ruttner, F. 1935. Die Cyanophyceen der deutschen limnologishen Sunda Expedition, etc. Arch. Hydrol., suppl. 14: 371-481 (as cited in Merola et al, 1981).
Gogarten, J.P. et al. 1989. Evolution of vacuolar H+-ATPase: implications for the origin of eukaryotes. Proc. NAS 68: 6661-85.
Gogarten, J.P., Doolittle, W. F., Lawrence, J. G. 2002. Prokaryotic Evolution in Light of Gene Transfer. Mol. Biol. Evol. 19: 2226-38 (uconn.edu).
Goloboff, P. A. 1993. Estimating character weights during tree search. Cladistics 9: 83–91.
Goloboff, P. A. 1999. Analyzing large data sets in reasonable times: solutions for composite optima. Cladistics 15: 415-428.
Goloboff, P.A., Catalano, S.A., Mirande, J.M., Szumik, C.A., Arias, J.S., Kallersjo, M., Farris, J.S. 2009. Phylogenetic analysis of 73, 060 taxa corroborates major eukaryotic groups. Cladistics 25: 211-30.
Goloboff, Pablo; Farris, James; Källersjö, Mari; Oxelman, Bengt; Ramiacuterez, Maria; Szumik, Claudia. 2003. Improvements to resampling measures of group support. Cladistics 19: 324–332.
Goodfellow, M., O'Donnell, A. (eds). 1993. Handbook of New Bacterial Systematics. Academic Press, New York.
Gouy, M., Li, W. H. 1989. Molecular phylogeny of the kingdoms Animalia, Plantae, and Fungi. Mol. Biol. Evol. 6: 109–122.
Griffin, J.L. 1988. Fine structure and taxonomic position of the giant ameboid flagellate *Pelomyxa*

palustris. J. Protozool. 35: 300-315.

Gupta, R.S. 1998a. Protein phylogenies and signature sequences: a reappraisal of evolutionary relationships archeobacteria, eubacteria, and eukaryotes. Microbiol. Mol. Biol. Rev. 62: 1435-1491 (ncbi.nlm. nhi.gov).

Gupta, R.S. 1998b. Life's 3rd domain: an established fact or an endangered paradigm. Theor. Popul. Biol. 54: 91-104.

Gupta, R.S. 2000. The natural evolutionary relationships among prokaryotes. Crit. Rev. Microbiol. 26: 111-131.

Gupta, R.S. 2003. Evolutionary relationships among photosynthetic bacteria. Photosynth. Res. 76: 173–183.

Gupta, R.S. 2011. Origin of diderm (Gram-negative) bacteria: antibiotic selection pressure rather than endosymbiosis likely led to the evolution of bacterial cells with two membranes. Antonie van Leeuwenhoek 100: 171-182.

Gupta R., Shami A. 2011. Molecular signatures for Crenarchaeota and Thaumarchaeota. Antonie van Leeuwenhoek 99:133–157.

Gupta, R.S., Singh, B. 1994. Phylogenetic analysis of 70 kD HSP sequences suggests a chimeric origin for the eukaryotic cell nucleus. Current Biol. 4: 1104-1114.

Guy, Lionel; Ettema, Thijs. 2011. The archeobacterial 'TACK' superphylum and the origin of eukaryotes. Trends in Microbiology 19: 580-87.

Hackett, J.D., Yoon, H.S., Li, S., Reyes-Prieto, A., Rummele, S.E., Bhattacharya, D. 2007. Phylogenomic analysis supports the monophyly of crytophytes and haptophytes and the association of 'Rhizaria' with chromalveolates. Mol. Biol. Evol. 8: 1702-13 (dblab.rutgers.edu).

Haeckel. E.H. 1866. Generelle Morphologie der Organismen. Reimer, Berlin (as cited in Lipscomb, 1991).

Haeckel, E. 1878. Das Protistenreich. Gunther, Leipzig (as cited in Lipscomb, 1991).

Haeckel, E. 1994. Systematische Phylogenie, Vol. 1. Reimer, Berlin (as cited in Lipscomb, 1991).

Haeckel, E. 1904. The Wonders of Life. Harper, New York and London (as cited in Lipscomb, 1991).

Hampl, V., Hug, L., Leigh, J.W., Dacks, J.B., Lang, B.F., Simpson, A.G.B., Roger, A.L. 2009. Phylogenomic analyses support the monophyly of Excavata and resolve relationships among eukaryote super-groups. Proc. NAS USA 106: 3859-64 (pnas.org).

Hartman, H., Fedorov, A. 2002. The origin of the eukaryotic cell: a genomic investigation. Proc. Natl. Acad. Sci. U S A. 99: 1420-25.

Heath, I. Brent. 1980. Variant Mitoses in Lower Eukaryotes: Indicators of the Evolution of Mitosis? International Review of Cytology 64: 1–80.

Hennig, Willi. 1966. Phylogenetic Systematics. Transl. D.D. Davis and Rainer Zangerl. U. of Ill. Press (books.google).

Ho, Chi-Chun; Lau, Susanna; Woo, Patrick. 2013. Romance of the three domains: how cladistics transformed the classification of cellular organisms. Protein Cell 4: 664–676 (springer.com).

Hogg, John. 1860. On the Distinctions of a Plant and an Animal, and on a Fourth Kingdom of Nature. Edinburgh New Philosphical Journal 12 (new series): 216-25.

Holt, J.G., Krieg, N.R., Sneath, P.H.A., Staley, J.T., Williams, S.T. (eds.). 1994. Bergey's Manual of Determinative Bacteriology, 9th ed. Williams & Wilkins, Baltimore.

Horaninow, P. 1834. Primae Lineae Systematis Naturae. St. Petersburg (as cited in Ragan, 1997).

Horaninoff, P. 1843. Tetractys Naturae. Weinhoberianis, St. Petersburg (as cited in Ragan, 1997).

Hori, H, Osawa, S. 1987. Origin and evolution of organisms as deduced from 5S rRNA sequences. Mol. Biol. Evol. 4: 445-472.

Hori, H. Stow, Y, Inoue, I, and Chihara M. 1990. Origins of organelles and algae evolution deduced from 5S rRNA sequences, pp. 557-559, in Dardon, P., Gianinazzi-Pearson, V., Grenier, A.M., Margulis, L., Smith, D.C. (eds.), Endocytology IV, INSA, Paris.

Horike et al. 2009. Gene (Science Direct).

Huber, H. et al. 2002. A new phylum of Archeota represented by a nanosized hyperthermophilic symbiote. Nature 417: 63–67.

Iwabe, N. et al. Evolutionary relationship of Archeobacteria, Eubacteria, and eukaryotes inferred from phylogenetic trees of duplicated genes. Proc. NAS 86: 9355-59.

Jain, Ravi; Rivera, Maria; Lake, James. 1999. Horizontal gene transfer among genomes: the complexity hypothesis. Proc. Natl. Acad. Sci. U. S. A. 96: 3801-6 (pnas.org).

Janouskovec J, Horák A, Oborník M, Lukes J, Keeling P.J. 2010. A common red algal origin of the apicomplexan, dinoflagellate, and heterokont plastids. Proc. Natl. Acad. Sci. USA 107: 10949-10954.

Javaux, Emmanuelle. 2007. The Early Eukaryotic Fossil Record, Chp. 1, in Eukaryotic Membranes and Cytoskeleton: Origins and Evolution, edited by Gaspar Jekely. Landes Bioscience/Springer.

Jeffrey, C. 1971. Thallophytes and kingdoms: a critique. Kew Bull. 25: 291-299.

Jeffrey, C. 1982. Kingdoms, codes, and classification. Kew Bull. 37: 403-416.

Joseph, R. 2010. The origin of eukaryotes: Archeotes, bacteria, viruses, and horizontal gene transfer. Journal of Cosmology 10: 3418-3445 (cosmology.com).

Jin, Guohua; Nakhleh, Luay; Snir, Sagi; and Tuller, Tamir. 2007. Inferring Phylogenetic Networks by the Maximum Parsimony Criterion: a Case Study. Mol. Biol. Evol. 24:324–337 (cs.rice.edu; oxfordjournals.org).

Karling, J.S. 1968. The Plasmodiophorales. Hafner, New York.

Karpov, S.A. 1999. Flagellate phylogeny, pp. 336-360, in The Flagellates (Leadbeater, B.S.C., Green, J.C., eds). Systematics Association Special Volume 59, Taylor and Francis, London and NY.

Karpov, S.A. 2000. Ultrastructure of the aloricate bicosoecid *Pseudobodo tremulans,* with revision of the order Bicosoecida. Protistology 1: 101-109 (protistology.imfo.ru).

Karpov, S.A. 2010. Flagellar apparatus structure of *Thaumatomonas* (Thaumatomonadida) and thaumatomonad relationships. Protistology 6: 236–244 (protistology.imfo.ru).

Katz, Laura. 2002. Lateral gene transfers and the evolution of eukaryotes: theories and data. Int. J. Syst. Evol. Microbiol. 52: 1893–1900 (science.smith.edu).

Keeling, Patrick; Burger, Gertraud; Durnford, Dion; Lang, Franz; Lee, Robert W.; Pearlman, Ronald; Roger, Andrew; Gray, Michael. 2005. The tree of eukaryotes. Trends Ecol. Evol. 20: 670-76.

Kessel, Edward (ed.). 1955. A Century of Progress in the Natural Sciences. California Academy of Sciences, San Francisco (Archive.Org).

Kitching, Ian et al. 1998. Cladistics. Oxford U. Press.

Kim, Eunsoo, Graham, Linda. 2008. EF2 Analysis Challenges the Monophyly of Archeoplastida and Chromalveolata. PloS One 3 (plosone.org).

Kim, E., Simpson, A.G.B., Graham, L. 2006. Evolutionary Relationships of Apusomonads. Mol. Biol. Evol. 23: 2455-66 (oxfordjournals.org).

Kim, Eunsoo; Harrison, James; Sudek, Sebastian; Jones, Meredith; Wilcox, Heather; Richards, Thomas; Worden, Alexandra; Archibald, John. 2011. Newly identified and diverse plastid-bearing branch on the eukaryotic tree of life. Proc. NAS USA 108: 1496–1500 (pnas.org).

Klaveness, D., Shalchian-Tabrizi, K., Thomsen, H.A., Eikrem, W., Jakobsen, K.S. 2005. *Telonema antarcticum* sp. nov., a common marine phagotrophic flagellate. Int. J. Syst. Evol. Microbiol. 55: 2595-2604 (sgmjournals.org).

Klein, R.M. 1970. Relationships between blue-green and red algae. Ann. NY Acad. Sci. 175: 623-33.

Kurland, C.G., Canback, B., Berg, O.G. 2003. Horizontal gene transfer: a critical view. Proc. Natl. Acad. Sci. U. S. A. 100: 9658-62 (pnas.org).

Kuroiwa, T. et al. 1994. Comparison of Ultrastructures between the Ultra-small Eukaryote *Cyanidioschyzon merloae* and *Cyanidium caldarium*. Cytologia 59: 149-58.

Kussakin, O. G., Starobogatoff, Y.I. 1973. On the very highest categories of the organic world, pp. 215–226, Problemy Évolyutsii, Nauka, Sibirskoe Otdelenie, Novosibirsk 3: 95-103 (in Russian).

Kussakin, O.G., Drozdoff, A.L. 1994 (vol.1), 1998 (vol. 2), Phylums of the Living World. Nauka, St. Petersberg (in Russian with English summaries)(books.google).

Kuske, Cheryl, Barns, Susan, and Busch, Joseph. 1997. Diverse uncultivated bacterial groups from soils of the arid southwestern United States that are present in many geographic regions. Applied and Environmental Microbiology. 63: 3614–3621.

Lake, J.A. 1983. An alternative to archeobacterial dogma. Nature 319: 626.

Lake, J.A., Henderson, E., Oakes, M., Clark, M.W. 1984. Eocytes: a new ribosome structure indicates a new kingdom with a close relationship to eukaryotes. Proc. NAS USA 81: 3786-3790.

Lake, J.A. 1986. Mapping evolution with 3-D ribosome structure. Syst. Appl. Microbiol. 7: 131-136.

Lake, J.A.1988. Origin of the eukaryotic nucleus determined by rate-invariant analysis of rRNA sequences. Nature 331: 184-186.

Lake, J. 2009. Evidence for an early prokaryotic endosymbiosis. Nature 460: 967-971 (see also the Lake Lab website).

Lee, J.J., Hunter, S.H., Bovee, E.C. (eds.). 1985 (1st ed.), 2000 (2nd ed.). Illustrated Guide to Protozoa. Allen Press, Lawrence, Kansas.

Leedale, G.F. 1974. How many are the kingdoms of organisms? Taxon 23: 261-70.

Lenaers, Guy; Scholin, Christopher; Bhaud, Yvonne; Saint-Hilaire, Danielle; Herzog, Michel. 1991. A molecular phylogeny of dinoflagellate protists (Pyrrhophyta) inferred from the sequence of 24S rRNA divergent domains D1 and D8. Journal of Molecular Evolution 32: 53-63 (springer.com).

Lewis, L. A. & McCourt, R.M. 2004. Green algae and the origin of land plants. Am. J. Bot. 91: 1535-1556.

Lienau, K., DeSalle, R., Rosenfeld, J., Planet, P. 2006. Reciprocal Illumination in the Gene Content Tree of Life. Syst. Biol. 55: 441-53 (researchgate.net).

Lienau, E.K., DeSalle, R., Allard, M., Brown, E.W., Swofford, D., Rosenfeld, J.A., Sarkare, I. N., and Planet, P.J. 2011. The mega-matrix tree of life: using genome-scale horizontal gene transfer and sequence evolution data as information about the vertical history of life. Cladistics 27: 417-27 (amnh.org).

Linder, D.H. 1940. Evolution of Basidiomycetes and its relation to the terminology of the basidium. Mycologia 32: 419-37 (as cited in Demoulin, 1974).

Linneus (von Linné), C. 1735. Systema Naturae. 1st ed. Lugduni Batavorum, Stockholm (as cited in Ragan, 1997).

Linneus, C. 1767. Mundus Invisibilis, pp. 385-408, in Amenitatis Academicae, vol. 7, Erlangen as cited in Ragan, 1997).

Lipscomb, Diana. 1985. The Eukaryotic Kingdoms. Cladistics 1: 127-40.

Lipscomb, D. 1989. Relationships among the eukaryotes, pp. 161-178, in The Hiearchy of Life, B. Fernholm, K. Bremer, & H. Jornvall (eds.), Elsevier, New York.

Lipscomb, D. 1991. Broad classification: the kingdoms and the protozoa, pp. 81-136, in Parasitic Protozoa,Vol. 1, 2nd ed. J.P. Kreier & J.R. Baker (eds.). Academic Press, San Diego.

Lopez-García, P., Moreira, D. 2004. The Syntrophy Hypothesis for the Origin of Eukaryotes. Cellular Origin, Life in Extreme Habitats, and Astrobiology 4: 131-146 (as cited in Katz, 2002).

Luttke, A. 1991. On the origin of chloroplasts and rhodoplasts: protein sequence composition. Endocyobiosis Cell Res. 8: 75-82.

Luzzatto, Michele; Palestrini, Claudia; & D'Entrèves, P.P. 2000. Hologenesis: the last and lost theory of evolutionary change. Italian Journal of Zoology 67: 129-138 (tandfonline.com).

Lutzoni, François; Vilgalys, Rytas.1998. Integration of morphological and molecular data sets in estimating fungal phylogenies. Can. J. Bot. 73 (Suppl. 1): S649-S659.

Maddison, W. D. 1997. Gene trees in species trees. Syst. Biol. 46: 523-36.

Maggs, C.A., Verbruggen, H., de Clerck, O. 2007. Molecular systematics of red algae (Chapter 6). In: Brodie, J., Lewis, J. (eds), Unravelling Red Algae. CRC Press, Boca Raton, Fl., (books.google).

Margulis, L. 1973. Five-Kingdom classification. In Evolutionary Biology, Vol. 7, pp. 45-78, 45-78, T. Dobzhansky, M.K. Hecht, and W.C. Steere (eds.), Plenum Publishing, NY.

Margulis, L, Schwartz, K. 1998. Five Kingdoms, 3rd ed. W.H. Freeman, NY.

Margulis, L., Melkonian, M., Corliss, J.O., Chapman, D.J. (eds.). 1990. Handbook of Protoctista. Jones and Bartlett, Boston.

Margulis, Lynn; Dolan, Michael; Guerrero, Ricardo. 2000. The chimeric eukaryote: Origin of the nucleus from the karyomastigont in amitochondriate protists. PNAS 97: 6954–6959 (pnas.org).

Margulis, L. 2010. Symbiogenesis. A new principle of evolution, rediscovery of Boris Mikhaylovich Kozo-Polyansky (1890–1957). Paleontological Journal 44: 1525-1539 (Abstract-springer.com).

Margush, T., McMorris, FR. 1981. Consensus n-trees. Bull. Math .Biol. 43: 239–244.

Martin, William and Müller, Miklos. 1998. The hydrogen hypothesis for the first eukaryote. Nature 392: 37-41 (researchgate.net).

McKenna, M.C. 1975. Towards a phylogenetic classification of Mammalia. In The Phylogeny of Primates, pp. 21-46. W. P.Luckett and F.S. Szalay (eds.). Plenum Publishing, NY.

Meneghini, G. 1839. Nuova specie di alga. Nuova Giornale de' Litterati 39: 67-68 (as cited in Merola et al, 1981).

Meneghini, G. 1841. Monographia Nostichinearum italicarum. Accad. Sci., Torino; Mat. Fis., ser. 2, 5: 1-143 (as cited in Merola et al, 1981).

Mereshkovsky, K. 1910. Theorie der zwei Plasmaarten als Grundlage der Symbiogenesis, einer neue Lehre von den Enstehung der Organismen. Biol. Zent. 30: 278-303, 321-47, 353-67.

Merola, A., Castaldo, R., De Luca, P., Gambardella, R., Musacchio, A., Taddei, R. 1981. Revision of *Cyanidium caldarium*: 3 species of acidophilic algae. Giorn. Bot. Ital. 115: 189-95.

Meyer, T.E., Cusanovich, M.A., Kamen, MD. 1986. Evidence against use of bacterial amino acid sequence data for construction of all-inclusive phylogenetic trees. Proc. NSA USA 83: 217-20.

Mindell, David. 1992.

Minge, M.A., Silberman, J.D., Orr, R.J.S., Cavalier-Smith, T., Shalchian-Tabrizi, K., Burki, F. Skjæveland, Å., Jakobsen, K.S. 2009. Evolutionary position of breviate amoebae and the primary eukaryote divergence. Proc. Biol. Sci. 276: 597–604 (nih.gov).

Möhn, E. 1984. System und Phylogenie der Lebewese. 2 vols. E. Schweizerbart'sche.

de Monet (Lamarck), J.B.J. 1806. Tableau du règne animal.

Moldowan et al., 1996. Abstract. Geol. 24.

Moreira D., Heyden S, von der Bass, D., Lopez-Garcia P., Chao E., and Cavalier-Smith T. 2006. Globoal eukaryote phylogeny: combined S and LSU rDNA trees support monophyly of Rhizaria, Retaria, and Excavata. Mol. Phyl. Evol. (sciencedirect.com).

Mulkidjanian, A.Y., Koonin, E.V, Makarova, K.S, Haselkorn, R., Galperin, M.Y. (year not given).

Origin and evolution of photosynthesis: clues from genome comparison (macromol.uni-osnabrueck.de).

Muller, Félix; Brissac, Terry; Le Bris, Nadine; Felbeck, Horst; Gros, Olivier. 2010. First description of giant Archeota (Thaumarchaeota) associated with putative eubacterial ectosymbiotes in a sulfidic marine habitat. Environmental Microbiology 12: 2371–2383.

von Munchhausen, O. 1765-76. Bibliotheca Botanico-Physico-Economica. Forsters, Hanover (as cited in Ragan, 1997).

Nagashima, H. et al. 1993. Several new strains of thermal alga *Cyanidioschyzon* as the most primitive eukaryotes, pp. 279-285, in Endocytobiology, Sato, V, S., Ishida, M., Ishakawa, H. (eds.), Tübingen U. Press.

Nelson, Gareth and Platnick, Norman. 1981. Systematics and Biogeography: Cladistics and Vicariance. Columbia University Press, New York (book-re.org).

Neushul, M. 1974. Botany. Hamilton Publishing Co., Santa Barbara, Cal. (as cited in Edwards, 1976).

Nikolaev, S.I., Berney, C., Fahrni, J.F., Bolivar, I., Polet, S., Mylnikov, A.P., Aleshin, V.V., Petrov, N.B., Pawlowski, J. 2004. The twilight of Heliozoa and rise of Rhizaria, an emerging supergroup of amoeboid eukaryotes. Proc. Natl. Acad. Sci. U.S.A. 101: 8066–8071 (nih.gov).

Nixon, K.C. 1999. The parsimony ratchet, a new method for rapid parsimony analysis. Cladistics 15: 407-414.

Not, Fabrice; Valentin, Klaus; Romari, Khadidja; Lovejoy, Conny; Massana, Ramon; Töbe, Kerstin; Vaulot, Daniel; Medlin, Linda. 2007. Picobiliphytes: A Marine Picoplanktonic Algal Group with Unknown Affinities to Other Eukaryotes. Science 315: 253-55 (Abstract-sciencemag.org).

Novak, F.A. 1930. Systematika Botanika. J.R. Vilimek, Praha.

Nozaki H., Iseki M., Hasegawa M., Misawa K., Nakada T., Sasaki N., Watanabe M. 2007. Phylogeny of primary photosynthetic eukaryotes as deduced from slowly evolving nuclear genes. Mol. Biol. Evol. 24: 1592-1595.

Okamoto, N., Chantangsi, C., Horák, A., Leander, B.S., Keeling, P.J. 2009. Molecular phylogeny and description of the Novel Katablepharid *Roombia truncata* gen. et sp.nov., and Establishment of the Hacrobia taxon nov. PLoS One. 4, e7080 (nih.gov).

Olendzenski, L. et al. 2000. Horizontal transfer of archeal genes into Deinococcaceae. J. Mol. Evol. 51: 587-99.

Olsen, G.J., Matsuda, H., Hagstrom, R., Overbeek, R. 1994. FastDNAml: a tool for construction of phylogentic trees of DNA sequences using maximum likelihood. CA-BIOS 10: 41-48.

Owen, Richard. 1859. Paleontology, or a systematic Summary of extinct Animals and their geological Relations. 2nd ed. 1861. A. & C. Black, Edinburgh (as cited in Whittaker, 1969).

Pace, N.R. 2006. Time for a change. Nature 441, 289 (Abstract-nature.com).

Parfrey L., Barber E., Lasser E., Dunthorn M., Bhattacharya D., Patterson D.J., and Katz L. 2006. Evaluating support for the current classification of eukaryotic diversity. PSOL Genetics 2: 220-38.

Parfrey, L., Grant, J., Tekle, Y.I., Lasek-Nesselquist, E., Morrison, H.G., Sogin, M.L., Patterson, D.J., Katz, L.A. 2010. Broadly Sampled Muligene Analyses Yield Well-Resolved Eukaryotic tree of Life. Syst. Biol. 59: 518-533.

Parker, Sybil (ed.). 1982. Synopsis and Classification of Living Organisms. McGraw-Hill, New York.

Pascher, A. 1931. Systematische Ubersicht uber die Flagellaten. Bot. Zentrlb. Beih, Abt. 248: 317-32 (as cited in Dodson, 1971).

Patterson, D.J. 1999. The Diversity of Eukaryotes. Am. Nat. 65 (supplement) (asu.edu).

Pawlowski, Jan. 2013. The new micro-kingdoms of eukaryotes. BMC Biology 11: 40-42 (researchgate.net; europepmc.org).

Pennisi, Elizabeth. 2004. Evolutionary biology. The birth of the nucleus. Science 305: 766-768.

Pester, Michael; Schleper, Christa; Wagner, Michael. 2011. Thaumarchaeota: an emerging view of its phylogeny and ecophysiology. Current Opinion in Microbiology 14: 300–306. (researchgate.net).

Philip, G.K., Creevey, C.J., McInerney, J.O. 2005. Opisthokonta and Ecdysozoa May Not Be Clades: Stronger Support for the Grouping of Plant and Animal than for Animal and Fungi and Stronger Support for Coelomata than Ecdysozoa. Mol. Biol. Evol. 22: 1175–1184 (bioinfo.nuim.ie).

Philippe, H. & Adoutte, A. 1998. The molecular phylogeny of Eukaryota: solid facts and uncertainties. In Evolutionary Relationships Among Protozoa, pp. 25-56, G.H. Coombs, K. Vickerman, M.A. Sleigh, M.A. and Warren, A. (eds.), Systcs. Assc. Spl. Vol. Srs. 56, Kluwer Academic, Dordrecht, Netherlands.

Philippe, H., Germot, A., Moreira, D. 2000. The new phylogeny of eukaryotes. Curr. Opin. Genetics Devel.10: 596–601 (max2.ese.u-psud.fr).

Pisani, D., Cotton, J., McInerney, J. 2007. Supertrees Disentangle the Chimeral Origin of Eukaryotic Genomes. Mol. Biol. Evol. 24: 1752-60.

Pisani, Davide; Benton, Michael; Wilkinson, Mark. 2007. Congruence of Morphological and Molecular Phylogenies. Acta Biotheor. 55: 269-81 (paleo.gly. Bris.ac.uk.).

Poole, A.M., Jeffares, D.C., Penny, D. 1998. The path from the RNA world. J. Mol. Evol. 46: 1-17.

Porter, Susannah. 2006. The Proterozoic Fossil Record of Heterotrophic Eukaryotes, Chapter 1. In Neoproterozoic Geobiology and Paleobiology, edited by Shuhai Xiao and Alan Jay Kaufman. Springer, Netherlands (geol.ucsb.edu).

Porter, S.M., Meisterfeld, R., Knoll, A.H. 2003. Vase shaped microfossils from the Neoproterozoic Chuar Group, Grand Canyon: a classification guided by modern testate amoebae. J. Paleont. 77: 409-429.

Portier, P. 1918. Les Symbiotes. Masson, Paris (as cited in Dodson, 1971).

Qi, J., Wang, B., Hao, B. 2004. Whole Proteome Prokaryote Phylogeny Without Sequence Alignment: a K-String Composition Approach. J. Mol. Evol. 58:1-11.

Raff, E.C., Diaz, H.B., Hoyle, H.D., Hutchens, J.A., Kimble, M., Raff, R.A., Rudolph, J.E., & Subler, M.A. 1987. Origin of multiple gene families--are there both functional and regulatory constraints? In Development as an Evolutionary Process, pp. 203-238, R.A. and E.C. Raff, Alan Liss, New York.

Ragan, M.A. 1997. A 3rd Kingdom of Life: the history of an idea. Arch. Protistenkd. 148: 225-43.

Ragan, M.A., Chapman, D.J. 1978. A Biochemical Phylogeny of Protistans. Academic Press, New York.

Raikov, Igor. 1982. The Protozoan Nucleus: Morphology and Evolution (Cell Biology Monographs, Vol. 9), transl. Nicholas Bobrov and Marina Verkhovtseva, Springer-Verlag, Vienna and New York.

Raoult, Didier; Audic, Stephane; Robert, Catherine; Abergel, Chantal; Renesto, Patricia; Ogata, Hiroyuki. 2004 The 1.2-Megabase Genome Sequence of Mimivirus. Science 306: 1344-50 (researchgate.net).

Rappé, M. S., Giovannoni, S. J. 2003. The Uncultured Microbial Majority. Annual Review of Microbiology 57: 369–394.

Raup, D.M.. 1986. Biological Extinction in Earth History. Science (New Series) 231: 1528-1533 (johnboccio.com).

Richards, T., Cavalier-Smith, T. 2005. Myosin evolution and the primary divergence of eukaryotes. Nature 436: 1113-1118.

Rivera, Maria & Lake, James. 2004. The ring of life provides evidence for a genome fusion origin of eukaryotes. Nature 431 (molevol.de/nature.com)

Rodriguez-Ezpeleta, N. et al. 2007. Towards resolving the eukaryotic tree: the phylogenetic positions of jakobids and cercomonads. Curr. Biol. 17: 1420-25.

Sachs, J. 1874. Lehrbuch der Botanik, 4te umgearb. Engelmann, Leipzig (as cited in Demoulin, 1974).

Saunders, G. W., Bailey, J.C. 1997. Phylogenesis of pit-plug-associated features in the Rhodophyta: inferences from molecular systematic data. Can. J. Bot. 75: 1436-1447 (nrcresearchpress.com).

Saunders, G.W, Potter D., Andersen, R.A. 1997. Phylogenetic affinities of Sarcinochrysidales and Chrysomeridales (Heterokonta) based on analyses of molecular and combined data. J. Phycol. 33: 310-318.

Saunders, G.W., Hommersand, M.H. 2004. Assessing Supraordinal Red Algal Diversity and Taxonomy in the Context of Contemporary Systematic Data. Am. J. Bot. 91: 1494-1507 (botany.wisc.edu).

Savile, D.B.O. 1968. Possible Interelationships between Fungal Groups. In Fungi, vol. 3, pp. 649-75, G.C. Ainsworth and A.S. Sussman (eds.), Academic Press, NY and London (as cited in Demoulin, 1974).

Shalchian-Tabrizi, K., Eikrem,W., Klaveness, D., Vaulot, D., Minge, M.A., Le Gall, F., Romari, K., Throndsen, J., Botnen, A., Massana, R., Thomsen, H.A., Jakobsen, K.S. 2006. *Telonemia*, a new protistan phylum with affinity to chromistan lineages. Proc. Biol. Sci. 273, 1833–1842 (nih.gov).

Schleper, C., Nicol, G.W. 2010. Ammonia-oxidizing archeobacteria--physiology, ecology, and evolution. Adv Microb Physiol. 57: 1-41.

Schuh, Randall. 2000. Biological Systematics: Principles and Applications, Cornell U. Press (books.google).

Seckbach, J. 1987. Evolution of eukaryotic cells via bridge algae, the cyanidia connection. Ann. NY Acad. Sci. 503: 424-37.

Seckbach, J. 1994. The 1st eukaryotic cells-acid hot-spring algae. J. Biol. Physics 20: 335-345.

Skophammer, R.G., Servin, J.A., Herbold, C.W., Lake, J.A. 2007. Evidence for a Gram-positive, eubacterial root of the tree of life. Mol. Biol. Evol. 24: 1761-1768.

Sogin, M.L., Edman, U., Elwood, H. 1989. A single kingdom of eukaryotes. In Heirarchy of Life, pp. 133-43, B. Fernholm, F. Bremer, H. Jornvall (eds.), Excerpta Medica, Amsterdam.

Spang, Anja; Saw, Jimmy; Jørgensen, Steffen; Zaremba-Niedzwiedzka, Katarzyna; Martijn, Joran; Lind, Anders; van Eijk, Roel; Schleper, Christa; Guy, Lionel; & Ettema, Thijs. 2015. Complex Archeota that bridge the gap between prokaryotes and eukaryotes. Nature 521: 173–179 (Abstract-nature.com).

Stanier, R.Y., Ingraham, J.L., Wheelis, M.L., Painter, P.R. 1986. The Microbial World. Prentice-Hall, Englewood Cliffs, NJ.

Starobogatoff, Y.I. 1986. On the number of kingdoms of eukaryotic organisms. Trudy Zool. Inst. 144: 4-25 (in Russian).

Stiller, J.W., Harrell, L. 2005. The largest subunit of RNA polymerase II from Glaucocystophyta: functional constraint and short-branch exclusion in deep eukaryotic phylogeny. BMC Evol. Biol. 5: 71 (biomedcentral.com).

Takemura, M. 2001. Poxviruses and the origin of the eukaryotic nucleus. Journal of Molecular Evolution 52: 419-425.

Tekle YI, Grant J, Cole JC, Nerad TA, Patterson DJ, Anderson OR, Katz LA. 2008. Phylogenetic placement of diverse amoebae inferred from multigene analysis and assessment of the stability of clades within 'Amoebozoa' upon removal of varying fast rate classes of SSU-rDNA. Molecular Phylogenetics and Evolution 47: 339–352.

Theophrastus. 200s BC. Enquiry into Plants, 10 vols. (9 extant) (transl. A. Hort, 1916) (as cited in Jones, Samuel & Luchsinger, Arlene, 1979, Plant Systematics, McGraw-Hill).

Tilden, J. 1898. Observations on some west American thermal algae. Bot. Gaz. 26: 89-105 (as cited in Merola et al, 1981).

Tilden, J. 1910. Minnesota Algae, Vol. 1 (the Myxophyceae). U. Minn. Press, Minneapolis.

Tillyard, R.J. 1921. A new classification of the order Perlaria. Can. Entomol. 53: 35–43.

Valas, R. E., Bourne, P. E. 2011. The origin of a derived superkingdom: how a Gram-positive bacterium crossed the desert to become an archeote. Biology Direct 6: 16 (ncbi.nlm.nhi.gov).

Vada, A.R. 1952. Philema Organyeskogo Mira. Botanyeski Zhyrnal 37: 639-47.

Van den Hoek, C., Mann, D.G., Jahns, H.M. 1995. Algae: an Introduction to Phycology. Cambridge U. Press.

Villarreal, L. P. & DeFilippis, V. R. 2000. A hypothesis for DNA viruses as the origin of eukaryotic replication proteins. Journal of Virology 74: 7079–7084.

Walker G., Simpson A.G.B., Edgcomb V., Sogin M.L., and Patterson D.J. 2001. Ultrastructural studies of *Mastigameba punctophora* and *simplex* and *Mastigella commutans* and assesssment of hypotheses of relatedness of pelobiotes (Protista). Eur. J. Protistol. 37: 25-49.

Wallin, I.E. 1927. Symbiosis and the Origin of Species. Williams and Wilkins, Baltimore (as cited in Dodson, 1971).

Walton, L. B. 1930. Studies concerning organisms occurring in water supplies with particular reference to those found in Ohio. Bull. Ohio Biol. Surv. 24: 1-86 (as cited in Whittaker, 1957).

Wessner, D. R. 2010a. The Origins of Viruses. Nature Education 3: 37 (nature.com).

Wessner, D. R. 2010b. Discovery of the Giant Mimivirus. Nature Education 3: 61 (nature.com).

Wheelis, M.L., Kandler, O, & Woese, C.R. 1992. On the nature of global classification. Proc. NAS USA 89: 2930- 2934.

Whittaker, R.H. 1957. The kingdoms of the living world. Ecology 38: 536-538.

Whittaker, R.H. 1959. On the broad classification of organisms. Quart. Rev. Biol. 34: 210-226.

Whittaker, R.H. 1969. New concepts of kingdoms. Science 163: 150-160.

Whittaker, R. Margulis, L. 1978. Protistan classification and the kingdoms of organisms. BioSystems 10: 3-18.

Wilkinson, Mark. 1994. Common cladistic information and its consensus representation: reduced Adams and reduced cladistic consensus trees and profiles. Syst. Biol. 43: 343-368.

Wilkinson, Mark. 1995. More on reduced consensus methods. Syst. Biol. 44: 436-440.

Williams, D. 1991. Phylogenetic relationships among Chromista: a review and preliminary analysis. Cladistics 7: 141-156.

Williams, D., Ebach, M. 2008. Foundations of Systematics and Biogeography. Springer, New York (amazon.com).

Williams TA, Foster PG, Cox CJ, Embley TM. 2013. An archeote origin of eukaryotes supports only two primary domains of life. Nature 504: 231–236 (researchgate.net).

Williams TA, Embley TM. 2014. Archeote 'dark matter' and the origin of eukaryotes. Genome Biol. Evol. 6: 474–481 (researchgate.net).

Wilson, T.B., Cassin, J. 1863. On a third kingdom of organized beings. Proc. Acad. Nat. Sci. Phila.15: 113-21 (jstor.org).

Witzgany, Günther. 2015. Supplemental data to main article "The Agents of Natural Genome Editing"

(researchgate.net).
Woese, C.R. 1987. Bacterial evolution. Microbiol. Rev. 51: 221-271.
Woese, Carl. 1998. The universal ancestor. Proc Natl Acad Sci U S A 95: 6854–6859 (nih.gov).
Woese, C.R. & Fox, G. E. 1977. The concept of cellular evolution. J. Mol. Evol. 10: 1-6.
Woese, C.R., Debrunner-Vossbrinck, B.A., Oyaizu, A., Stackebrandt, S., Ludwig, W. 1985. Gram-positive bacteria: possible photosynthetic ancestry. Science 229: 762-765.
Woese, C.R., Olsen, G.V. 1986. Archeobacterial phylogeny: perspectives on the urkingdoms. Syst. Appl. Microbiol. 7: 161-177.
Woese, C.R., Kandler, O., & Wheelis, M.L. 1990. Towards a natural system of organisms: proposal for the domains Archea, Bacteria, and Eucarya. Proc. NAS USA 87: 4576-4579.
Wolf, Y., Rogozin, I., Grishin, N., Tatusov, R., and Koonin, E. 2001. Genome trees constructed using five different approaches suggest new major bacterial clades. BMC Evol. Biol. 1: 8 (nih.gov).
Wolf, Y. I., Rogozin, I. B., and Koonin, E. V. 2004. Coelomata and not ecdysozoa: evidence from genome-wide phylogenetic analysis. Genome Res. 14: 29–36.
Wyss, A.R., Novacek, M.J., & McKenna, M.C. 1987. Amino acid sequence versus morphological data and the interordinal relationships of mammals. Mol. Biol. Evol. 4: 9-116.
Yang, S., Doolittle, R. F., and Bourne, P. E. 2005. Phylogeny determined by protein domain content. Proc. Natl. Acad. Sci. USA 102: 373–378.
Yoon, Hwan Su: Hackett, Jeremiah; Pinto, Gabriele; and Bhattacharya, Debashish. 2002. The single, ancient origin of chromistan plastids. PNAS 99: 15507–15512.
Yong, Ed. 2013. Giant viruses open Pandora's box (nature.com).
Yoon HS, Grant J, Tekle YI, Wu M, Chaon BC, Cole JC, Logsdon JM, Patterson DJ, Bhattacharya D, Katz LA. 2008. Broadly sampled multigene trees of eukaryotes. BMC Evol. Biol. 8: 14 (biomedcentral.com).
Zillig, W. et al. 1989. Did eukaryotes originate by a fusion event? Endocytobiosis Cell Res. 6: 1-25.

Part 2

Chapter 10 Bacteria

The 4th sort of creatures...which moved through the 3 former sorts, were incredibly small, and so small in my eye that I judged, that if 100 of them lay one by another, they would not equal the length of a grain of course sand; and according to this estimate, ten hundred thousand of them could not equal the dimensions of a grain of such course sand. There was discover'd by me a fifth sort, which had near the thickness of the former, but they were almost twice as long. (The first time bacteria were observed.) - Antonie van Leeuwenhoek, Letter to H. Oldenburg, Oct. 9, 1676, quoted in The Collected Letters of Antonie van Leeuwenhoek (1957), Vol. 2, p. 95 – todayinsci.com.

The classification of bacteria has had a checkered and relatively short past. They were "invented" by celebrated Dutch microscopist Anton Leeuwenhoek (1683) who considered them animalcules, and in the Linnean system of 1735 they were placed as specia dubia in Vermes. Probably the first definite, named description of what we now know as bacteria was by Otto Friedrich Mueller in 1773, the genera being *Monas* and *Vibrio* (Buchanan, 1925). They were regarded as infusorians (the full name being Animalcula Infusoria, established by Ledermüller in 1760-63, refering to life forms able to produce dessication-resistant stages and which can be reactivated by an infusion of water to hay or pepper contaminated with such resting stages) by German naturalist, zoologist, geologist, and microscopist C. G. Ehrenberg in 1838, who first named them "Bakterien" (the English cognate was coined in 1847) and established 2 families for them: Monadina, including only *Monas*, and Vibrionia, including *Vibrio, Bacterium, Spirocheta, Spirillum, Spiradiscus,* and possibly several others. They were switched to Plantae by German biologist Ferdinand Cohn in 1872, who was one of the founders of modern bacteriology and microbiology, and who realized that blue algae were related to bacteria and set up the order Schizosporeae to accommodate both groups with the latter as family Bacteriaceae, which contained 6 genera in 4 tribes--Spherobacteria (*Micrococcus*), Microbacteria (*Bacterium*), Desmobacteria (*Bacillus* and *Vibrio*), and Spirobacteria (*Spirillum* and *Spirochete*), and later integrated them as Schizophyta in 1875, dividing them into Gleogenae (17 genera) and Nematogenae (25 genera). In the first taxonomic tree (Haeckel, 1871) they were in Moneres, along with amebas, in Protista. In 1877, McNab recognized 5 families for Cyanophyceae and 1, Bacteriaceae, for Schizomycetes. Trevisan, in 1879, divided bacteria into 2 tribes: the unicellular Bacterieae, with 2 subtribes and 5 genera, and the pluricellular Vibrionieae, with 3 subtribes and 12 genera.

In 1883, Danish physician Hans Christian Gram accidentally discovered the basic bacterial cell wall difference using a staining method he had just invented, one that would take his name and become famous, while working on a deceased pneumonia patient in a Berlin hospital. The discovery was published by Friedlander, a colleague of Gram and head of the lab where Gram made the discovery, in 1883 in an article titled Die Mikrokokken der Pneumonie in the Fortschritte der Medicin 1: 715, and by Gram himself 4 months later in the article Ueber die isolirte Farbung der Schizomyceten in Schnitt- und Trockenpreparaten for the Fortschritte der Medicin 2: 185 in March of 1884 and in the proceedings of the Congrès périodique international des sciences medicales, 8me Session, Section de pathologie générale et d'anatomie pathologique, which took place in Copenhagen, Aug. 10-16, of the same year and was titled Ueber die Farbung der Schizomyceten in Schnittenpreparaten, p.116-117 (Austrian, 1959, 1960). The method involves the use of crystal or gentian violet (the primary stain), iodine (the mordant), alcohol and/or acetone (the decolourizer), and safranin or carbol fuchsin (the counterstain or secondary stain). The Gram negatives don't retain the purple or blue colour of the 1st dye so go red or pink with the reddish counterstain because they have a thin monolayer of murein (10% of the cell wall), while the Gram positives do retain it with the same counterstain because they have thick, multilayered murien (50-90% of the cell wall). But the Gram negative-positive dichotomy was not recognized taxonomically until about 90 years later, this by Gibbons and Murray.

Also in 1883, they were placed in Fungi in Eichler's Thallophyta as Schizomycetes. In the same year, Zopf's arrangement appeared, which had 3 groups: spherical (*Leuconostoc*), genera with coccus and rod stages, and genera with spiral stages. Schroeter, in 1886, recognized Coccobacteria, Eubacteria (rods), and Desmobacteria (long filaments, usually with sheaths). In the same year, Flugge saw bacteria as 2 groups: spherical and cylindrical. In 1887, Maggi recognized the same 4 tribes as Cohn's 1872 arrangement but in the reverse order and with 11 genera. In 1888, Hansgirg divided Schizomycetes into 3 orders: Desmobacteria (filamentous; 3 families, Clado-, Creno-, and Leptothricaceae), Eubacteria (2 families, Bacteriaceae and Myconostocaceae), and Spherobacteria (1 family, Mycococcaceae).

Trevisan in 1887, Toni and Trevisan in 1889, and Ludwig in 1892 all recognized 3 groups: a coccoid, a bacillary, and a filamentous one. The 1st and 2nd schemes included Trichogenae, Baculogenae, and Coccogenae. The 1st had 26 genera in 1 family, but the 2nd was probably the most elaborate of its time, with 650 species described. It also had only 1 family and the 3 taxons were subfamilies. The 3rd author had no formal names for the 3 taxons and included 18 genera.

Matsuschita in 1902, Flugge in 1907, Heim in 1911, and Lohnis in 1913 all recognized 3 groups still based on simple cell morphology: Coccaceae, Bacillaceae, and Spirillaceae. Migula in 1890, Sternberg in 1892, and Fischer in 1897 and 1903 all recognized the same 3 groups but with a filamentous taxon, Trichobacterinae, added or separated out, the last author placing the 3 basic groups into the order Haplobacterinae, therefore establishing 2 orders, the 1st with 3 families and the 2nd with 1. Benecke, in 1912, had the same orders except the 1st had 6 families and the 2nd was named Desmobacterinae. Baumgartner's classification of 1890 had a monomorphic division, with 3 genera representing coccoid, bacillary, and helical forms, respectively, and a pleomorphic division, also with 3 genera. Kendall, in 1902, also recognized the same 3 basic morphological groups but added or separated out Chlamydobacteriaceae and Beggiatoaceae. In 1909, Conn recognized the same 3 basic groups also, adding a group designated as "higher". Migula, in 1900, modified his taxonomy placing the 3 basic groups as families into the order Eubacteria and adding or separating out Thiobacteria, which had 2 families: Beggiatoaceae and Rhodobacteriaceae. Engler in his 1912 edition of his syllabus followed Migula but had 6 families in Eubacteria.

Chester's 1901 scheme included Coccaceae, Bacillaceae, Chlamydobacteriaceae, Beggiatoaceae, and Mycobacteriaceae. In 1909, Orla-Jensen established 2 orders: Cephalotrichinae (7 families) and Peritrichinae (presumably with only 1 family). In 1917, the SAB (Society of American Bacteriologists) Committee's Preliminary Report (Winslow et al) divided Class Schizomycetes into 4 orders: Myxobacteriales, Thiobacteriales, Chlamydobacteriales, and Eubacteriales, with a 5th group being 4 genera considered intermediate between bacteria and protozoans: *Spirocheta, Cristospira, Saprospira*, and *Treponema*. In the same year, R. E. Buchanan, who was part of the committee, arranged bacteria into 6 orders: Eubacteriales, Chlamydobacteriales, Actinomycetales, Thiobacteriales, Myxobacteriales, and Spirochetales. In 1920, the Final Report of the SAB Committee (also by Winslow et al) was published, and 1923 saw the 1st edition of the famous and classic standard Manual of Determinative Bacteriology (Bergey et al) in which Winslow et al were generally followed. In 1921, Hilda Hempl-Heller established Bacteria as a phylum and divided it into 2 classes: Eubacterieae (3 orders: Eubacteriales, Thiobacteriales, and Chlamydobacteriales) and Myxobacterieae. Ernst Pribham (1929), bacteriology professor at Loyola in Chicago, presented a comprehensive taxonomy in 1928 at the annual meeting of the ASM (American Society for Microbiology), which included 1 class, Schizomycetes, 4 subclasses, 10 orders, 21 families, and 96 genera (Table 10-1).

Table 10-1. Pribham's Classification, 1928.

Protozoobacteria
  Spirochetales
    Spirochetaceae (2 gen.)
    Cristispiraceae (3 gen.)
Eubacteria
  Protobacteriales
    Nitrobacteriacae (5 gen.)
    Thiobacilli (1 gen.)
  Metabacteriales
    Pseudomonadaceae (3 tribes, 4 gen.)
    Bacteriaceae (4 tribes, 6 gen.)
    Micrococcaceae (2 tribes, 5 gen.)
Mycobacteria
  Bacteriomycetales
    Leptotrichaceae (2 tribes, 4 gen.)
    Bacteroidaceae (2 tribes, 6 gen.)
  Bacillomycetales
    Bacillaceae (2 sbfams., 2 tribes each, 4 subtribes, 9 gen.)
    Clostridaceae (2 sbfams., 2 tribes each, 7 gen.)
  Actinomycetales
    Mycobacteriaceae (3 tribes, 8 gen.),
    Actinomycetaceae (2 tribes, 2 gen.)
Algobacteria
  Desmobacteriales
    Spherotilaceae (1 gen.)
  Siderobacteriales
    Chlamydotrichaceae (2 tribes, 3 gen.)
    Siderocapsaceae (3 gen.)
  Thiobacteriales
    Rhodobacteriaceae (2 sbfams., 5 tribes, 16 gen.)
    Beggiatoaceae (3 gen.)
    Achromatiaceae (4 gen.)
  Myxobacteriales
    Polyangiaceae (2 gen.)
    Myxococcaceae (1 gen.)

Three of the subclasses grouped those families which had animal, fungal, or plant similarities and were deemed to be connected to those higher organisms. Eubacteria were considered as being "true bacteria." Gram positives occupy most of Mycobacteria and the Gram negatives were seen as polyphyletic.

Bacteria were first recognized as a kingdom by German zoologist Gunther Enderlein in 1925. Stanier and van Neil (1941) also recognized them as a kingdom, naming them Monera, which included 2 phyla, Myxophyta and Schizomycetae, the latter comprising classes Eubacteriae (3 orders), Myxobacteriae (1 order), and Spirochetae (1 order). Bisset's (1962) scheme had 1 class and 4 orders, Eubacteriales, Actinomycetales, Streptomycetales, and Flexibacteriales, but also a tree which comprised 6 taxons: *Spirillum-Vibrio*, Spirochetes, Trichobacteria, a pseudomonadoid group, a Cytophaga-Myxobacteria group, and Gram positives. Gibbons & Murray (1978), divided them into 4 phyla, Gracilicutes, Firmicutes, Tenericutes (Mollicutes), and Mendosicutes.

French physician and biologist André-Romain Prévot (Caratini, 1971), of the Institut Pasteur, designed a system that also excluded blue algae, was patterned after Pribham, had 4 subphyla, 8 classes, 19 orders, and 39 families, and is in Table 10-2.

Table 10-2. Prévot's System (the numbers in parentheses refer to the family totals).

sbph. Eubacteriales
 cl. Asporulales
  Micrococcales (2)
  Bacteriales (6)
  Spirillales (2)
 cl. Sporulales
  Bacillales (2)
  Clostridiales (2)
  Plectridales (2)
  Sporovibrionales (1)
sbph. Mycobacteriales (with pseudo-mycelium)
 cl. Actinomycetales
  Actinobacteriales (2)
  Mycobacteriales (1)
 cl. Myxobacteriales
  Myxococcales (1)
  Angiobacteriales (3)
  Asporangiales (1)
 cl. Azotobacteriales
  Azotobacteriales (1)
sbph. Algobacteriales
 cl. Siderobacteriales
  Chlamydobacteriales (3)
  Caulobacteriales (1)
 cl. Thiobacteriales (Sulfobacteriales)
  Rhodobacteriales (3)
  Chlorobacteriales (2)
  Beggiatoales (2)
sbph. Protozoobacteriales
 cl. Spirochetales
  Spirochetales (2)

Möhn's (1984) classification is in Table 10-3.

Table 10-3. Uberreich Prokaryonta by Möhn .

Oberreich Archeobacteria
 Reich Archeobacteriobiota

- Unterreich Methanobacteriobiota
    - Stamm Methanococcacea
    - Stamm Methanospirillacea
    - Stamm Methanosarcinacea
    - Stamm Methanobacteriacea
- Unterreich Halobacteriobiota
    - Stamm Halobacteriacea
    - Stamm Halococcacea
- Unterreich Caldariabiota
    - Stamm Sulfolobacea
    - Stamm Thermoplastacea

Oberreich Neobacteria

- Reich Bacteriobiota
    - Unterreich Eubacteria
        - Uberstamm Grambacteria
            - Stamm Streptococcacea
            - Stamm Clostridacea
            - Stamm Bacillacea
            - Stamm Mycoplasmatacea
            - Stamm Chlamydiacea
            - Stamm Micrococcacea
            - Stamm Actinomycetacea
        - Uberstamm Agramabacteria
            - Stamm Desulfovibrionacea
            - Stamm Rhodobacteriacea
            - Stamm Thiobacillacea
            - Stamm Azotobacteriacea
            - Stamm Pseudomonada
            - Stamm Photobacteriacea
            - Stamm Enterobacteriacea
    - Unterreich Spirochetae
        - Stamm Spirochetacea
    - Unterreich Myxobacteriacea
        - Stamm Myxobacteriacea
- Reich Cyanobiota
    - Stamm Cyanophyta
    - Stamm Prochlorophyta

He also designated Spirochetes as evolved from Clostridacea and Cyanobacteria as evolved from Rhodobacteriacea.

The 2nd 4-volume edition of Bergey's by Garrity and Holt from 2002 has 23 eubacterial phylums but are not classified into higher groups (Garrity, Bell, and Lilburn, 2004).

The following are various molecular taxonomies for eubacteria (totalling 20 (with Brown et al considered as 2 and Horike et al considered as 4)(figures in parentheses are bootstrap values).

Table 10-4. Molecular Taxonomies for Eubacteria.

Woese, 1987

- *Thermotoga*
- Supergroup 1
  - Chloroflexi
  - Supergroup 2
    - Deinothermi
    - Supergroup 3
      - Supergroup 4
        - Spirochetes
        - Chlorobia+Cytophagales (Bacteroidetes)
        - Proteinimurus
      - Sporobacteria nom. nov.
        - Mollifirmicutes+Cyanobacteria
        - Proteobacteria Stackebrandt et al 1986

Kuske et al, 1997

- *Aquifex*
- SG 1
  - Planctobacteria
  - SG 2
    - SG 3
      - Bacteroidetes
      - SG 4
        - Cyanobacteria+Mollifirmicutes
        - Proteobacteria

Gupta, 1998

- Posibacteria
  - Mollicutes
  - Actinobacteria (high G-C)
  - Endospora+Thermotogae
- Deinothermi
- Didermata
  - Chloroflexi
  - Cyanobacteria
  - Spirochetes
  - Chlamydiae
  - Aquificales
  - Proteobacteria
    - Delta+-Epsilon
    - Alpha

Beta
Gamma

Brown et al, 2001

a) (23 data sets) (CI .47, RI .59)

Spirochetes
Exoflagellata (100)
 Chlamydiales
 SG 2 (97)
  Chlorobium-Porphyromonas (Bacteroidales)
  SG 3 (99)
   Deinococcus-Actinobacteria
   SG 4 (78)
    Clostridium
    SG 5 (77)
     Mollicutes+Bacilli-Clostridia (100)
     SG 6 (83)
      Aquifex-Thermotoga (98)
      Cyanobacteria-Proteobacteria (78)

b) (14 data sets, (9 (HGT) data sets removed)) (CI .49, RI .63)

Aquifex
SG 1
 Thermotoga
 SG 2
  SG 3 (60)
   SG 4 (51)
    (Chlamydiales-Spirochetes+*Chlorobium-Porphyromonas*)
    Actinobacteria
   Clostridium + (Mollicutes+Bacilli-Clostridia)
  SG 5
   Cyanobacteria+*Deinococcus* (80)
   Proteobacteria (86)

Wolf et al, 2001

Parasitica
 Mollicutes
 Chlamydiae-Spirochetes
 Rickettsiae
SG 2
 Aquifex-Thermotoga
 Proteobacteria
 Bacilli

Actinomycetes-*Deinococcus*-Cyanobacteria

Brochier and Philippe, 2002

Planctomyceteales
Mureinomura
  Bacteroidaceae
  Metamureinomura
    Spirochetes
    Aquificales-Thermotogales+Fusobacteria
    Gamma-Proteobacteria
    Beta-Proteobacteria
    Alpha-Proteobacteria
    Delta-Proteobacteria
    Epsilon-Proteobacteria
    Mollicutes-Endospora + Chloroflexi
    Actinobacteria
    Cyanobacteria
    Thermales-Deinococci+Chlorobia

Daubin et al, 2002 (using supertrees)

Spirochetes
SG 1
  Chlamydiales
  SG 2
    Aquifex+Thermotoga
    Sporobacteria
      Cyanobacteria+Mollicutes-Endospora
      SG 4
        Deino-Thermus+Actinobacteria
        Proteobacteria

Rappé and Giovannoni, 2003

Aquificae+*Desulfobacterium*
SG 1
  SG 2
    OP1
    SG 3
      Dictyoglomus
      Thermotogae+*Coprothermobacter*
  SG 4
    Thermodesulfobacteria
    SG 5
      Deinococcus-Thermus
      SG 6

Chloroflexi
SG 7
  Spirochetes+(Synergistes+Caldiserica)
  SG 9
    SG 10
      Poribacteria
      Proteinimurus s.l.
        Planctobacteria+OP3
        Chlamydiae+(*Lentisphera*+Veruccomicrobia)
    SG 12
      SG 13
        Fusobacteria
        Cyanobacteria+Mollicutes-Endospora
      Actinobacteria+Armatimonada
  SG 14
    SG 15
      (Bacteroidetes+Chlorobia)+*Caldithrix*
      Fibrobacteria+Gemmatimonada
    Proteobacteria+(*Chrysiogenes*+Desulfobacteria)
    Elusimicrobia +Acidobacteria

Battistuzzi, Feijao, and Hedges, 2004 (using proteins)

Aquifex
SG 1 (>95)
  Thermotoga
  SG 2 (41)
    SG 3 (45)
      Mollicutes-Endospora (66)+*Fusobacterium* (84)
      Terrabacteria (>95)
        Cyanobacteria+*Deinococcus* (92)
        Actinobacteria (>95)
    SG 4 (70)
      Chlorobium
      SG 5 (60)
        Chlamydiae-Spirochetes (82)
        Proteobacteria (>95)
          Epsilon
          Alpha+(Beta+Gamma) (>95)

Qi, Wang, and Hao, 2004

Spirochetes+Chlamydiae
Proteobacteria
SG 1
  SG 2
    Aquificae+Thermotogae

Fusobacteria+Mollicutes-Endospora
SG 3
Chlorobia
SG 4
Deinobacteria
Cyanobacteria
Actinobacteria

Francesca Ciccarelli et al, 2006

Mollicutes-Endospora
SG 1
SG 2
SG 3
Actinobacteria
Planctobacteria+Spirochetes
SG 4
Chlamydiae
SG 5
Fibrobacteria
Chlorobia+Bacteroidetes
SG 6
SG 7
Fusobacteria+(*Aquifex*+*Thermotoga*)
Cyanobacteria+DeinoThermi
Proteobacteria
Protoprotei
Metaprotei
Deltaprotei
Acidobacteria

Lienau et al, 2006

Mollicutes 1
SG 1
Mollicutes 2
SG 2
Proteinimurus
SG 3
Bacteroidetes
SG 4
Clostridia
SG 5
SG 6
Deltaprotei 1+Epsilonprotei
Thermotogae+(*Geobacter*+Actinobacteria)

- SG 7
  - Aquificae
  - SG 8
    - Cyanobacteria
    - SG 9
      - *Deinococcus* + *Rhodopirellula*
      - SG10
        - Deltaprotei 2
        - Neoprotei
          - Gammaprotei+Betaprotei
          - Alphaprotei

Pisani et al, 2007 (using supertrees)

- *Aquifex*
- Supergroup 1
  - SG 2
    - Cyanobacteria
    - SG 3
      - SG 4
        - Chlorobia+Bacteroidetes
        - Spirochetes+Chlamydiae
      - SG 5
        - SG 6
          - *Pirellula*
          - Deino-Thermi + Actinobacteria
        - Proteobacteria

Fabia Battistuzzi and S. Blair Hedges, 2008

- Thermotogae
- SG 1
  - Aquificae
  - SG 2
    - Fusobacteria
    - SG 3
      - Hydrobacteria
        - SG 4
          - Spirochetes+Planctobacteria
          - Chlamydiae+Chlorobia-Bacteroidetes
        - Proteobacteria
          - Epsilon-Bacteria
          - Neoprotei
            - Solibacteres+Delta-bacteria
            - Alpha+ Beta-Gammabacteria
      - Terrabacteria
        - SG 9

- Mollicutes-Endospora
- Deino-Thermi
- SG 10
  - Cyanobacteria+Chloroflexi
  - Actinobacteria

Yarza et al, 2008, 2010

- Thermotogales
- SG1
  - Thermodesulfobacteria
  - SG 2
    - Synergistes
    - SG 3
      - Caldiserica
      - SG 4
        - Aquificales
        - SG 5
          - Mollicutes-Endospora
          - SG 6
            - Actinobacteria+Chloroflexi
            - SG 7
              - Nitrospirae
              - SG 8
                - Deinococcales
                - SG 9
                  - SG 10
                    - SG 11
                      - SG 12
                        - *Caldithrix*
                        - Gemmatimonada+Fibrobacteria
                      - SG 13
                        - *Phycisphera*+Plancobacteria
                        - Chlamydiae + (*Lentisphera*+Verrucomicrobia)
                    - Chlorobia-Bacteroidetes
                  - SG 14
                    - Thiobacteria
                      - Deltaprotei
                      - *Desulfovibrio*+(*Desulfurella*+Epsilonprotei)
                    - SG 15
                      - SG 16
                        - Acidobacteria
                        - SG 17
                          - *Chrysiogenes*+Deferribacteria
                          - SG 18
                            - Spirochetes
                            - Fusobacteria+Elusimicrobia+Armatimonada

- Proteobacteria
  - Zetaprotei (*Mariprofundus*)
  - Rhodobacteria
    - Alphaprotei
    - Chromatibacteria
      - Gammaprotei
      - Betaprotei

Horike et al, 2009

a) concatenated tree with thermophilic metabacteria as outgroup

- *Thermotoga*
- SG 1
  - *Aquifex*
  - SG 2
    - Chloroflexi
    - SG 3
      - SG 4
        - SG 5
          - Chlamydiae
          - Bacteroidales+Chlorobia
        - Spirochetes+*Fusobacterium*
      - SG 6
        - Planctobacteria+Proteobacteria
        - Terrabacteria
          - Actinobacteria+*Deinococcus*
          - Mollicutes-Endospora + Cyanobacteria

b) supertree with thermophilic metabacteria as outgroup

- *Thermotoga*
- SG 1
  - *Aquifex*-Chloroflexi
  - SG 2
    - SG 3
      - Bacteroidales+Chlorobia+*Fusobacterium*
      - Spirochetes+(Planctobacteria+Proteobacteria)
    - Terrabacteria
      - Actinobacteria+*Deinococcus*
      - (Chlamydiae+Mollicutes-Endospora)+Cyanobacteria

c) concatenated tree with thermophilic metabacteria as outgroup

- *Thermotoga*
- SG 1
  - *Aquifex*

Chloroflexi
SG 2
SG 3
Chlamydiae+(Bacteroidales+Chlorobia)
Spirochetes+*Fusobacterium*
SG 4
Planctobacteria+Proteobacteria
Terrabacteria
Actinobacteria+*Deinococcus*
Mollicutes-Endospora+Cyanobacteria

d) concatenated tree with mesophilic metabacteria as outgroup

Thermotoga-Aquifex
SG 1
SG 2
Fusobacterium+Mollicutes-Endospora
Chloroflexi
SG 3
SG 4
SG 5
Chlamydiae+Spirochetes
Proteobacteria
Bacteroidales+Chlorobia
SG 7
Planctobacteria
Terrabacteria
Actinobacteria+*Deinococcus*
Cyanobacteria

Qi, Wang, and Hao (2004) use a different molecular approach, one without sequence alignment, the composition vector method, which had essentially the same results as Wolf et al, Battistuzzi and Hedges, and Ciccarelli et al. In Wolf, the 1st supergroup (which I dub Parasitica) has mostly small genome size and is parasitic, while the 2nd has large genome size or is free-living with small genome size (*Thermotoga* and *Aquifex*), and the 5 approaches used are i) presence-absence of genomes in clusters of orthologous genes; ii) conservation of local gene order (gene pairs) among prokaryotic genomes; iii) parameters of identity distribution for probable orthologs; iv) analysis of concatenated alignments of ribosomal proteins; v) comparison of trees constructed for multiple protein families.

In the 20 molecular taxonomies above, 6 using RNA, 11 using proteins, 2 using genomes (Table 10-5), and Wolf et al (2001), Chlorobia-Saprospirae show up 12 times out of the 13 times the 2 are included, Spirochetes+Chlorobia-Bacteroidetes+Proteinomura appear 4 times, Spirochetes is an independent lineage in 5 of the trees, Proteinimurus appears 4 times out of 7, Cyanobacteria with Mollifirmicutes or Actinobacteria are found 9 times, Proteobacteria+Cyanobacteria-Mollifirmicutes are in 3 of the trees, Deino-Thermi group with Mollifirmicutes 3 times but are an independent lineage also 3 times, *Thermotoga* groups with *Aquifex* 6 times (probably a case of LGT as suspected by Gupta) but are independent also 6 times, Proteobacteria s.l. are in 15 (Proteobacteria s.s. are in another 4), Posibacteria are in 3 of 13, and Chloroflexi are independent in 6 of 10. We see from these that relationships in Eubacteria (which, like Metabacteria, can be considered a single kingdom) are unresolved, and, although most classifications agree on the 3 domains and a Euryarcheota-Crenarcheota dichotomy in Metabacteria, among eubacteria they show there is considerable disagreement between RNA and proteins, as represented principally by Woese (1987) for the former and Battistuzzi et al (2004) for the latter.

Table 10-5.

RNA

Woese (1987)
Kuske et al (1997)
Brochier and Philippe (2002)
Daubin et al (2002, NJ and ML versions)
Rappé and Giovannoni (2003)
Yarza et al (2008, 2010)

proteins

Gupta (1998)
Brown et al (2001, 23- and 14-data set versions)
Battistuzzi et al (2004)
Qi et al (2004)
Ciccarelli et al (2006)
Horike et al (2009, 4 versions)
Battistuzzi and Hedges (2008)

genomes

Lienau et al (2006)
Pisani et al (2007)

Togabacteria contains 10 or 11 genera (*Thermosipho, Fervidobacteria, Thermotoga, Geotoga, Petrotoga, Marinitoga, Thermococcoides, Kosmotoga, Oceanotoga, Defluvitoga, Mesotoga*) and about 30 species. Most of the genera are sulfur, sulfate, and/or thiosulfate reducers, and all species have ether lipids like metabacteria, a result of LGT. The first 3 are isloated from geothermally heated marine sediments, tidal springs, or hot springs. Many are petrophilic (*Geotoga, Petrotoga, Oceanotoga, Kosmotoga*, and *Thermococcoides*), occuring in oil fields (garciajeanlouis.org; Yixiao Feng et al (bioteke.com); Acton, 2011; LPSN (List of Prokaryotic Names in Standard Nomenclature). The group stains Gram negative, the toga is sometimes confused as an outer membrane, the wall is thick, and the taxon is sometimes considered Gram positive. It has both class 1 and 2 C DNA polymerase. Gram negatives have class 1 only. Cyanobacteria has class 3. The molecular taxonomy for 8 of them presented here is the most representative of the 5 and is the 16S analysis by DiPippo et al (2009).

Table 10-6.

phyl. Togabacteria Cavalier-Smith, 1992 (including *Aquifex*), orthog. emend., syn. class

Thermotogae Reysenbach, 2002, order Thermotogales Reysenbach, 2002
fam. Thermotogaceae Reysenbach, 2002
sbf. Thermotogoideae
tribe Thermotogeae
Thermotoga Stetter and Huber, 1986
tribe Fervidobactereae
Thermosipho Huber et al, 1989
Fervidobacterium Patel et al, 1985
fam. Petrotogaceae
sbf. Kosmotogoideae
Kosmotoga DiPippo et al, 2009
sbf. Petrotogoideae
tribe Petrotogeae
Petrotoga Davey et al, 1993
Geotoga Davey et al, 1993
tribe Marinitogeae
Marinitoga Wery et al, 2001

The other analyses are by DiPippo et al (2009, the 23S analysis), Feng et al (Acton, 2011), Nanoura et al (2011), and Gupta and Bhandari (2011). There is only 1 family recognized so the divisions here are my own. *Thermococcoides* Feng et al, 2010 is in Thermotogaceae as sister group derived from Thermotogoideae in Yixiao Feng et al but was subsequently placed in *Kosmotoga* by Nunoura et al (2011). *Oceanotoga* was named by Jaysinghearchchi et al in 2011, *Defluvitoga* by Ben Hania et al in 2012, and *Mesotoga* (for mesophilic species) by Nesbø et al in 2013.

Deinothermi are aerobic, immotile, nonsporulating, and heterotrophic, and have an atypical wall profile--thick as in Firmicutes but with an outer membrane as in Gracilicutes. The phylum is usually refered to as *Deinococcus-Thermus* and the class as Deinococci Garrity & Holt 2002. Thermi's genera are *Thermus* Brock and Freeze 1969*, Marinithermus* Sako et al 2003*, Meiothermus* Nobre et al 1996*, Oceanothermus, Rhabdothermus* Steinsbu et al 2011, and *Vulcanithermus* Miroshnichenko et al 2003 and are seen as only 1 family, Thermaceae da Costa & Rainey 2002. They are thermophilic aerobic rods that reduce nitrate to nitrite and their major quinone is MK-8. They are isolated from hydrothermal environments. *Thermus acquaticus*, discovered by Thomas Brock in Yellowstone in 1969, has become the source of the Taq polymerase, a DNA enzyme so heat resistant it can withstand the cycling temperatures required for PCR (polymerase chain reaction) now used for DNA sequencing as well as legal and industrial applications. Deinobacteria are aerobic rods or coccuses and contain 2 families (in order Deinococcales Rainey et al, 1997): Deinobacteriaceae (or Deinococcacee Brooks & Murray, 1981)(*Deinococcus* Brooks & Murray, 1981, *Deinobacterium* Oyaizu et al, 1987) and Trueperaceae Rainey et al, 2005 (*Truepera* da Costa et al, 2005). They are highly radio-resistant, able to tolerate as much as 3 mln. rads of ionizing radiation, while other bacteria are killed by only 100 rads and humans by only 500. *Deinococcus,* also discovered by Brooks (in 1981), was isolated from atomic reactors, *Deinobacterium,* from a Finnish paper mill in 2011, and *Truepera* from a geothermal zone on San Miguel Isl. (westernmost of California's Channel Isls. in the Pacific in Santa Barbara County) in 2005. (garciajeanlouis.net; Margulis & Schwartz, 1998; Taxonomicon).

Hydrogenobacteria (Aquificae) are aerobic, thermophilic, flagellate or immobile, often sulfur- or hydrogen-oxydizing rods found in terrestrial or marine environments, and comprise 9 or 10 genera in 2 or 3 families: Hydrogenobacteriaceae (Aquificaceae Reysenbach 2002, *Aquifex, Calderobacterium* [sometimes considered part of *Hydrogenobacter*], *Hydrogenobacter, Hydrogenivirga,*

Hydrogenobaculum, Thermocrinis) and Hydrogenothermaceae Eder and Huber 2003) (*Hydrogenothermus, Persephonella, Sulfurihydrogenobium, Venenivibrio*). They are isolated from hot springs, sulfur pools, and hydrothermal vents. The first genera discovered were *Hydrogenobacter* in 1984 by Kawasumi et al and *Aquifex* in 1992 by Huber and Stetter. Sometimes Desulfobacteria are included. (garciajeanlouis.net; Taxonomicon; LPSN).

The Saprospirae-Chlorobia group is probably an artifact; the genetic relatedness of organisms would show up in the phenotype. Flexirubins, gliding motility, MK-6 and MK-7, and aerobiosis occur in Flavobacteria but are absent in Bacteroidetes, which are unpigmented, anaerobic, immotile, and have MK-8 to MK-12, except for family Marinilabaceae (considered part of Bacteroidetes), which has MK-7, flexirubin-like pigments, gliding motility, and facultative anaerobiosis. Within subphylum Flavobacteria there are 2 classes: Flavobacteriae (1 order), which has MK-6, and Cytophagae (orders Marinilabales, Sphingobacteriales, and Cytophagales), which has MK-7. Subphylum Bacteroidetes has 1 order. The usual taxonomy recognizes 4 classes: Flavobacteria, Cytophagae, Sphingobacteria, and Bacteroidetes (garciajeanlouis.net).

Photosynthesis occurs in 7 groups: Chloroflexi, Chlorobia, Heliobacteria, *Chloracidobacterium* Bryant et al 2007 (in Acidobacteria), Rhodobacteria, Chromatia, and Cyanobacteria. Battistuzzi, Feijao, and Hedges (2004) provide 3 possible explanations for this wide distribution: 1) it evolved in 1 lineage and spread through lateral gene transfer (LGT), 2) the common ancestor of these groups had this metabolism and genetic machinery (and it was lost in several lineages), and 3) there was a combination of the 2. They assume the 2nd. They also imply that the presence of pigments such as carotenoids, which are photoprotective compounds, is an indication of photosynthetic ancestry. Carotenoids, aside from phototrophs, are present in Thermi, Deinobacteria, Myxobacteria, Methylomonada, Flavobacteria, and Enterobacteria, the last 4 showing up as related to phototrophic taxons in molecular taxonomies. And Gupta states that the similarities in component parts and overall charge transfer mechanisms in different reaction centers indicate that they all originated from a common ancestor.

The first phototroph has been variously proposed as *Heliobacillus* (by Vermaas in 1994 and Gupta in 2003), *Chlorobium* (by Buttner et al in 1992), *Chloroflexus* (by Pierson in 1994), and purple bacterial (by Xiong et al in 2000)(Armen Mulkidjanian, et al, Origin and evolution of photosynthesis: clues from genome comparison-macromol.uni-osnabrueck.de). Mulkidjanian et al state the arguments in favour of purple bacteria do not appear valid, and that there seems to be some support for each of the other candidates. However, they do not give a reference for Buttner et al, but it is presumably "Photosynthetic reaction center genes in green sulfur bacteria and in photosystem 1 are related "(1992. Proc. Nat. Acad. Sci. USA 89: 8135-8139), but this does not mention such an origin, and the Beverly Pierson reference is the wrong one. I did find an article by her and colleagues from the year cited, which does make the suggestion that *Chloroflexus* might be the original phototroph: "On the basis of morphology and physiology it appears that *Chloroflexus* may be more similar than other extant phototrophs to the early microfossils (Olsen and Pierson, 1987; Pierson and Olsen, 1989) although others have suggested these Precambrian filamentous microfossils could be Cyanobacteria (Awramik, 1992)"(Pierson et al, 1994, p. 35). Mulkidjanian and colleagues hold to the LGT option and say that the original phototroph was an anoxygenic ancestor of Cyanobacteria because Tice and Lowe in 2004 presented geological evidence that the Buck Reef Chert, 250-400-m-thick rock running along the South African coast, was produced by phototrophic microbial communities c. 3.4 bya. and defined the inhabitants as partially filamentous phototrophs, which, according to the carbon isotopic composition, used the Calvin cycle to fix CO_2.

Mulkidjanian is also noted for the Zinc World theory of the origin of life, which is an extension of Wächtershäuser's Iron-Sulfur World, and which was further expanded by Mulkidjanian and colleagues with a study to reconstruct the "hatcheries" of the first cells combining geochemical analysis with phylogenomic scrutiny of the inorganic ion requirements of universal components of modern cells. It indicated that the ionic composition conducive to the origin of cells is compatible with emissions of vapour-dominated zones of inland geothermal systems, so the precellular stages of evolution may have taken place in shallow "Darwin ponds". But all theories for the origin of life are controvesial.

Photobacteria are not to be confused with *Photobacterium*, which is a genus in Vibrionaceae containing some species that are luminescent and symbiotes in specialized luminous organs in marine fish. It was established by Hauser in 1885.

Chloroflexi, previously with only 1 family and 2 genera (*Chloroflexus* and *Chloronema*), have 2 orders, Chloroflexales (12 species) and Herpetosiphonales (1 gen.). Five other classes are dubiously attached to the phylum based only on molecules. Two of these classes, Aerolineae (5 gen.) and Caldilineae (1 gen.) are isolated from digestors, Dehalococcoidetes (1 genus) are isolated from underground water contaminated with chlorinated solvants, *Thermomicrobium* is found in a Yellowstone hot spring, *Spherobacterium*, attached to the previous genus as class Thermomicrobia, is Gram positive and was previously assigned to Actinobacteria, where it probably belongs, and Ktedonobacteria (3 gen.) are mycelial, sporulating, and Gram positive (garciajeanlouis.net), and probably belong in Actinobateria also. For Chloroflexales the taxonomy is usually as follows:

Chloroflexales Gupta et al 2013
  Chloroflexaceae Gupta et al 2013
    Chloroflexus Pierson and Castenholz 1974
    Chloronema Dubinina and Gorlenko 1975
    Heliothrix Pierson et al 1986
    Roseiflexus Hanada et al 2002
  Oscillochloridaceae Keppen et al 2000
    Oscillochloris Gorlenko and Pivovarova 1989 (reassigned from *Oscillatoria cerulescens* Gicklhorn 1921 in Cyanophyceae), emend. Keppen et al 2000
Herpetosiphonales Gupta et al 2013
  Herpetosiphonaceae Gupta et al 2013
    Herpetosiphon Holt & Lewin 1968

For Chlorobia the usually recognized taxonomy is as follows:

Chlorobiales Gibbons and Murray 1978
  Chlorobiaceae Copeland 1956
    Chlorobium Nadson 1906
    Pelodictyon (Szafer 1911) Lauterborn 1913
    Prosthecochloris Gorlenko 1970
    Ancalochloris Gorlenko and Lebedeva 1971
    Chloroherpeton Gibson et al 1985
    Chlorobaculum Imhoff 2003
Ignavibacteriales Iino et al 2010
    Ignavibacterium Iino et al 2010

Heliobacteria contain 1 family, Heliobacteriaceae Madigan and Asao 2010, and 4 genera: *Heliobacterium* Gest and Favinger 1985, *Heliobacillus* Beer-Romero and Gest 1987, *Heliophilum* Ormerod et al 1996, *Heliorestis* Bryantseva et al 2000.

Chromatiales Imhoff 2005, the purple sulfurs, include 3 families, Chromatiaceae Bavendamm

1924, being the major and original family, with the type genus, *Chromatium* Perty 1852. The first was *Thiospirillum jenense* (Ehrenberg 1838) Migula 1900. There are 10 genera traditionally recognized.

Rhodospirillales Pfennig and Trüper 1971, the purple non-sulfurs, have 1 family, Rhodospirillaceae Pfennig and Trüper 1971, and 6 genera, with the type genus, *Rhodospirillum*, going back to 1907 when it was named by Molisch along with *Rhodobacter* and *Rhodopseudomonas* in the same year. There are 6 genera traditionally recognized.

Modern classification for Cyanobacteria began with the Geitler system in 1932, which was based on morphology and specimen collection and had 3 orders: Chroococcales, Chemesiphonales, and Hormogonales. It included 1300 species in 145 genera (Schaechter, 2009).

Drouet's system (1981), which was based on herbarium specimens and morphology, simplified the previous system including only 62 species in 24 genera, which was not considered appropriate by taxonomists but was welcomed by biochemists and physiologists because of its ease of use, and several names/laboratory strains, such as *Anacystis nidulans* and *Escherichia coli*, are derived from this system.

Bourelly made a critical reevaluation of the Geitlerian system in 1985 in his "Les algues d'eau douce III," 2nd ed., based primarily on morphology and reproductive characteristics.

The taxonomy most used today is that of Rippka and colleagues (1979), which includes five orders: Chroococcales, Pleurocapsales, Oscillatoriales, Nostocales, and Stigonematales. It is based on gross morphology, fine structure, mode of reproduction, physiology, chemistry, and sometimes genetics based on specimens from collections and culture.

The classification of Anagnostidis and Komarek, based on gross morphology, fine structure, mode of reproduction, and other criteria and presented in a series of articles from 1985 to 1990 in the journal Algology Studies and another series in the journal Archives of Hydrobiology, also from 1985 to 1990, included 4 orders: Chroococcales, Oscillatoriales, Nostocales, and Stigonematales. Pleurocapsales were merged with Chroococcales.

Molecular analyses based on 16S rRNA by Giovannoni and colleagues in 1988 and Turner and colleagues in 1989 included 10 orders. So Chroococcales and Oscillatoriales are polyphyletic. These results, presented below, correspond mostly to the phenotype -- for example, *Gleobacter* is probably the most primitive, with no filaments, thylakoids, hormogoniums, akinetes, nor heterocysts, and reproducing by binary fission instead of sporulation, and Nostocales + Stigonematales standing out mainly by the possession of heterocysts.

Table 10-7. Evolutionary Taxonomy for Cyanobacteria.

- Gleobacteria
- Metacyanobacteria nom. nov.
  - Pseudanabenales
  - Neocyanobacteria nom. nov.
    - Oscillatoriales 1
    - Cenocyanobacteria nom. nov.
      - Prochlorales 1
      - Argacyanobacteria nom. nov.
        - Synechococcales+Prochlorales 2
        - Anacyanobacteria nom. nov.
          - Chroococcales 4 + Pleurocapsales
          - Hormogoneae
            - Oscillatoriales 2

Heterocysta nom. nov. (Nostocales+ Stigonematales)

Proteobacteria is divided up in molecular taxonomies as shown earlier, and Alphaprotei has the synapomorphy of co-enzyme Q-10 (Goodfellow and O'Donnell, 1993, p. 229; jeanlouisgarcia.net), the Beta-Gamma group has the synapomorphies of Q-8, sulfur compound oxidation, and chemolithotrophy, and together they have the synapomorphy of ubiquinones (Goodfellow and O'Donnell, loc. cit.; jeanlouisgarcia). Deinobacteria also has ubiquinones and Aquificae also has chemolithotrophy, and a genus of Flavobacteria and one in Gammaprotei have Q 10, but are perhaps cases of LGT (except for Aquificae); Deinobacteria does group with Proteobacteria (but with *Thermus*) in Lienau et al (2006) and is close to it in Pisani et al (2007)(see earlier).

In essence, the Gamma-Beta group comprises Chromatia, Thiobacteria (sulfide- or sulfur-oxidizers), Siderobacteria (iron-oxidizers), Methylomonada (methanol-oxidizers), Nitrobacteria (nitrogen-oxidizers), Pseudomonada, and Enterobacteria; the Alpha group is made up of Rhodobacteria and Caulobacteria; the Delta group contains Desulfobacteria and Myxobacteria; and the Epsilon group has only Campylobacteria as the major group.

The carotenoid spirilloxanthin occurs in Myxobacteria and Methylomonada and various other genera of Proteobacteria such as *Acidiphilium, Rhizobium, Bradyrhizobium*, and *Sphingomonas*, and is possibly a synapomorphy for this heterogeneous phylum, and occurs as well in *Terriglobus*, which is in Acidobacteria, which, like Desulfobacteria, has iron-reducing genera. In Lienau et al, Epsilonprotei does not group with Proteobacteria. As Gupta (1998, p. 1461) points out, the heterogeneity of the group has been cause for concern.

The phylum can be arranged into subphylum Ubiquinona, containing classes Q-8 (Enterobacteria, Pseudomonada, Chromatia, and Clatobacteria (the chemolithoautotrophic proteobacteria, which are Siderobacteria, Thiobacteria, Methylomonada, and Nitrobacteria) and Q-10 (Rhizobia, Caulobacteria, and Rhodobacteria), subphylum Myxospora, and subphylum Desulfobacteria-Campylobacteria.

There are 4 genera in Siderobacteria: *Gallionella, Sideroxydans, Acidovorax*, and *Aquabacterium*. All are in Beta-Proteobacteria. *Siderocapsa, Spherotilus, Crenothrix*, and *Metallogenium* are ambiguously iron-oxidizing; *Acidithiobacillus* and *Thiomonas*, both in Beta-Protobacteria also, are possible members; and "*Ferritrophicum*" and "*Ferrovum*", both in Beta-Proteobacteria as well, are candidate genera (hence the quotation marks). Nitrobacteria, also have 2 families; one is for the nitrite-oxidizers (*Nitrobacter, Nitrococcus, Nitrospina*) and the other for the ammonia-oxidizers (*Nitrosococcus, Nitrosolobus (Nitrospira), Nitrosomonas*, *Nitrosospira*). The Nitrobacteria taxon and its 2 families do not show up as monophyletic in molecular analyses, nor do Methylomonada (the methylotrophs), which contain some 30 genera.

Thiobacteria include 2 families, which are traditionally recognized, Sulfomonada (Thiobacilliaceae) (as tribe Thiobacilleae Pribram, 1929) and Beggiatoaceae Kendall, 1902. The former includes the mostly non-filamentous, polar-flagellated *Thiobacillus, Thiobacterium, Thiospira, Thiovulum, Thiomicrospira, Thiosphera, "Thiodendron", Thermothrix, Macromonas, Sulfurivirga*, and *Thioalkalimicrobium*. Only *Thiobacterium* is nonmotile and only *Thermothrix* is filamentous. The *Beggiatoa* family comprises the mostly gliding, nonflagellar, and filamentous *Achromatium, Beggiatoa, Leucothrix*, which is very similar to blue algae, *Thioploca, Thiothrix*, and "*Thiospirillopsis*." The first 5 are gliders and the only non-filamentous member is *Achromatium*. Thiotrichaceae Garrity et al 2005 (in Gamma-Protei) are based on molecular data but include most (10) of these 16 genera, including members of both groups--the 4 (recognized) gliders, 5 polar flagellates, and *Thiobacterium,* and add *Thiomargarita*, which is the largest bacterium known, some being as large as .7 mm and can be seen with the naked eye, and looks like a string of pearls (hence the name, *margarita*, being Greek for "pearl") even without the sulfur inclusions, which reflect incident light, which gives it luster, as it occurs as coccoids in chains (see Thiomargarita namibiensis-Wikipedia).

Both *Micromonas* and *Achromatium* have calcium carbonate, and the latter is known for the deposition of moonmilk, which is a speleothem (secondary cave formation) that, when wet, is soft and spongy, sometimes even viscous or liquid, and generally very white and when dry forms crusts on the walls and floors of caves, sometimes stalactites and stalagmites (dripstones). The moon element comes from the medieval notion that the moon's rays condensed on Earth, so it was believed the pure white mineral was petrified moonlight. It was alternatively believed elves left it in caves because of its healing powers. It was sold by pharmacies and was prescribed until the 19th century. As it is simply calcite, it definitely cured acalcinosis and probably cardialgia by neutralizing the acid and it definitely had no adverse effects. I imagine this alternate view was also because caves would be too dark to have moonlight. Other microorganisms precipitate the calcium crystals, including actinomycetes and algae. The word is of German origin (Mondmilch) and was coined by 16th century Swiss naturalist Conrad Gesner, who used the name Lac Lunae in his Latin text of 1555 concerning the beautiful Swiss Alps. A few years ago a discovery in a Basque cave became world famous, a 1000-foot-long river consisting of moonmilk. (ShowCaves.Com).

The 8th edition of Bergey's Manual from 1974 had 2 Endospora orders, Bacillales and Clostridiales, and Actinobacteria contained the orders Corynebacteriales (Caryophanaceae, Corynebacteriaceae, Brevibacteriaceae, and Propionibacteriaceae), Eubacteriales (1 family), Actinomycetales (1 family), and Mycobacteriales (Mycobacteriaceae, Frankiaceae, Actinoplanaceae, Dermatophilaceae, Nocardiaceae, Streptomycetaceae, and Micromonosporaceae).

The 9th edition from 2000 contains the Micrococci group, the Endospora group, 2 Non-Sporulating Rods groups, Mycobacteria (1 genus), and 8 Actinomycetes groups: Nocardioforms, the *Frankia* group, Actinoplanetes, Streptomycetes, Maduromycetes, and 2 thermophilic groups, the latter of which also has endospores as opposed to the usual actinospore of Actinobacteria (Actinomycetes), and a *Glycomyces* group.

In Garrity et al, there are 5 classes in Endospora (which are usually egregiously refered to as Firmicutes; the latter name definitely includes also Actinobacteria, so can't be restricted to Endospora) and 6 in Actinobacteria (5 in the Garcia website presentation, which is essentially the same) (which they strangely call subclasses and the orders are consequently called suborders) based only on molecular analysis, but for Endospora there are usually 3 classes recognized, Bacilli, Clostridia, and Mollicutes. A taxonomy by Matthew Wolf et al (2004) using PGK amino acids and ML (maximum likelihood) is as follows:

- Bacilli 1
- SG 1
  - Clostridia 1
  - SG 2
    - Clostridia 2
    - SG 3
      - *Thermotoga*
      - SG 4
        - Mollicutes
        - Bacilli 2 (including *Fusobacterium*)

In this, then, Clostridia are paraphyletic, Bacilli are polyphyletic, but Mollicutes are monophyletic.

For Actinobacteria, Garrity and colleagues and Garcia and colleagues use the taxonomy presented in Stackebrandt et al (1997) and from which they take the inadvisable ranking, but they do not represent it properly because they omit the ranks between family and order (suborder as they call it). Here is the phylogeny in tabulated form down to order:

Table 10-8. Actinobacteria-Evolutionary Classification (SG=supergroup).

Rubrobacteria (4 ords., 5 fams., 6 gen.)
SG 1
  Coriobacteria (1 fam., 8 or 13 gen.)-Spherobacteria (1 gen.)
  SG 2
    Acidomicrobia (1 or 2 fams., 5 or 6 gen.)
    SG 3
      Bifidobacteria (1 fam., 7 gen.)
      SG 4
        SG 5
          SG 6
            Glycomycetes (1 fam., 3 gen.)+Maduromycetes (3 fams., 27 gen.)
            Propionobacteria (2 fams.)
          SG 7
            Actinomycetales (1 fam., 7 gen.)+Micrococcales (16 fams., c.100 gen.)
            Corynebacteria (6 fams., 17 gen.)
        SG 8
          Streptomycetes (1 fam., 10 gen.)
          SG 9
            Pseudonocardiae (1 fam., 32 gen.)
            Frankiae (6 fams., 13 gen.)+Actinoplanetes (1 fam., 31 gen.)

Most of the groups have less than 90 % bootstrap support, so they might have weak support. In any case, we must keep in mind that molecular results are no more necessarily correct than phenotypic ones and are only working hypotheses. But many of the groups do correspond to the phenotype. The Streptosporangaceae family is Maduromycetes with several genera added, the Mycobacteriaceae family is Mycobacteria (1 genus) with a genus added, Corynebacteria are recognized but expanded, the Micromonosporaceae family is Actinoplanetes with several genera added, and the Micrococci group is greatly expanded. Also, Corynebacteriaceae, Brevibacteriaceae, Propionibacteriaceae, Actinomycetaceae, the Nocardioform subgroups, Glycomycetaceae, and the endosporulating actinomycete Thermonosporae group are recognized but somewhat differently circumscribed. And it does recognize a large clade (SG 4) distinguished by the possession in some or most of B-type murein, A1gamma murein, MK-10, 11, or 12, exospores, and mycelia. Heliobacteria are a family in Clostridia, unlike earlier editions of Bergey's, where it was misplaced among Gram negatives so it could be with the other eubacterial phototrophs.

Rubrobacteria are immotile, nonsporulating, aerobic, catalase positive rods and have the A1gamma murein type (in 2 genera) and 2 genera are pigmented. *Rubrobacterium* is highly radio-resistant (as is *Arthrobacter*, placed in the Micrococci family), so Deinobacteria probably belong with or close to Gram positives, and DeinoThermi are, in fact, with or close to Posibacteria in Yuri Wolf et al, Daubin et al, Pisani et al, Qi, Wang, and Hao, and Battistuzzi and Hedges as we saw earlier.

Acidomicrobacteria are aerobic, catalase positive, nonsporulating, usually immotile, acidophilic, thermophilic, iron-oxidizing and -reducing rods found in geothermal sites in Yellowstone (*Ferrithrix*), a Welsh mine (*Ferrimicrobium*), a Japanese solfatara (*Aciditerrimonas*), Japanese estuary sediment

(*Illumatobacterium*), and various hot habitats (the type genus).

Coriobacteria are not only immotile and nonsporulating, but also obligately anaerobic, catalase negative, and unpigmented, and have MK-6 as the major quinone--the other classes have MK-8-MK-12--so the group appears to be the most primitive.

Wolfgang Ludwig questions the status of Bifidobacteria as a class within the Actinobacteria phylum (Garrity et al). Bifidobacteria are fermentative, nonmotile, nonsporulating, catalase negative rods, but *Bifidobacterium* is coryneform and has the V configuration and palisades like many in the coryneform group.

Since then, 4 new orders were added to SG 4: Catenulispora (2 fams., 2 gen.), Actinopolyspora (1 gen.), Kineospora (1 fam., 5 gen.), and Jiangellales (1 fam., 2 gen.).

Also added was class Nitriliruptoria (discovered in 2009), containing 2 genera and considered deeply branching. They are nonmotile, nonsporulating, aerobic, catalase-positive rods isolated from soda lake sediment in Russia (*Nitriliruptor*) and a sea cucumber (*Euzebya*). The former genus is alkaliphilic.

Until recently, metabacteria were classified separately as part of the Gram-Negative, Chemolithotrophic Bacteria (*Sulfolobus* Brock et al., 1972), the Methane-Producing Bacteria (*Methanobacterium* Kluyver and van Niel 1936, etc.), and as part of the Gram-Negative Aerobic Rods and Cocci (Halobacteriaceae) (Parker, 1982). Since then, more sophisticated classifications have been done and many new species have been discovered. Unfortunately, internal arrangements have not undergone cladistic analysis based on classical evidence, but for methanogens 3 orders can be recognized. Methanobacteriales is distinguished by the use of formate and H2+CO2 as catabolic substrates and pseudomurein in the wall, Methanococcales by the use of methanol and trimethyalamine as catabolic substrates, lysis by detergents or hypotonic shock, and absence of pseudomurein, and Methanosetales (*Methanosarcina* and *Methanothrix* [*Methanoseta]*) by the use of acetate as catabolic substrate (Bergey's Manual, 9th ed., Holt et al, 2000). New genera *Microarcheum, Parvarcheum*, and *Aciduliprofundum* are of uncertain position.

Methanogens have walls of pseudomurein, protein, glycoprotein, or heteropolysaccharides, and now contain 35 genera arranged as follows (garciajeanlouis.net):

cl. Methanobacteria (1 ord., 2 fams., 6 gen.)
cl. Methanococci (1 ord., 2 fams., 4 gen.)
cl. Methanomicrobia (3 ord., 8 fams., 24 gen.)
cl. Methanopyri (1 gen.)

They inhabit sewage sediments, animal intestines, bogs, swamps, and estuaries. They are responsible for marsh gas, which accounts for 90% of natural gas, and produce over 2 bln. tons of methane per year.

For Halobacteria there were 1 order, 3 families, and 6 genera Halobacteriaceae (*Haloarcula* and *Halobacterium*) possess PGS (phosphatidyl glycerol sulfate) and TGD (trigalactosyldiacylglycerol), Halococcaceae (*Halococcus* and *Haloferax*) contain SDGD1, Natronaceae (*Natronobacterium* and *Natronococcus*), the alkaliphilic family, has a pH range of 8.5-11 (Bergey's Manual, 9th ed.). But many genera were added for a total of 44, and there is only 1 order and 1 family recognized. There is also the new group Nanohaloarcheota.

Halobacterium Schoop 1935 ex Elazari-Volcani 1957 was also named *Flavobacterium* Elazari-Volcani 1940 and *Halobacter* Anderson 1954.

Halobacteria dwell in hypersaline habitats such as salt lakes, salt ponds, soda lakes, salt flats, marine salterns, and salted fish and meats. Its salt concentrations are up to 5.5 M NaCl (32%), which is at the saturation limit. The walls contain glycoproteins and heteropolysaccharides.

In Sulfobacteria there is superorder Sulfolobi, which contain 1 order, Sulfolobales, and has optimal growth below pH 4, a relatively low temperature range (45-96 C.), and is aerobic, coccoid, and thermoacidophilic; and superorder Thermoprotei, which contains Thermoproteales, Desulfococcales, and Thermococcales, has optimal growth above pH 4 and a relatively high temperature range (70-110 C.), and is anaerobic and coccoid, disc-shaped, rod-shaped, or filamentous (Bergey's Manual, 9th ed.).

The Garcia arrangement has 32 genera in 2 classes: Thermoprotei and Thermococci (although the latter is placed in Euryarcheota). Thermoprotei have 5 orders, 7 families, and 28 gen, while Thermococci have 1 order, 1 family, and 3 genera, and 1 unclassified genus.

Sulfobacteria inhabit Icelandic geothermal sources, Yellowstone geysers and hot springs, carbon mine refuse, and submarine volcanic eruption fluids. *Sulfolobus* and Thermoproteales have glycoproteinaceous walls arranged in a hexagonal pattern.

The orders and families are based on descriptions in Bergey's Manual 9th ed. but which does not use ranks for them, with the new groups based on Wikipedia.

Thermoplasmata had only 1 genus, but 4 new ones have been found: *Picrophilus* Schleper et al 1996, *Ferroplasma* Golyshina et al 2000, *Thermogymnomonas* Itoh et al 2007, and *Acidiplasma* Golyshina et al 2008. They branch deeply in Woese (1997) and Barns et al (1996), where they group with methanogens. *Thermoplasma* Darland et al 1970, *Ferroplasma,* and *Thermogymnomonas* have no walls, but *Picrophilus* has an S-layer wall. Derived features in common with Sulfobacteria are sulfur metabolism, and those with methanogens are histone-like proteins. *Thermoplasma* is found in self-heating solfatara fields and acidic coal refuse piles. *Acidiplasma* is found in hydrothermal pools and chalcopyrite ores. The current arrangement for the group is as follows:

Thermogymnomonas
Unnamed
  Thermoplasma
  Unnamed
    Picrophilus
    Ferroplasmaceae
      Acidiplasma
      Ferroplasma

Archeoglobi also had only 1 genus with 2 new ones being discovered. *Ferroglobus* Hafenbradl et al 1997 and *Geoglobus* Kashefi et al 2002, like *Archeoglobus* Stetter 1988, were isolated from near or in hydrothermal vents, but, unlike *Archeoglobus*, which is a sulfate reducer, they reduce iron. The group branches deeply in molecular classifications where it is in Euryarcheota.

The following gives the taxon totals for bacterial phylums.

Table 10-9. Taxon Totals for Bacterial Phylums.

	cl.	ords.	fams.	gen.	sp.
Endospora	2	9	c. 40	c. 400	c. 2500
Proteobacteria	4	36	c. 80	c. 460	c. 2000
Saprospirae	3	5	15	c. 200	c. 2000
Actinobacteria	4	10	c. 60	c. 300	c. 1500
Cyanobacteria	2	11	c. 20	c. 150	c. 1000

Mollicutes	1	5	6	11	c. 350
Spirochetes	1	1	4	15	200+
Togabacteria	1	1	1	11-12	30
DeinoThermi	1	2	3	10	28
Aquificae	1	2	3	14	21
Chlorobia	1	1	1	6	21
Chlamydiae	1	1	1	6	17
Acidobacteria	2	3	3	12	16
Chloroflexi	2	3	3	9	14
Planctobacteria	1	1	1	11	10

The figures are taken from the LPSN except for species. The number of species for Proteobacteria is said to be 1600 by Springer Reference (springerreference.com) but nearly 2000 by Margulis and Chapman (2009). The figure for Posibacteria in all is reported as c. 2500 by Earth Life (earthlife.net), but this is probably confused with Firmicutes 1 (i.e., Clostridia and Mollicutes). The number of families for Cyanobacteria are from Parker (1982). The number of species for Spirochetes was taken from Paster and Dewhirst (2000), for Togabacteria from Battistuzzi et al (2004), for Acidobacteria from garciajeanlouis.net, and for Aquificae, Chloroflexi, Planctobacteria, Chlamydieae, DeinoThermi, Mollicutes, and Chlorobia from Garrity et al (2004). Flavobacteriaceae (in Flavobacteria) and Bacillaceae (in Clostridia) are the largest families of bacteria with about 100 genera each.

About a dozen small taxons have been recently proposed (some discovered much earlier) as separate and independent phylums by some (e.g., Bergey's [Garrity et al, 2004]) but are in all probabilty allied to other groups as can be seen in some of the taxonomies presented earlier. These taxons are Nitrospirae (4 gen., 7 sp., discovered '86), Gemmatimonada (1 sp.), Fibrobacteria (1 gen., 2 sp.), Lentisphera (2 gen., 2sp.), Verrucomicrobia (5 gen., 30 sp.), Chrysiogenetes (3 gen., 3 sp.), Deferribacteria (6 gen., 7 sp., discovered '97), Fusobacteria (2 fams., 10 gen., 25 sp.), Elusimicrobia (1 sp.), Armatimonada (2 gen., 2 sp.), Synergistetes (11 gen., 17 sp.), Caldiserica (1 sp., found in 2009), Dictyoglomi (1 species, discovered in '85), and Thermodesulfomicrobia (4 ge., 8 sp., probably part of Aquificae)(stats taken from garciajeanlouis.net).

A convenience classification for prokaryotes is presented below.

Table 10-10. Convenience Classification for Prokaryotes.

- sbk. Eubacteria
  - spph. Gracilicutes
    - phyl. Heterotropha tax. nov.
      - cl. Neganaerobia tax. nov.
        - ord. Chlamydiae-Rickettsiae
        - ord. Spirochetes
        - ord. Bacteroidetes
        - ord. Spirillobacteria (Campylobacteria)
        - ord. Enterobacteria
        - ord. Desulfobacteria
      - cl. Negaerobia tax. nov.
        - ord. Planctobacteria
        - ord. Caulobacteria
        - ord. Pseudomonada

- ord. Flavobacteria
- ord. Myxospora
- phyl. Autotropha tax. nov.
  - cl. Clatobacteria tax. nov.
    - ord. Hydrogenobacteria (Aquificae)
    - ord. Siderobacteria
    - ord. Nitrobacteria
    - ord. Methylobacteria
    - ord. Thiobacteria
      - Beggiatoae
      - Sulfomonada (Thiobacilli)
  - cl. Photobacteria
    - sbcl. Anoxyphotobacteria
      - spord. Chlorobacteria
        - ord. Chloroflexi
        - ord. Acidobacteria
        - ord. Chlorobia
      - spord. Porphyrobacteria
        - ord. Rhodobacteria
        - ord. Chromatia
    - sbcl. Cyanobacteria
- spph. Mollicutes
- spph. Firmicutes
  - phyl. Deinobacteria
  - phyl. Thermi
  - phyl. Togabacteria
  - phyl. Eufirmicutes
    - sbph. Endospora
      - cl. Bacilli
      - cl. Clostridia
      - cl. Heliobacteria
    - sbph. Metafirmicutes
      - cl. Micrococci
      - cl. Corynebacteria
      - cl. Actinobacteria
- sbk. Metabacteria (Mendosicutes)

Anaerobia are parasitic (except for Desulfobacteria) and unpigmented (except for Enterobacteria), while Aerobia are free-living (except for a few pseudomonads and a few myxosporans) and pigmented (except for Planctobacteria and Caulobacteria). The protein-walled groups are the first in each class. Exceptions in Autotropha are Chloroflexi, Acidobacteria, and Rhodobacteria, which are heterotrophic.

Acton, Ashton Q. (ed.) 2011. Issues in Life Science, p. 850. Scholarly Editions. (books.google).

Austrian, Robert. 1959. Concerning Friedländer, Gram, and the Etiology of Lobar Pneumonia, an Historical Note. Trans. Am. Clin. Climatol. Assoc. 71: 142–149 (nih.gov).

Austrian, Robert. 1960. The Gram Stain and the Etiology of Lobar Pneumonia, an Historical Note. Bacteriol. Rev. 24: 261–265 (nih.gov).

Battistuzzi, F. U., Feijao, A., Hedges, S. B. 2004. A genomic timescale of prokaryote evolution: insights into the origin of methanogenesis, phototrophy, and the colonization of land. BMC Evolutionary Biology 4: 44.

Battistuzzi, F. U., Hedges, S. B. 2008. A Major Clade of Prokaryotes with Ancient Adaptations to Life on Land. Molecular Biology and Evolution 26: 335–343 (mbe.oxfordjournals.org).

Bisset, K.A. 1952. Bacteria, 1st ed. Livingston, London; 2nd ed. 1962.

Brochier, Céline & Philippe, Hervé. 2002. Phylogeny: a nonhyperthermophilic ancestor for Bacteria. Nature 417: 244 (nature.com).

Brown, J.R, Douady, C. J., Italia, M.J., Marshall, W.E., Stanhope, M.J. 2001. Universal trees based on large combined protein sequence data sets. Nature Genetic 28: 281-85 (nature.com).

Buchanan, Robert. 1925. General Systematic Bacteriology. Williams and Wilkins (archive.org).

Caratini, Roger. 1971. La vie des plantes, vol. 10, Encyclopédie Bordas, Bordas (after Manuel de classification et détermination des bactéries anaérobies, 1958, Masson, Paris and Bactéries in Précis de Botanique, 1963, Masson, Paris)

Ciccarelli, F. D., Doerks, T., Von Mering, C., Creevey, C.J., Snel, B., Bork, P. 2006. Toward Automatic Reconstruction of a Highly Resolved Tree of Life. Science 311: 1283–1287 (binf.bio.uu.nl).

Daubin, Vincent, Gouy, Manolo, Perrière, Guy. 2002. A Phylogenomic Approach to Bacterial Phylogeny: Evidence of a Core of Genes Sharing a Common History. Genome Res. 12: 1080-90 (genome.cshlp.org).

DiPippo, J.L., Nesbø, C.L., Dahle, H., Doolittle, W.F., Birkland, N.K., Noll, K.M. 2009. *Kosmotoga olearia* gen. nov., sp. nov., a thermophilic, anaerobic heterotroph isolated from an oil production fluid. Int J Syst Evol Microbiol. 59 (Pt 12): 2991-3000 (ijs-sgmjournals.org).

Drouet, F. 1981. Summary of the classification of blue-green algae. Beih. Nova Hedwiga 66: 135-209.

Enderlein, G. 1925. Bakterien-Cyclogenie. de Gruyter, Berlin.

Feng, Yixiao; Cheng, Lei; Zhang, Xiaoxia; Li, Xia; Deng, Yu; and Zhang, Hui. 2010. *Thermococcoides shengliensis* gen. nov., sp. nov. representing a novel genus of the order Thermotogales from oil-production fluid (bioteke.com).

Garrity, George, Bell, Julia, Lilburn, Timothy. 2004. Taxonomic Outline of Prokaryotes (uni-muenster.de).

Gibbons, N.E. & Murray, R.E. (eds.). 1978. Bergey's Manual of Determinative Bacteriology, 9th ed. Williams & Wilkins, Baltimore.

Giovannoni, S.J., Turner, S., Olsen, G.J., Barns, S., Lane, D.J., Pace, N.R. 1988. Evolutionary Relationships among Cyanobacteria and green chloroplasts, J. Bacteriol. 170: 3584-92 (jb.asm.org).

Goodfellow, M. and O'Donnell, A., eds. 1993. Handbook of New Bacterial Systematics. Acad. Press.

Gupta, Radhey. 1998. Protein phylogenies and signature sequences: a reappraisal of evolutionary relationships archeobacteria, eubacteria, and eukaryotes. Microbiol. Mol. Biol. Rev. 62: 1435-1491 (researchgate.net).

Gupta, Radhey. 2003. Evolutionary relationships among photosynthetic bacteria. Photosynthesis Research 76: 173–183 (researchgate.net).

Gupta, RS, Bhandari, V. 2011. Phylogeny and molecular signatures for the phylum Thermotogae and its subgroups. Antonie Van Leeuwenhoek 100: 1-34 (Abstract)(science.mcmaster.ca) .

Holt et al. 2000. Manual of Deteminative Bacteriology, 9th ed. Lippincott, Wilkins, and Wiliams.

Horike et al. 2009. Gene (Science Direct).

Kuske, Cheryl, Barns, Susan, and Busch, Joseph. 1997. Diverse uncultivated bacterial groups from soils of the arid southwestern United States that are present in many geographic regions. Applied and Environmental Microbiology. 63: 3614–3621 (aem.asm.org; researchgate.net).

Lienau, Kurt; DeSalle, Robert; Rosenfeld, Jeffrey; and Planet, Paul. 2006. Reciprocal Illumination in the Gene Content Tree of Life. Syst. Biol. 55: 441-53 (researchgate.net).

Margulis, Lynn, and Chapman, Michael. 2010. Kingdoms and Domains (books.google).

Möhn, E. 1984. System und Phylogenie der Lebewese. 2 vols. E. Schweizerbart'sche.

Nunoura, T., Hirai, M., Imachi, H., Miyazaki, M., Makita, H., Hirayama, H., Furushima, Y., Yamamoto, H., Takai, K. 2010. *Kosmotoga arenicorallina* sp. nov., a thermophilic and obligately anaerobic heterotroph isolated from a shallow hydrothermal system occurring within a coral reef, southern part of the Yaeyama Archipelago, Japan, reclassification of *Thermococcoides shengliensis* as *Kosmotoga shengliensis* comb. nov., and emended description of the genus *Kosmotoga*. Archives of Microbiology 192: 811-819 (Abstract)(ncbi.nlm.nih.gov).

Parker, Sybil (ed.). 1982. Synopsis and Classification of Living Organisms. McGraw-Hill, New York.

Paster, Bruce and Dewhirst, Floyd. 2000. Phylogenetic Foundation of Spirochetes. J. Mol. Microbiol. Biotechnol. 2: 341-344 (horizonpress).

Pierson, Beverly; Valdez, Diane; Larsen, Mark; Morgan, Elizabeth; Mack, E. Erin. 1994. Chloroflexus-like organisms from marine and hypersaline environments: distribution and diversity. Photosynthesis Research 41: 35-52 (springer.com).

Pisani, Davide; Cotton, James; and McInerney, James. 2007. Supertrees Disentangle the Chimerical Origin of Eukaryotic Genomes. Mol. Biol. Evol. 24: 1752-60 (academic.uprm.edu).

Pribham, Ernst. 1929. A contribution to the classification of microorganisms. J. Bacteriol. 18: 361-394 (jb.asm.org).

Qi, J., Wang, B., Hao, B. 2004. Whole Proteome Prokaryote Phylogeny Without Sequence Alignment: a K-String Composition Approach. J. Mol. Evol. 58:1-11 (pnas.org).

Rappé, M. S., Giovannoni, S. J. 2003. The Uncultured Microbial Majority. Annual Review of Microbiology 57: 369–394.

Rippka et al. 1979. J. Gen. Microbiol. 111: 1-61 (as cited in Holt et al, 2000).

Schaechter, Miselio (ed.). 2009. Encyclopedia of Microbiology, 3rd ed, vol. 1 (books.google).

Stackebrandt, Erko; Rainey, Fred; and Ward-Rainey, Naomi. 1997. Proposal for a New Hierarchic Classification System, Actinobacteria classis nov. IJSB 47: 479-91 (ijs.sgmjournals.org).

Stanier, R.Y. & van Neil, C.B. 1941. The main outlines of bacterial classification. J. Bacteriol. 42: 437-466.

Turner, Sean; Burger-Wiersma, Tineke; Giovannoni, Stephen; Mur, Luuc; Pace, Norman. 1989. The relationships of a prochlorophyte *Prochlorothrix hollandica* to green chloroplasts. Nature 337: 380-85 (Abstract)(researchgate.net).

Woese, Carl. 1987. Bacterial evolution. Microbiol. Rev. 51: 221-271.

Wolf, Yuri; Rogozin, Igor; Grishin, Nick; Tatusov, Roman; and Koonin, Eugene. 2001. Genome trees constructed using five different approaches suggest new major bacterial clades. BMC Evol. Biol. 1: 8 (nih.gov).

Wolf M, Müller T, Dandekar T, Pollack JD. 2004. Phylogeny of Firmicutes with special reference to Mycoplasma (Mollicutes) as inferred from phosphoglycerate kinase amino acid sequence data. Int. J. Syst. Evol. Microbiol. 54 (Pt 3): 871–5 (ijs.microbiologyresearch.org).

Yarza, P., Richter, M., Peplies, J. R., Euzeby, J., Amann, R., Schleifer, K. H., Ludwig, W., Glöckner, F. O., Rosselló-Móra, R. 2008. The All-Species Living Tree Project: a 16S rRNA-based phylogenetic tree of all sequenced type strains. Systematic and Applied Microbiology 31: 241–250 (researchgate.net).

Yarza, Pablo; Ludwig, Wolfgang; Rosselló-Móra, Ramon. 2010. Update of the All-Species Living Tree Project based on 16S and 23S rRNA sequence analyses. Systematic and Applied Microbiology 33: 291–299 (researchgate.net).

Chapter 11 Red Algae

It has been quite widely accepted that the Rhodophyta are one of the most ancient lineages of eukaryotes. - Van Den Hoek et al, 1995, p. 95.

Algae, known by Theophrastus, the Father of Botany, and the other ancients, the word coined in 1551, but known only from seaweeds and red tides, were organized by Linneus into 4 genera, the filamentous *Conferva*, the membranous *Ulva*, the thalloid *Fucus*, and the gelatinous *Tremella*. The few unicelluar algae then known, such as *Volvox*, he placed in Vermes and then *Chaoticum*. They were probably first recognized as a class in 1750 by Karl August von de Bergen (his 21st class out of 22) (Adanson, 1763). Lamouroux in 1813 distinguished the orders Fucacées, Floridées, and Ulvacées. Agardh (1824) recognized 6 orders: Diatomeae, Nostichinae, Confervoidea, Ulvaceae, Florideae, Fucoideae. Harvey (1836) established 3 subclasses: Chlorospermae (including blue algae and xanthophyceans which would both be separated out later in the century), Melanospermae (brown algae), and Rhodospermae. Haeckel classified them as Archephyta (blue algae and most green algae), Characeae (stoneworts), Florideae (red algae), and Fucoidae (brown algae) in Plantae while placing Diatomae, Myxocystoda (Noctilucae), and Flagellata (*Peridium*, *Volvox*, and *Euglena*) in Protista. And Eichler (1883) divided them into the familiar Cyanophyceae, Rhodophyceae, Chlorophyceae, and Pheophyceae in his Thallophyta, which, of course, included also funguses, but the phylum was 1st proposed by Endlicher (1836), with phytoflagellates claimed also by many protozoologists.

Red algae were first recognized as a kingdom by Jeffrey in 1971.

Here is a list of synonyms for the group:

Rhodospermae Harvey 1836
Heterocarpeae Kützing 1843
Rhodophyceae Ruprecht 1851
Phycoerythrinophycées Marchard 1895
Rhodophyta Wettstein 1901
Rhodophycophyta Papenfuss 1946
Mesoprotista Dougherty, Gordon, & Allen 1957
Aconta Christiansen 1962
Rhodobiota Jeffrey 1971
Rhodobionta Edwards 1976
Biliphyta Cavalier-Smith 1981
Rhodymeniontes Starobogatoff 1986
Rhodobiontes Kussakin and Drozdov 1994

Prerhodophyceans

The earliest descriptions of prerhodophyceans, which are blue-green, unicellular, thermoacidophilic algae, were by Meneghini (1839; 1841 [in *Coccochloris* (=*Anaphorathece*, Cyanobacteria)]) and Tilden (1898 [in *Protococcus*, a green alga]; 1910 [in *Pleurocapsa*, a cyanobacterium]), but these descriptions were invalid as they referred to mixed populations. The 1st valid description was by Galdieri in 1899 as *Pleurococcus sulphurarius*, and set up by Merola et al (1981) as a new genus, *Galdieria sulphuraria,* and referred to red algae and the other prerhodophyceans. They established the new class Cyanidiophyceae and the families Cyanidiaceae, for

Cyanidium and *Cyanidioschyzon*, and Galdieriaceae for *Galdieria*. The 2nd valid description was of *Cyanidium caldarium* by Geitler and Ruttner (1935) but seen as a synonym with *Pleurocapsa caldaria*. *Cyanidioschyzon merolae* was discovered as part of *Cyanidium caldarium* by Tilden (1898) but recognized and named as such (because of its longitudinal fission) only in 1978 by De Luca et al, the 2 being previously designated as *Cyanidium caldarium* forma A and B, respectively. The other species are *Galdieria maxima, Galdieria partita, Galdieria daedala, and Cyanidium maximum*. *Cyanidioschyzon* is obligately autotrophic, while the other 2 genera are autotrophic but capable of heterotrophy.

Glaucophyceans

Glaucobiota are unicellular algae with chlorophyll a, phycobilins, beta-carotene, xanthophylls, a proteinaceous or cellulosic wall, lamellar cristas, 2 anisokontic flagella with mastigonemes, an MLS (multi-layered structure), cruciate rootlets, open mitosis, and cyanelles (cyanobacterial endosymbiotes). They are blue-green, sometimes form in loose colonies, are found in freshwater, and are limited in distribution. The genera are *Glaucocystis* Itzigsohn in Rabenhorst, 1868 (8 sp.), *Gloeochaete* Lagerheim, 1883 (1 sp.), and *Cyanophora* Korshikov, 1924 (3 sp.). *Glaucocystis* was variously assigned to Cyanophyceae, Chlorophyceae, and Rhodophyceae. *Gleochete* was first assigned to Cyanophyceae, then Chlorophyceae, and finally to Rhodophyceae. As mentioned in Chapter 6, *Glaucosphera* was once included but was found to belong in red algae proper. Listed in Margulis et al as being of uncertain affiliation with glaucophyceans and as genera inquirendae are the coccoid *Archeopsis* Skuja 1954 (=*Gleocapsa* Kützing 1843, *Palmoglea* Kützing, 1849) and *Glaucocystopsis* Bourelly 1960, the mastigotes *Peliaina* Pascher 1929 and *Strobilomonas* Schiller 1954 (which might be cryptomonads), and the capsalean *Cyanoptyche* Pascher 1929 and *Chalarodora* Pascher 1929.

There is 1 order, Glaucocystales Bessey, 1907 and 1 family, Glaucocystaceae G. S. West, 1904. They were raised to class rank in 1901 by Bohlin and phylum rank in 1954 by Skuja. Kies and Kremer, in 1986, established 2 orders, Cyanophorales and Gleochetales.

The first report of a genus living with a bacterial endosymbiote was for the thecameban *Paulinella* by Lauterborn in 1895, who did not know whether the endosymbiote was a cyanobacterium or a plastid, recognized the similarities between them, and suggested, before Mereschkowsky's 1905 endosymbiosis theory, that chromatophores (plastids) were symbiotically associated with the cell they live in. The blue-green inclusions were called chromatophores. After thorough investigations Geitler and Pascher concluded these 2 genera and *Paulinella* were symbioses between heterotrophic host cells and modified, autotrophic, bacterial endosymbiotes functioning like chloroplasts and were named "cyanelles" by Pascher in 1929, who named the whole complex "endocyanome" and called this type of endosymbiosis "endocyanosis". The cyanelles of *Cyanophora* and *Glaucocystis* were raised to the rank of bacterial taxons—*Cyanocyta* and *Skujapelta*, respectively—by Hall and Claus in 1963, who created for them 2 new families. Cyanelles, which have limited biochemical and genetic ability and therefore resemble plastids, have thin cell walls which were shown to be a murein layer. (Kies and Kremer, 1990).

Rhodobiota

The cladistic classification for red algae by Gabrielson and Garbary (1985, 1987), but with the 3 Porphyridiales families included, is presented in Table 11-1.

Table 11-1. Kingdom Rhodobiota.

Stylonematophyceae Yoon et al 2006
- Stylonematales K.M. Drew 1956 (1 family)

Parastylonemata nom. nov.
- Porphyridiophyceae Kylin 1937
  - Porphyridiales Kylin 1937 (without peripheral thylakoids; GB association with mitochondrion) (1 family)
- Paraporphyridia nom. nov.
  - Rhodellophyceae Cavalier-Smith 1998 (GB association with the nucleus)
    - Rhodellales Yoon et al 2006 (2 gen.)
  - Metarhodophycidae Magne 1989 (syn. Metarhodophytina Saunders and Hommersand 2004)
    - Erythropeltales (1 family)
    - Eurhodophycidae Magne 1989 (syn. Eurhodophytina Saunders and Hommersand 2004) (GB association with mitochondrion)
      - Compsopogonales (1 genus)
      - Neorhodophyceae nom.nov.
        - Rhodochetales Skuja 1939 (4 families)
        - Macrorhodophyceae Cavalier-Smith 1998
          - Bangiophyceae Wettstein 1901
            - Bangiales Engler 1892 (1 fam., 2 gen.)
          - Acrophyceae nom. nov.
            - Acrochetiales Feldmann 1953 (1 fam.)
            - Nemaliophycidae nom. nov.
              - Nemaliales Schmitz in Engler 1892 (7 fams.)
              - Batrachospermales Pueschel & K.M. Cole 1982 (1 fam.)
              - Corallinophycidae Le Gall & Saunders 2007
                - Corallinales Silva & Johansen 1986 (2-4 fams.)
              - Hildenbrandiophycidae Saunders and Hommersand 2004 emend.
                - Hildenbrandiales Pueschel & Cole 1982 (1 fam.)
                - Gelidiales Kylin 1928 (3 fams.)
                - Palmariales Guiry & Irvine 1978 (1 fam.)
              - Gigartinophycidae nom. nov.
                - Gigartinales Schmitz in Engler 1892 (50 fams.)
                - Bonnesmaisoniarae nom. nov.
                  - Bonnemaisoniales Feldmann 1952 (1 fam.)
                  - Rhodymeniarae nom. nov.
                    - Rhodymeniales Schmitz in Engler 1892 (2 fams.)
                    - Ceramiales Oltmanns 1904

(4 fams.)

They used 15 orders and 37 traits, Wagner parsimony, branch swapping, Adams consensus, and the Physis program. There were 62 steps, 6 MPTs, and a CI of .60.

Recognized in addition by Hommersand and Saunders (2004) based on molecular classifications are:

Ahnfeltiales Maggs & Pueschel 1989 (1 family)
Halymeniales Saunders & Kraft 1996
Gracilariales Fredericq & Hommersand 1989
Nemastomatales Kylin 1925
Plocamiales Saunders & Kraft 1994

They recognize 30 orders in all, but 26 are recognized here, with the exclusion of Prerhodophyceae and unnecessary higher level status for some families. Dixon (1982) recognized 10 orders and 67 families. While he recognized 5 orders in Florideophycidae (Cenorhodobiota), Pueschel and Cole in '82 distinguished 11, and Gabriel and Garbary in '86 also distinguished 11, adding 1 and taking out 1 (Gabrielson et al, 1993). Chadefaud and Emberger (1960) had recognized 11 orders as follows:

sbph. Proto-Floridées
    Bangiales (Porphyridiaceae and Bangiaceae)
sbph. Floridées
    Eo-floridées (Acrochetales and Eu-Nemaliales)
    Méso-floridées (the remaining Nemaliales, plus Chetangiales, Gelidiales, Gigartinales, Cryptonemiales, and Rhodymeniales)
    Meta-floridées (Bonnemaisoniales and Ceramiales)

The latter subphylum contains 99% of the genera.

Magne (1989) established Archeorhodophycidae, Metarhodophycidae, and Eurhodophycidae. "Florideae" is properly a synonym for angiosperms and is a misnomer for red algae so I use the name Neorhodophyceae.

Van den Hoek et al (1995) proposed the following taxonomy:

Table 11-2.

Bangiophycidae
    Rhodochetales
    Erythropeltidales+Bangiales (erect filaments or parenchymatous fronds in the gameophyte phase, which exhibit apical cell division and lack pit plugs)
Floridophycidae (pit plugs with cap membranes and 2-layered plug caps, trichogynes on the carpogoniums, gonimocarp, tetrasporangia)
    Acrochetales+Palmariales
    unnamed 1 (carpogonia at tips of specialized carpogonial branches)
        Nemaliales
        Batrachospermales
        unnamed 2 (formation of auxillary cells)

Corallinales
unnamed 3 (1 plug cap layer)
  Gelidiales
  unnamed 4 (loss of cap layer)
    Gigartales
    unnamed 5
      Rhodymeniales
      Ceramiales

In this, Porphyridiales is considered to be 3 independent lineages possibly derived from Rhodochetales, Erythropeltidales, and Bangiales. The 1-layered cap is also present in Hildebrandales so in Saunders and Hommersand it is considered primitive.

Although uncertain, there is some fossil evidence indicating the great age of Rhodobiota--extinct fossil genera *Eosphaera* Barghoorn, 1965, from Schreiber Beach, Ont., which is comparable to *Porphyridium* Nägeli, 1849, and *Huroniospora* Barghoorn, 1965, which is comparable to *Rhodella* L.V. Evans, 1970, discovered in the chert of the Gunflint Formation in the Province of Ontario, going back c. 2 bln. yrs. to the Precambrian (Tappan, 1980). The 4 have binary fission and budding in a mucilaginous sheath, with the latter 2 having tolerance for a wide range of temperatures and salinities. According to Javaux (2007), as mentioned in Chp. 7, the earliest red alga is *Bangomorpha* from the Middle Proterozoic 1.2 bya.

Other red algal fossils include especially the corallines because they tend to preserve well as they are encrusted with calcium carbonate crystals and go back to the Mesozoic, and the calcified Solenosporaceae, allied to Corallinales, go back to the Cambrian (Gabrielson et al, 1990). Corallines number about 30 genera (Dawson, 1966). Like coral animals (see Chapter 18), they extract calcium and carbonate from sea water and produce and deposit limestone (calcium carbonate), and are the second most important contributors to coral-reef formation (Coral Reef Facts-coral-reef-info.com). They can also build reefs, called algal reefs, independent of coral reefs, in cooler and deeper waters than the usual shallow, warm, tropical or subtropical waters of coral reefs (Stafford-Deitsch, 1991). They were formally classified in Cryptonemiales. There are also calcareous green (11 major genera in Siphonales and Dasycladales) and brown algae (1 major genus, *Padina* in Dictyotales) but these are mostly lightly calcified, while corallines are mostly heavily calcified (Dawson, 1966).

About 50 genera and some 100 species are heterotrophic and these are "floridophyceans" and are parasitic in other red algae and most are related to their hosts, often in the same family, and are called adelphoparasites. Parasites unrelated to their hosts are called alloparasites.

The largest family is Rhodomelaceae, in Ceramiales, with 100 genera and 500 species.

The kingdom is predominantly marine and benthic (see Chp. 17). It is the only marine kingdom besides Retaria. In it we find purple laver (*Porphyra* C. Agardh, 1824, Porphyraceae Kützing, 1843= Bangiaceae Reinbold 1891), coral weed (*Corallina* Linneus, 1758, Corallinaceae Lamouroux, 1812, Corallinales), Irish moss or carragheen (*Chondrus* Stackhouse, 1797, Gigartinaceae Kützing, 1843, Gigartinales), dulse (*Palmaria* Stackhouse, 1801, Palmariaceae Guiry, 1974, Palmariales), threadweed (*Nemalion* Duby, 1830, Nemaliaceae (Farlow) De Toni and Levi, 1886, Nemaliales), and pitcherweed (*Ceramium* A.W. Roth, 1797, Ceramiaceae Dumortier, 1822, Ceramiales), among other large and attractive forms.

Chadefaud, Marius; Emberger, Louis. 1960. Traité de botanique systématique, tome 1. Masson, Paris (as cited in La vie des plantes, Roger Caratini, ed., 1971).
Dawson, E. Yale. 1966. Marine Botany: an Introduction. Holt, Rinehart, and Winston.

Dixon, P.S. 1982. Rhodophycota. In Sybil Parker (ed.), Synopsis and Classification of Living Organisms, McGraw-Hill, New York.

Gabrielson, P.W., Sommerfeld, M. R., Townsend, R. A., and Tyler, P. L.1990. Phylum Rhodophyta (Chp. 8). In Handbook of Protoctista, Lynn Margulis et al eds. Jones & Bartlett.

Gabrielson P. W., Garbary, J. D. 1985. Systematics of the red algae (Rhodophyta). Critical Reviews in Plant Sciences 3: 325-366.

Gabrielson, P. W., Garbary, D. J. 1987. A cladistic analysis of Rhodophyta: Florideophycidean orders. British Phycological Journal 22: 125–138.

Javaux, Emmanuelle. 2007. The Early Eukaryotic Fossil Record. In Eukaryotic Membranes and Cytoskeleton: Origins and Evolution, Gaspar Jekely (ed.), Chp. 1. Landes Bioscience/Springer.

Kies, Ludwig; Kremer, Bruno. 1990. Glaucocystophyta. In Handbook of Protoctista, Margulis et al, (eds.), Chp. 12. Jones and Bartlett, Boston.

Magne, F. 1989. Classification et phylogénie des Rhodophycées. Cryptogamie Algologie 10: 111-13.

Saunders, G.W. and Hommersand, M.H. 2004. Assessing Supraordinal Red Algal Diversity and Taxonomy in the Context of Contemporary Systematic Data. Am. J. Bot. 91: 1494-1507.

Stafford-Deitsch, Jeremy. 1991. Reef: a Safari Through the Coral World. Holt, Rinehart, and Winston.

Tappan, Helen. 1980. The Paleobiolgy of Plant Protistans. Freeman.

Van den Hoek, C., Mann, D.G., Jahns, H.M. 1995. Algae: an Introduction to Phycology. Cambridge U.

Chapter 12 Plantae

The limits of the plant kingdom are those which will include the two groups Chlorophyceae (green algae) and Embryophyta (higher plants). - H.F. Copeland, 1938 (first to delimit the plant kingdom phylogenetically).

Historical Overview

Theophrastus, in his "'Historia plantarum" (c. 372 – c. 278 B.C.), distinguished between trees, shrubs, undershrubs, and herbs; biennial and perennial; centripetal and centrifugal; polypetalous and gamopetalous; superior and inferior ovaries; and flowering and nonflowering (Lawrence, 1960; Core, 1955); modes of generation, localities, sizes, and practical uses such as foods, juices, herbs, etc. (Wikipedia).

Theophrastus (New World Encyclopedia; IEP [Internet Encyclopedia of Philosophy]; Wikipedia) was a Greek scientist and a favourite student of Aristotle, founded botany in Greece, and devised the first known taxonomy of plants. Most of our information on him comes from Diogenes Laertius' "Lives and Opinions of Eminent Philosophers," written 400 years later in the 3rd century AD. Theophrastus' given name was Tyrtamos, but was changed by Aristotle to Euphrastos, meaning "eloquent" and then to Theophrastus, meaning "divine speech". He inherited Aristotle's works on botany and continued to expand them. In Athens he was in charge of the first botanical garden. In his will, Aristotle made him guardian of his children and bequeathed to him his library and the originals of his works.

He first studied under Leucippus or Alcippus on Lesbos, then maybe under Plato, and then under Aristotle, who appointed him his successor as leader of the Lyceum when he moved to Chalcis around 313 B.C. Aristotle and Theophrastus lived several years on Lesbos, the latter's place of birth, and it is probably there they started research into natural science, Aristotle on animals and Theophrastus on plants.

Theophrastus also recognized the role of sex in the reproduction of some higher plants, though this last discovery was lost in later ages. Some of his names survive into modern times, such as carpos for fruit, and pericarpion for seed vessel.

He presided over the Peripatetic School for 35 years. Under his guidance it flourished, and had at one period more than 2000 students. He was highly esteemed by 3 kings, Philip of Macedonia, Cassander, and Ptolemy I Soter. In his will, he left his house and garden to the Lyceum as a permanent seat of instruction. He famously said, "life is ruled by fortune, not wisdom."

He also conducted the first recorded experiment to show that the Mediterranean was formed by the inflowing Atlantic by releasing a message in a bottle. In 314 B.C. he noted that tourmaline becomes charged when heated, the first known reference to pyroelectricity.

He wrote 227 books on a wide variety of subjects, including philosophy, natural science, politics, rhetoric, and medicine, totaling nearly 233, 000 lines of text, which is probably on average about 200 pages each. However, the only (nearly) complete works still existing today are 2 large tomes on botany, "De historia plantarum" ("A History of Plants" or "Enquiry into Plants"), and "De causis plantarum" ("On the Causes of Plants" or "About the Reasons of Vegetable Growth"), and The Characters. We also possess, in fragments, a "History of Physics," "On Stones," "On Sensation," and "Airopta" (on metaphysics).

His 2 botanical works are extant probably because Pope Nicholas V ordered them translated into Latin in the mid-15th century. For several centuries they became an indispensable guideline for the teaching and understanding of botany. They were of a general nature, only casually referring to particular species. He also recorded what he knew of the foreign species of India, Persia, Bactria, Syria, Egypt, and Libya.

"A History of Plants" consists of 9 books (originally 10) discussing plant anatomy, trees, germination of seeds, the difference between wild and cultivated trees, perennials, wild and cultivated vegetables, cereals, and saps and medicine. The 1st book deals with the parts of plants; the 2nd with reproduction and times and manner of sowing; the 3rd, 4th, and 5th are devoted to trees, their types, locations, and practical applications; the 6th with shrubs and spiny plants; the 7th with herbs; the 8th with plants that produce edible seeds; and the 9th deals with plants that produce useful juices, gums, resins, etc.

"Reasons of Vegetable Growth" consists of 6 books (originally 8) discussing the growth and propagation of plants; the effect of environmental changes on the growth of plants; how various types of cultivation affect plants; propagation of cereals; artificial and unnatural influences on plants; plant disease and death; and the odour and taste of plants.

"Enquiry into Plants" was first published in a Latin translation by Theodore Gaza, in Treviso, in 1483. In its original Greek it first appeared from the press of Aldus Manutius in Venice in 1495–98, from a 3rd-rate manuscript, which, like the majority of the manuscripts that were sent to printers' workshops in the 15th and 16th centuries, has disappeared. Wimmer identified 2 manuscripts of first quality, the Codex Urbinas in the Vatican Library, which was not made known to J. G. Schneider, who made the first modern critical edition in 1818–21, and the excerpts in the Codex Parisiensis in the Bibliothèque Nationale de France. The English translation is by Athur Hort from 1916.

Parasara in the "Vrikshayur Veda" (science of plants and plant life) in Ancient India knew of the cell (rasakosa) and chlorophyll (ranjakena pakyamanat) and recognized several plant families (ganas), including the mustard, bean, and squash families (Core, 1955, p. 27, from Majumdar, 1946). The descriptions of plant morphology were quite detailed. Chlorophyll was rediscovered and named by Joseph Pelletier and Joseph Caventou in 1816.

In the 13th century, scholar and bishop Albertus Magnus (Albert von Bollstadt, "Doctor Universalis"), the Aristotle of the Middle Ages, who had as one of his pupils Thomas Aquinas (the Angelic Doctor), and whose writings would fill 21 folio volumes, had already recognized monocots and dicots in his "De vegetabilis," dividing plants into Leafless (Cryptogams in part) and Leafy (mostly phanerogams with some cryptogams) and the latter into Corticate (monocots) and Tunicate (dicots).

In his most famous work, "De plantis," from 1583, Italian botanist and physician to Pope Clement VIII (after 1592) Andrea Cesalpino classified plants according to habit, dividing them into trees and herbs, but further divided them according to type of fruit and seed, ovary position, presence or absence of bulbs, type of sap, and the number of locules in the ovary. He was essentially a philosopher of the Aristotelian school so he sought a classification system based on reason instead of utility, the latter approach being that of other herbalists. He believed that leaves provide protection for buds, flowers, and fruit, and that plants had a nutritive soul, and treated the pith of dicots as homologous to the vertebral column, and that the organs of fructification were more important than habit. His work was not appreciated by his contemporaries but did influence later taxonomists such as Ray, Tournefort, and Linneus. The following is his classification as presented by Sachs.

Table 12-1. Cesalpino Classification.

Arboreae (Arbores et frutices).

I. Corde ex apice seminis (e.g., *Quercus, Fagus, Ulmus, Tilia, Laurus, Prunus*).

II. Corde ex basi seminis (e.g., *Ficus, Cactus, Morus, Rosa, Vitis, Salix*, Coniferae).

Herbaceae (Suffrutices et herbae).

III. Solitariis seminibus (e.g. *Valeriana, Daphne, Urtica, Cyperus*, Gramineae).
IV. Solitariis pericarpiis (e.g. Cucurbitaceae, Solaneae, *Asparagus, Ruscus, Arum*).
V. Solitariis vasculis (e.g. various Leguminosae, Caryophylleae, Gentianeae).
VI. Binis seminibus (Umbelliferae).
VII. Binis conceptaculis (e.g. *Mercurialis, Poterium, Galium, Orobanche, Hyoscyamus, Nicotiana*, Cruciferae).
VIII. Triplici principio (ovary) non bulbosae (e.g. *Thalictrum, Euphorbia, Convolvulus, Viola*).
IX. Triplici principio bulbosae (large-flowered monocots).
X. Quaternis seminibus (Borragineae and Labiatae).
XI. Pluribus seminibus, anthemides (some Compositae).
XII. Pluribus seminibus, cichoraceae aut acanaceae (some Compositae, *Eryngium*, and *Scabiosa*).
XIII. Pluribus seminibus, flore communi (e.g. *Ranunculus, Alisma, Sanicula, Geranium, Linum*).
XIV. Pluribus folliculis (e.g. *Oxalis, Gossypium, Aristolochia, Capparis, Nymphaea, Veratrum*).
XV. Flore fructuque carentes (Filices, Equiseta, Musci (including corals), Fungi).

The examples appended by Sachs are meant to show that, with the exception of the 6th and 10th classes, none completely represents a natural group (Sachs included the 15th class as a natural group). Most of them are a collection of heterogeneous plants, and the distinction of dicots and monocots, is largely omitted; the ninth class does contain only monocots but not all of them. Yet, as Sachs also points out, Cesalpino intended it to be a system based on natural affinities. We find much the same intention-result contradiction in Darwinian era, pre-cladistic classifications meant to be evolutionary. Cesalpino prepared an herbarium of 768 well-mounted plants that is still in existence, one of the oldest known.

Swiss Protestant herbalist Gaspard (Kaspar) Bauhin (Core, 1955; Lawrence, 1960) recognized the distinction between genus and species, which was lacking in the earlier herbalists, and introduced binomial nomenclature, although sometimes it was trinomial or quadrinomial. The work by his brother Jean (Johann), "Historia plantarum universalis," published posthumously in 1650-51, nearly 40 years after his decease, which contained descriptions of 5000 plants and had 3500 figures, inspired him to do a similar compilation, collecting all that had been previously written on plants, using his herbarium of 4000 specimens he had collected in his travels, and drawing up a concordance of all the names given by various authors to the same species, on which he spent several years but, like his brother, who was 19 years older, was unable to complete. There were, however, the publication of 3 preliminary volumes: "Phytopinax" in 1596, "Prodromus theatri botanici" in 1620, and, most importantly, "Pinax theatri botanici" in 1623. Not unusual for the times he placed duckweeds among the cryptogams, water ferns among mosses, and united corals and sponges with seaweeds. "Pinax," which means register, converted chaos into order in nomenclature. Of all the synonyms, Bauhin chose one name for each species as the most suitable based on his 40 years of research, which earned him the title "legislateur en botanique."

Cambridge naturalist, philosopher, and cleric-theologian John Ray (Sachs, 1890; Core, 1955; Lawrence, 1960) divided plants, based on relationship of form and gross morphology, specifically fruit type and leaf and floral features, into herbs and trees, with the former subdivided into Imperfectae (4 classes: Plantae submarinae (chiefly polyps and *Fucus*), Fungi, Musci (Confervae, mosses, and lycopods). Capillares (ferns, *Lemna, Equisetum*) and Perfectae (seed plants), further subdivided into dicots (19 classes) and monocots (3 classes). Trees were also divided into monocots (1 class) and dicots (6 classes), making 33 classes in all. This was in the last edition/revision of his "Methodus planatarum" in 1704, which accounted for nearly 18,000 species. "Historia plantarum" was published in 3 volumes between 1686 and 1704. The Historia was immediately accepted as the standard botanical work of the day and would go on to influence Jussieu and Candolle.

Table 12-2. John Ray's Taxonomy.

A. Plantae gemmis carentes (herbae)

(a) Imperfectae.

I. Plantae submarinae (chiefly polypes and *Fucus*).
II. Fungi.
III. Musci (Confervae, mosses, lycopods).
IV. Capillares (ferns, *Lemna, Equisetum*).

(b) Perfectae.

Dicotyledones

V. Apetalae.
VI. Planipetalae lactescentes.
VII. Discoideae semine papposo.
VIII. Corymbiferae.
IX. Capitalae (vi-ix are Compositae).
X. Semine nudo solitario (Valerianeae, *Mirabilis, Thesium*, etc.).
XI. Umbelliferae.
XII. Stellatae.
XIII. Asperifoliae.
XIV. Verticillatae (Labiatae).
XV. Semine nudo polyspermo (*Ranunculus, Rosa, Alisma*).
XVI. Pomiferae (Cucurbitaceae).
XVII. Bacciferae (*Rubus, Smilax, Bryonia, Solanum, Menyanthes*).
XVIII. Multisiliquae (*Sedum*, Helleboreae, *Butomus, Asclepias*).
XIX. Vasculiferae monopetalae (various).
XX. Vasculiferae dipetalae (various).
XXI. Tetrapetalae siliquosae (Cruciferae, *Ruta, Monotropa*).
XXII. Leguminosae.
XXIII. Pentapetalae Vasculiferae enangiospermae (various).

Monocotyledones

XXIV. Graminifoliae floriferae vasculo tricapsulari (Liliaceae, Orchideae, Zingiberaceae).
XXV. Stamineae (Grasses).
XXVI. Anomalae incertae sedis.

B. Plantae gemmiferae (arbores).

(a) Monocotyledones

XXVII. Arbores arundinaceae (Palms, *Dracaena*).

(b) Dicotyledones.

XXVIII. Arbores fructu a flore remoto seu apetalae (Coniferae and various others).
XXIX. Arbores fructu umbilicato (various).
XXX. Arbores fructu non umbilicato (various).
XXXI. Arbores fructu sicco (various).
XXXII. Arbores siliquosae (woody Papilionaceae).
XXXIII. Arbores anomalae (*Ficus*).

Of these classes Sachs says only the Fungi, Capillares, Stellatae, Labiatae, Pomiferae, Tetrapetalae, Siliquosae, Leguminosae, Floriferae, and Stamineae can pass as wholly or approximately natural groups, but Fungi and Capillares, as they were circumscribed then, are not natural either. Ray's chief merit is that he to some extent recognized natural affinities in their broader features; the systematic separation of the smaller groups was little advanced by him.

Joseph Pitton de Tournefort (Core, 1955; Lawrence, 1960; Wikipedia), like others before him, divided plants into trees and herbs, but with each of these categories being subdivided according to such characteristics as presence or absence of petals, petals simple or compound (akin to polypetalous and gamopetalous), and flowers regular or irregular, and included 22 classes. This system, published in "Eléments de botanique, ou Méthode pour reconnaître les Plantes" (the Latin translation being "Institutiones rei herbariae," published in 1700 and 1719) in 1694, which described around 7,000 plant species and 673 genera, was widely adopted throughout Europe, prevailing in France until supplanted by Jussieu's, and in the rest of western Europe until superseded by that of Linneus, and nearly all classifications published for the next 50 years were based on it. Unlike Ray, it did not distinguish between cryptogams and phanerogams nor between dicots and monocots, but it contained easy identification and most botanists at the time had little interest in a natural system.

Tournefort's family intended that he enter the Church, but he pursued botany instead. After 2 years collecting, he studied medicine at Montpellier, where he had Pierre Magnol as professor, after whom the magnolia was named, and was appointed professor of botany at the Jardin du Roy in Paris in 1683. He was made a member of the Académie on the nomination of Jean-Pierre Bignon, after who *Bignonia* was named.

He is known as the father of the genus concept, although this was, of course, invented before him, but, whereas Bauhin described only the species, Tournefort described only the genera, thereby shifting the center of gravity. *Acer* (maple), *Betula* (birch), *Bignonia* (catalpa), *Castanea* (chestnut), *Fagus* (beech), *Magnolia* (magnolia), *Populus* (poplar), *Quercus* (oak), *Salix* (willow), *Ulmus* (elm), and *Verbena* (verbena) are all said to be named by him, but Brummit credits Linneus except for *Castanea*, which he credits to Miller. The word "herbarium" was apparently his invention; previously, herbariums had been called by a variety of names, such as hortus siccus (dry garden). He wrote other books including "Histoire des plantes" qui naissent aux environs de Paris in 1698. He was killed by a cart or carriage in Paris; the road on which he died now bears his name.

The Linnean system for plants was 1st presented in the 2nd edition of his "Hortus uplandicus" in 1732 and was expanded and served as the basis for "Genera plantarum" from 1737. The latter had descriptions of 935 genera and was published in 5 editions plus 2 supplements called mantissae for a total of 1336 genera. The "Species plantarum" from 1753 is considered as a starting point in modern nomenclature.

Table 12-3. The Linnean System for Plants.

- Publicae
  - Monoclinia
    - Diffinitas
      - Indefferentismus
        - Classis 1. Monandria: 1 stamen
        - Classis 2. Diandria: 2 stamens
        - Classis 3. Triandria: 3 stamens
        - Classis 4. Tetrandria: 4 stamens
        - Classis 5. Pentandria: 5 stamens
        - Classis 6. Hexandria: 6 stamens
        - Classis 7. Heptandria: 7 stamens
        - Classis 8. Octandria: 8 stamens
        - Classis 9. Enneandria: 9 stamens
        - Classis 10. Decandria: 10 stamens Classis 11. Dodecandria: 12 stamens
        - Classis 12. Icosandria: 20 (or more) stamens, perigynous
        - Classis 13. Polyandria: many stamens, inserted on the receptacle
      - Subordinatio
        - Classis 14. Didynamia: flowers with 4 stamens, 2 long and 2 short
        - Classis 15. Tetradynamia: flowers with 6 stamens, 4 long and 2 short
    - Affinitas
      - Classis 16. Monadelphia; flowers with the anthers separate, but the filaments united, at least at the base
      - Classis 17. Diadelphia; flowers with the stamens united in two separate groups
      - Classis 18. Polyadelphia; flowers with the stamens united in several separate groups
      - Classis 19. Syngenesia; flowers with 5 stamens, the anthers united at their edges
      - Classis 20. Gynandria; flowers with the stamens united to the pistils
  - Diclinia
    - Classis 21. Monoecia: monoecious plants
    - Classis 22. Dioecia: diecious plants
    - Classis 23. Polygamia: polygamodiecious plants
- Clandestina
  - Classis 24. Cryptogamia: ferns, fungi, algae, and bryophytes.

Swedish botanist, physician, and zoologist, Carolus Linneus (Carl von Linné)(1707-78) is known as the father of modern taxonomy, and considered one of the fathers of modern ecology, has been called *Princeps botanicorum* (Prince of Botanists), "The Pliny of the North," and "The Second Adam".

His father was an amateur botanist and Lutheran minister and his mother a rector's daughter. He received most of his higher education at Uppsala University, but first at Lund, where he enrolled in 1727, and latterly at Harderwijk, and began giving lectures in botany at Uppsala in 1730. He lived abroad between 1735 and 1738, where he studied and published the first edition of his Systema Naturae in the Low Countries. He then returned to Sweden, where he became professor of medicine and botany at Uppsala. In the 1740s, he was sent on several journeys through Sweden to find and classify plants and animals. In the 1750s and '60s, he continued to collect and classify animals, plants, and minerals, and published several volumes. He named many of the major plant genera in 1753.

Linneus had 17 students who were the most promising and committed, one of whom was Pehr Kalm, who made an expedition to North America, and Linneus named *Kalmia* after him, which is in the Ericaceae family, and which contains the mountain laurel, bog laurel, sand myrtle, and Alpine azalea, among others. Perhaps his most famous apostle was Carl Peter Thunberg, after which Linneus named *Thunbergia*, which is in the Acanthaceae family, and which contains the thunbergia and clock vine, among others.

Michel Adanson, a French naturalist of Scottish descent, in his "Familles des plantes" (1763), recognized 58 families, the first 5 being cryptogams (Bissus, Champignons, Fucus, Epatiques, and Fougères). In this work he developed his system, which, in its adherence to natural botanical relations, was based on that of Joseph Pitton de Tournefort, and had been anticipated to some extent nearly a century before by John Ray. The success of this work was hindered by its innovations in the use of terms, which were opposed by the defenders of the popular sexual system of Linneus.

Adanson (1727 – 1806), after leaving the Collège Sainte-Barbe, was employed in the herbarium cabinets of R. A. F. de Réaumur (who invented the octogesimal temperature scale, based on 0 and 80 degrees for the freezing and boiling points of water, in 1730) and celebrated botanist Bernard de Jussieu, as well as in the Jardin des Plantes in Paris.

After his return to Paris from Senegal in 1754 he made use of a small portion of the information he had collected there in his "Histoire naturelle du Sénégal" from 1757. Sales of the work were slow, and after the publisher's bankruptcy and the reimbursement to subscribers, Adanson estimated the cost of the book to him had been 5,000 French pounds, beginning the penury in which he lived the rest of his life. This work has a special interest from the essay on shells, printed at the end, where Adanson proposed his universal method, a system of classification distinct from those of Leclerc (comte de Buffon) and Linneus. This system for all organized beings was founded on the consideration of each individual organ. As each organ gave birth to new relations, he established a corresponding number of arbitrary arrangements. Those beings possessing the greatest number of similar organs were referred to one great division, and the relationship was considered more remote in proportion to the dissimilarity of organs.

In 1774, Adanson submitted to the consideration of the French Academy of Sciences an enormous work, extending to all known life forms and substances. It consisted of 27 large volumes of manuscript, which displayed the general relations of all these matters and their distribution; 150 volumes more, occupied with the alphabetical arrangement of 40,000 species; a vocabulary containing 200,000 words with their explanations; and a number of detached memoirs, 40,000 figures, and 30,000 specimens of the 3 kingdoms of nature recognized at the time: plant, animal, and mineral. The committee to which the inspection of this monumental work was entrusted found it too long so strongly recommended Adanson to leave out what was merely compilation, but he rejected this advice, and the work, at which he continued to labour, was never published.

Antoine-Laurent de Jussieu's system of 1789 is often said to be the first natural system of classification for plants, instead of Adanson's, and is as follows:

Table 12-4. Jussieu's System.

Acotyledones I. (contains the first 6 families: Fungi, Algae, Hepaticae, Musci, Filices, Naiades)
Monocotyledones

Stamina hypogyna II.
perigyna III.
epigyna IV.
Dicotyledones
Apetalae
Stamina epigyna V.
perigyna VI.
hypogyna VII.
Monopetalae
Corolla hypogyna VIII.
perigyna IX.
epigyna antheris connatis X.
distinctis XI.
Polypetalae
Stamina epigyna XII.
hypogyna XIII.
perigyna XIV.
Diclines irregularis XV.

Antoine-Laurent, nephew of Bernard, Antoine, and Joseph, sons of Christophe de Jussieu (or Dejussieu), an apothecary of some repute, who published "Nouveau traité de la thériaque" (1708) (a theriac is an ancient medicine made of dozens of ingredients), was born, like his brothers, in Lyons. In 1765 he went to Paris, where he first studied medicine, graduating in 1770. In the same year he was appointed professor (a post he occupied till 1826) and demonstrator at the Jardin du Roi in place of Lemonnier, after which he devoted himself entirely to botany. In 1804 he was made professor of botany in the medical faculty in Paris, where he lectured until 1826. His memoir on the classification of Ranunculaceae in 1773 led to his election to the Academie des sciences. He adopted his uncle Bernard's ideas concerning the natural system, expanded them, gave them a theoretical basis, and applied them to the different families. All our knowledge concerning the natural system of his uncle we owe to him; consequently, it is not possible to make a clear distinction between the work of the two men. As early as 1774, during his uncle's lifetime, appeared the treatise "Exposition d'un nouvel Ordre des Plantes, adopté dans les démonstrations du Jardin royal", in the "Mém. de l'Acad. des sciences" of 1774. His chief work, the result of many years' study on the same subject, was "Genera plantarum secundum ordines naturales disposita, juxta methodum in horto Regio Parisiensi exaratum anno 1774" in 1789, which greatly influenced scholars in France, especially Cuvier and de Candolle. At a later date it also affected the German and English scholars, who had been at first suspicious of it as a product of the French Revolution and were extreme adherents of the Linnean system. Even more vigorously than his Uncle Bernard, he upheld the theory of unequal value of the characteristics of plants. The characteristics are "weighed, not simply counted" ("pesés et non comptés"). Antoine-Laurent gave to the 3 main groups of the original classification of his uncle the names of Acotyledon, Monocotyledon, and Dicotyledon, and divided them into 15 classes, containing in all 100 families and clearly defined the characteristics of the families.

Antoine-Laurent also published "Principes de la méthode naturelle des végétaux" in Paris in 1824. He partly prepared a greatly desired 2nd edition of the "Genera plantarum", but it was never issued, only what had been left ready for print, an entirely rewritten "Introductio" for the 2nd edition, was published after his decease by his son Adrien-Henri, who also became a botanist, in Ann. des Sc. nat. in 1837.

Bernard de Jussieu (1699-1777) started medical practice in 1720 but stayed as a physician only for a short time, finding the work uncongenial, and gladly accepted his brother's invitation to Paris in 1722, when he succeeded Sébastien Vaillant as assitant demonstrator of plants in the Jardin du Roi. He devoted most of his energies to the royal garden, which his brother left almost entirely to him. In 1725 he brought out a new edition of Tournefort's "Histoire des Plantes des environs de Paris," in 2 vols., which was afterwards translated into English by John Martyn, the original work being incomplete, and which gained his admission into the Academie des sciences and for which he communicated several research papers.

Long before 1700s French biologist Abraham Trembley published his "Histoire des polypes d'eau douce," and before Peyssonel, Jussieu maintained these organisms were animals and not the flowers of marine plants, then the current notion, and to confirm his views he made 3 trips to the Normandy coast, and published his views in the "Mémoires" of 1742, and sought to classify them at this early date into genera. He also separated the whale from the fish and placed it among the mammals.

Very modest and retiring, he published very little, in spite of the wide range of his learning, caring little for the credit of new discoveries, so long as they were made public. The few botanical articles he published, between 1739 and 1942, treat of 3 water plants. But in 1758, after Louis XV made him superintendent of the royal garden at Trianon at Versailles near Paris, he arranged the plants in the garden according to his own classification scheme. This arrangement, sometimes called the Trianon System, is printed in his nephew's Genera, and formed the basis of that work. The genera and families are arranged according to consideration of all the characteristics, which are not regarded of equal value. On the decease of his brother Antoine, he could not be induced to succeed him in his office, and prevailed upon L. G. Lemonnier to assume the position.

Adrien-Henri de Jussieu (1797-1853), son of Antoine-Laurent, was born in Paris and in 1824 received a doctorate of medicine, presenting a treatise on Euphorbiaceae. When his father retired in 1826 he was made professor of agricultural botany at the Jardin des Plantes. In 1845 he was made professor of plant organography at the university. His textbook, "Cours élémentaire de botanique", went through numerous editions and translations. Besides a "Géographie botanique" in 1845, he also published monographs on several plant families, especially Malpighiaceae in 1843. He was also president of the Académie des ciences.

Antoine de Jussieu (1686-1758)(Wikipedia; Catholic Encyclopedia) went to Paris in 1708, succeeding the celebrated Tournefort at the Jardin des Plantes. His own original publications are not generally of marked importance, but he edited an edition of Tournefort's "Institutiones rei herbariae" in 3 vols. in 1719, and also a posthumous work of the Dominican Jacques Barrelier, "Plantae per Galliam, Hispaniam, et Italiam observatae, & co." in 1714, a large and not unimportant treatise. Barrelier had left numerous drawings of plants and the text for a large work; the text was destroyed in a fire after Barrelier's decease, but the drawings were saved. The work contains 334 botanical plates, in folio, with nearly 1400 figures. He also wrote "Traité des vertus des plantes" in 1771. He also practised medicine, chiefly devoting himself to the poor.

After making botanical journeys over a large part of France, in 1716 he explored the flora of the Pyrenean peninsula. From 1718 he also made use in his practice of quassia bark (*Cortex simarube*), the first of which had been sent in 1713 to the Jesuit priest Soleil in Paris from Cayenne. De Jussieu wrote an account of the bark in the "Mémoires" of the Academy for 1728, and Linneus named *Simaruba jussiei* after him. The "Mémoires" also contain research papers by de Jussieu on human anatomy, zoology, paleontology, and mineralogy.

Joseph de Jussieu, explorer and traveller, brother of Adrien-Henri and Antoine, born in Lyons (1704-1779). Highly educated in many directions and able to act as physician, botanist, engineer, and mathematician, he became a member of the scientific expedition sent by the Académie to Peru in 1735 to measure an arc of the meridian, after which he supported himself chiefly by the practice of medicine. His extended and arduous explorations in Peru took place mainly in the years 1747-50. The botanical results of these journeys were large, but the greater part of his manuscripts and collections was lost, and he finally returned to Paris in 1771, in poor health.

He sent the seed of the *Heliotropium peruvianum* to his brother Bernard, so that the introduction of this ornamental plant into Europe is due to him. He also undertook an investigation of the area in which the cinchona tree flourishes and of the first use of its bark by the Jesuits in South America.

The "Prodromus systematis naturalis regni vegetabilis" was a monumental undertaking by Genevan botanist Augustin de Candolle, having the intention of classifying and describing all known species of vascular plants (30, 000 at the time). The 1st volume appeared in 1824 and de Candolle went on to publish 5 others himself, with his son Alphonse continuing the effort, publishing 10 more tomes plus 4 index volumes. The last edition, "Théorie élémentaire," was in 1844. The system dominated plant taxonomy until 1860 even though there were 2 dozen other such systems presented between 1825 and 1845, and is as follows (the orders are the equivalent to modern families and there were 161):

Table 12-5. The Candollean System.

Vasculares
- Class Exogenae (vascular bundles in a ring) (dicots)
  - Diplochlamydeae (both calyx and corolla)
    - Thalamiflorae (polypetalous and hypogynous)(46 orders)
    - Calyciflorae (polypetalous or sympetalous, perigynous or epigynous)(38 orders)
    - Corolliflorae (sympetalous and hypogynous)(24 orders)
  - Monochlamydeae (calyx only)(20 orders, the last one is Coniferae)
- Class Endogenae (vascular bundles scattered)
  - Phanerogamae (flowers present) (22 orders)(monocots)
  - Cryptogamae (flowers absent, hidden, or unknown)(ferns, horsetails, quillworts, club mosses) (5 orders)

Cellulares (no cotyledons)
- Class Foliacae (leafy; sexuality known)(mosses and liverworts)(2 orders)
- Class Aphyllae (not leafy; sexuality unknown)(Algae, Fungi, Lichens)(4 orders)

Augustin Pyrame de Candolle (1778 – 1841)(Wikipedia; HowStuffWorks) was a Swiss botanist born in Geneva whose family originated in ancient Provence but relocated to Geneva at the end of the 16th century to escape religious persecution.

De Candolle originated the idea of "nature's war", which influenced Charles Darwin and the principle of natural selection. He recognized that multiple species may develop similar characteristics that did not appear in a common evolutionary ancestor; this was later termed analogy and is now called convergence.

During his work with plants, he noticed that leaf movements follow a near 24-hour cycle in constant light, suggesting that an internal biological clock exists. Though many scientists doubted Candolle's findings, experiments over a century later demonstrated that "the internal biological clock" indeed exists.

The world-wide system of biogeographical regions was originated by him in 1820; it was based on plant distribution, and he was primarily interested in documenting the nature and floral composition of the plant "formations" (which we call biomes today) of the regions and their relations to climate

(Cox, 2001)(more on this in Chapter 17).

He also contributed to agronomy, paleontology, medical botany, and economic botany.

In 1794, he began his scientific studies at the Collège Calvin under Jean Pierre Vaucher (for whom the genus *Vaucheria* is named), who later inspired Candolle to make botanical science the chief pursuit of his life. In 1796, he moved to Paris after receiving an invitation from French geologist Déodat Gratet de Dolomieu. His botanical career began with the help of René Desfontaines, who recommended Candolle for work in the herbarium of Charles Louis L'Héritier de Brutelle during the summer of 1798. The position elevated de Candolle's reputation and also led to valuable instruction from Desfontaines himself. Within a couple of years de Candolle had established his first genus, *Senebiera*, in 1799, and went on to document 100s of plant families and create a new natural plant classification system.

De Candolle's first works, "Plantarum historia succulentarum" (4 vols.) in 1799 and "Astragalogia" in 1802, brought him to the notice of Georges Cuvier and Jean-Baptiste de Monet, comte de Lamarck. De Candolle, with Cuvier's approval, acted as deputy at the Collège de France in 1802. Lamarck entrusted him with the publication of the 3rd edition of "Flore française" from 1803–1815. In the introduction to this work, Principes élémentaires de botanique, de Candolle proposed his plant classification. The premise of his method is that taxons do not fall along a linear scale; they are discrete, not continuous.

In 1804, de Candolle published his "Essai sur les propriétés médicales des plantes" and was granted a doctorate of medicine by the medical faculty of Paris. Two years later, he published "Synopsis plantarum in flora Gallica descriptarum." He then spent the next 6 summers making a botanical and agricultural survey of France at the request of the French government, which was published in 1813. In 1807 he was appointed professor of botany in the medical faculty of the University of Montpellier, where he would later become the first chairman of botany in 1810. While at Montpellier, he published his "Théorie élémentaire de la botanique" in 1813, which introduced the word "taxonomie". He moved back to Geneva in 1816 and in the following year was invited by the government of the Canton of Geneva to fill the newly created chairmanship of natural history.

He spent the rest of his life in an attempt to elaborate and complete his natural system of botanical classification. He published initial work in his "Regni vegetabillis systema naturale," but after 2 volumes he realized he could not complete the project on such a large scale. Consequently, he began his less extensive "Prodromus systematis naturalis regni vegetabilis" in 1824 but was able to finish only 7 volumes, or two-thirds of the project. Even so, he was able to characterize over 100 families, helping to lay the empirical basis for general botany.

In all, he published 180 articles and left 40 unfinished manuscripts. He also wrote biographics and poetry. Many of his research papers are considered classics in the field, and more than 300 plants have been dedicated in his memory. He is remembered in the genera *Candollea* and *Candolleodendron*, both plants, and Candollea, a scientific magazine that publishes articles on botanical taxonomy.

He was the first of 4 generations of botanists in the de Candolle dynasty. His son, Alphonse, eventually succeeded in his father's chair in botany and continued the Prodromus. Augustin's grandson, Casimir Pyrame, also contributed to the Prodromus through his detailed, extensive research and characterization of Piperaceae. Augustin's great-grandson, Richard Emile Augustin de Candolle, was also a botanist.

John Lindley proposed a Cellulares-Vasculares dichotomy in 1830 (An Introduction to the Natural System of Botany) and the following in 1833 (Nixus plantarum):

Table 12-6. Lindley's Classification.

Sexuales
- Vasculares
  - cl. Exogenae Angiospermae
    - sbcl. Polypetalae (7 cohorts)
    - sbcl. Incompletae (5 cohorts)
    - sbcl. Monopetalae (5 cohorts)
  - cl. Exogenae Gymnospermae
  - cl. Endogenae (5 cohorts)
- Evasculares
  - cl. Rhizanthae

Esexuales
- Filicales
- Lycopodales
- Muscales
- Charales
- Fungales (Fungi, Lichenes, Algae)

In 1845 (The Vegetable Kingdom (biodiversitylibrary.org) he presented this scheme:

asexual, flowerless
- Thallogens (Algales, Fungales, Lichenales)
- Acrogens (Muscales, Lycopodales, Filicales)

sexual, flowering
- Rhizogens (1, partly monocots)
- Endogens (11) (monocots)
- Dictyogenes (1, mostly monocots)
- Gymnogens (1)
- Exogens (dicots)
  - Diclinous (8)
  - Hypogynous (14)
  - Perigynous (10)
  - Epigynous (7)

The Stefan Endlicher taxonomy, presented in his "Genera plantarum" in 1836-40, was as follows:

Table 12-7. Endlicher's Taxonomy.

- region Thallophyta
  - section Protophyta
    - class Algae
    - class Lichenes
  - section Hysterophyta
    - class Fungi
      - Gymnomycetes
      - Hyphomycetes
      - Gasteromycetes

Pyrenomycetes
Hymenomycetes
region Cormophyta
section Acrobrya
cohort Anophyta
cl. Hepaticae
cl. Musci
cohort Protophyta
cl. Equiseta
cl. Filices
cl. Hydropterides
cl. Selagines
cl. Zamiae
cohort Hysterophyta
cl. Rhizantheae
sect. Amphybrya (11 classes)
sect. Acramphibrya
coh. Gymnospermae (1 class)
coh. Apetalae (6 classes)
coh. Gamopetalae (9 classes)
coh. Dialypetalae (22 classes)

Endlicher (1804–49) was an Austrian botanist, numismatist, cleric, and Sinologist. In 1828 he was appointed to the Austrian National Library to reorganize its manuscript collection, in 1836 was appointed keeper of the court cabinet of natural history, in 1840 became professor at the University of Vienna and director of its Botanical Garden, and was fundamental in establishing the Imperial Academy of Sciences (Akademie der Wissenschaften) but was not elected president.

The majority and the most valuable of his works are on botany. Foremost, besides "Genera plantarum," were "Grundzüge einer neuen Theorie der Pflanzenerzeugung" (Foundations of a new theory of plant breeding) in 1838 and "Die Medicinalpflanzen der österreichischen Pharmakopöe" (Medicinal plants in the Austrian pharmacopeia) in 1842. He established the botanical periodical Annalen des Wiener Museums der Naturgeschichte (Annals of the Viennese Museum of Natural History) in 1835, which continues.

German botanist August Eichler was botany professor at the Technical U. of Graz, director of the botanical garden of Graz, later received an appointment to the U. of Kiel, and was then directer of the herbarium at the U. of Berlin. His system (1883, 3rd ed., "Syllabus der Vorlesungen über Phanerogamenkunde" (Syllabus of lectures on phanerogam knowledge) was included in the 1970 edition of the Encyclopedia Britannica.

Table 12-8. Eichler's Taxonomy.

A. Cryptogamae
phylum I. Thallophyta
classis I. Algae
Cyanophyceae
Chlorophyceae
Pheophyceae

Rhodophyceae
classis II. Fungi
classis III. Lichenes
phylum II. Bryophyta
classis I. Hepaticae
classis II. Musci
phylum III. Pteridophyta
classis I. Equisetinae
classis II. Lycopodinae
classis III. Filicinae
B. Phanerogamae
phylum I. Gymnospermae
phylum II. Angiospermae
classis I. Monocotyleae
classis II. Dicotyleae
subclassis I. Choripetalae
subclassis II. Sympetalae

George Bentham and Joseph Hooker, in "Genera plantarum," 1862-83, included nearly 100,000 species, and their system is as follows (the cohort is equivalent to the modern order and the order to the modern family):

Table 12-9. Bentham and Hooker's System.

Cryptogamae
Thallophyta
Bryophyta
Pteridophyta
Phanerogamae
Dicots
Polypetalae
Series 1. Thalamiflorae (hypogynous)(6 cohorts, 34 orders)
Series 2. Disciflorae (receptacle expanded as a disc)(3 cohorts, 22 orders)
Series 3. Calyciflorae (receptacle cup-like; perigynous or epigynous) (5 cohorts, 27 orders)
Gamopetalae
Series 1. Inferae (3 cohorts, 9 orders)
Series 2. Heteromerae (3 cohorts, 12 orders)
Series 3. Bicarpellatae (4 cohorts, 24 orders)
Monochlamydeae (simple perianth, i.e., apetalous)(8 series, 17 orders)
Gymnospermae (Gnetacae, Coniferae, Cycadaceae)
Monocots (7 series, 34 orders)

This system was largely patterned after Eichler's taxonomy, but also and not surprisingly, after De Candolle's, as Bentham was a close friend and colleague of his, but the text differs in that every genus was studied anew from material in the British and Continental herbariums. It was adopted from the outset by the Anglophone world.

German botanists Adolf Engler and Karl Prantl published a system, appearing in a monumental 20-volume work, "Die Natürlichen Pflanzenfamilien," 1897-1915, with later editions by Engler and Gilg in 1924 and Engler and Diels in 1936, that was based mostly on Eichler's and was widely adopted well into the late 20th century. Like Eichler it was supposed to be phylogenetic but was largely artificial. It was originally written by Adolf Engler as part of a guide to the botanical gardens at Breslau U. It divided the plant kingdom into 14 phyla (Abteilungen). He also published "Syllabus der Pflanzenfamilien," 1st ed., in 1892.

Because of its great detail it was quickly adopted as a valuable reference work in most countries except the British Empire (where the Bentham and Hooker system continued to be used); a vast literature developed based on it, and plant specimens in many herbariums around the world were arranged according to the "Engler-Prantl sequence."

"Genera siphonogamarum ad Englerian conscripta" was a numberd synopsis by Dalla Torre and Harms in 1900-07, with an extensive bibliography and synonymy, of Engler's system based on his various works especially as summarized in the Syllabus of 1892 and successive editions. The 12th edition of the Syllabus was edited by Melchior in 1964. And Wettstein's and Rendle's systems were patterned after it.

Engler wrote many books and did notable work also on phytogeography. He received the Linnean Medal in 1913. The International Association for Plant Taxonomy established the Engler Medal in his honour in 1986, to be awarded for outstanding contributions to plant taxonomy. The journal Englera, published by the Berlin-Dahlem Botanical Garden is named after him, as well as many genera.

The Englerian system is as follows:

Table 12-10. Engler's Taxonomy.

Schizophyta
Myxomycetes
Flagellatae
Dinoflagellatae
Heterocontae
Bacillariophyta
Conjugatae
Chlorophyceae
Charophyta
Pheophyceae
Rhodophyceae
Eumycetes
Embryophyta Archegoniatae (bryophytes and pteridophytes)
Embryophyta Siphonogama (seed plants).
- sbph. Gymnospermae (7 classes, 15 families)
- sbph. Angiospermae
  - cl. Monocotyledoneae (11 orders, 45 fams.)
  - cl. Dicotyledoneae
    - sbcl. Archichlamydeae (33 orders, 210 families)
    - sbcl. Metachlamydeae (Sympetalae)(11 orders, 57 families)

Tippo (1942) had recognized 2 plant subkingdoms, Thallophyta (10 phylums) and Embryophyta

(2 phylums, Bryophyta and Tracheophyta (vasculars, aka, Stelophyta), the latter containing classes Psilopsida, Lycopsida, Sphenopsida, and Pteropsida), and Bold (1973) 3 subkingdoms, Prokaryonta, Chloronta, and Achloronta. But Tippo's arrangement was more of a synthesis instead of original as his scheme for non-vasculars followed Smith (1938) and for vasculars that of Eames (1936).

Table 12-11. Tippo's Taxonomy.

sbk. Thallophyta (9 phylums)
sbk. Embryophyta
 phyl. Bryophyta (3 classes)
 phyl. Tracheophyta
  sbph. Psilopsida (2 orders)
  sbph. Lycopsida (5 orders)
  sbph. Sphenopsida (3 orders)
  sbph. Pteropsida
   cl. Filicineae (Cenopteridales*, Ophioglossales, Marattiales, Filicales)
   cl. Gymnospermae
    sbcl. Cycadophytae (3 orders)
    sbcl. Coniferophytae (4 orders)
   cl. Angiospsermae
    sbcl. Dicotyledoneae
    sbcl. Monocotyledoneae

A phylogenetic plant kingdom (green algae + embryophytes) was recognized by the following:

Plantae (Chlorophyta) Copeland 1938, 1947, 1956
Plantae Leedale 1974
Chlorophyta Cavalier-Smith 1978
Viridiplantae Cavalier-Smith 1981
Chlorobionta Möhn 1984
Plantae Lipscomb 1985, 1989, 1991
Plantae Starobogatoff 1986
Plantae Corliss 1994, 1995
Kussakin and Drozdoff 1994, 1998
Viridiplantae Holt and Uidica 2007

A chloroplast kingdom was recognized as monophyletic by the following:

Plantae Takhtazhan 1973
Plantae Leedale 1974
Phyta (Plantae) Bold et al 1987

The following table presents a cladistic classification for plants.

Table 12-12. Plantae Phylogeny (based on Bremer, 1985; Mishler and Churchill, 1985; Doyle and Donoghue, 1992; Mishler at al, 1994; Doyle, 1998; and Kenrick and Crane, 1997, etc.)

All the groups are status novus except Spermatophyta; synapomorphies are in parentheses as are the totals; *indicates extinct groups.

sbk. Primiphyta nom. nov.
infk. Loxophyta nom. nov. (syn. Loxophyceae)(Pedinomonadales)(3 or 4 gen.)
infk. Paraloxophyta nom. nov.
hpph. Pyramiphyta (Pyramimonadales)(2 or 3 fams., 3 or 5 gen.)
hpph. Parapyramiphyta nom. nov.
spph. Mamiellophyta (Mamiellales)(3 fams., 7 gen.)
spph. Paramamiellophyta nom. nov.
phyl. Pycnococcophyta (Pycnococcales)(syn. Pseudoscourfieldiales)(1-3 fams., 2-5 gen.)
phyl. Ulvophyta nom. nov. (14 ords., 70 fams., 400 gen., 4000 sp.)
sbk. Noviphyta nom. nov. (unilateral rootlets, glycolate degradation by glycolate oxidase)
infk. Chlorokybophyta (1 sp.)
infk. Cenophyta nom. nov. (rosette complex)
pvk. Klebsormidiophyta (1 ord., 1 fam., 3 gen., 40 sp.)
pvk. Anaphyta nom. nov.
mck. Gamophyta (Zygnematales)(1 or 2 ords., 3 fams., 60 gen., 4600 sp.)
mck. Streptophyta Jeffrey 1967 nom. nud., stat. nov. (cell plate)
nnk. Coleochetophyta nom. nov. (1 ord., 1 fam., 2 gen., 15 sp.)
nnk. Apicophyta nom. nov. (apical growth, C6-C3 phenylpropanoid-derived flavonoid compound, similarities in mitotic mechanisms)
pck. Charophyta Rabenhorst 1863 (stoneworts and musk grasses)(1 ord., 1 fam., 6 gen., 200 sp.)
pck. Embryophyta Endlicher 1836 (syn. Cormophyta Endlicher 1836, Metaphyta Haeckel 1894) (embryo, multicellular sporophytes, archegonia, lignins, meiospores with exine, sporangium, sporopollenin)
sbpck. Hepaticophyta (liverworts) (6 ords., c. 70 fams., c. 300 gen., 6000 sp.)
sbpck. Stomophyta nom. nov.(stomates, guard cells, collumate sporanguim, D-methionine)
infpck. Anthocerotophyta Stotler & Stotler-Crandall, 1977 (hornworts)(2 or 3 ord., 2-5 fams., 14 gen., 200 sp.)
infpck. Coscinophyta nom. nov. (Gk. "coscinos", sieve)(leptoid [nutrient-conducting cell]-sieve-element homology, axial gametophyte, terminal gametangiums, persistent and internally differentiated sporophytes)
pvpck. Muscophyta nom. nov.
phyl. Muscophyta (mosses)(18 ords., c. 90 fams., 800 gen., 8,000 sp.)
pvpck. Polysporangiophyta (multiple sporangia [sporangial branching], independent alternation of generations)
mcpck. Horneophyta* (protracheophytes)
phyl Horneophyta (3 gen.)
mcpck. Stelophyta Pia 1931 (syn. Tracheophyta Eames 1936) (stele [vascular cylinder], vascular system, water-conducting cells with annular-helical thickenings [tracheids], lignin)
grdtrph. Aglaophyta*
phyl. Aglaophyta (1 gen.)

grdtrph. Eustelophyta nom. nov. (thick lignified, decay-resistant wall layer in the tracheids, pitlets between thickenings or with pits, sterome)

hprtrph. Rhyniophyta* (4 gen.) (protracheophytes)

phyl. Rhyniophyta Banks 1958 stat. nov.

hprtrph. Cooksoniophyta nom. nov., stat. nov.

hprtrph. Neostelophyta nom. nov., stat. nov. (endarchy [centrifugal xylem maturation], terminal branching, lateral endogenous root branching pattern, terminal sporangia)

sptrph. Lycophyta Eichler 1883 stat. nov.

phyl. Lycophyta (club mosses and quillworts [Merlin's grass])(2 ords.[5 more extinct], 2 fams., 5 gen., 1000 sp.)

sptrph. Telomophyta (syn. Euphyllophyta Kenrick and Crane 1997) (telomation, longitudinal sporangial dehiscence, centrarch or mesarch xylem) stat. nov.

trph. Eophyllophyta stat. nov.

phyl. Eophyllophyta* (1 gen.)

trph. Metatelomophyta nom. nov., stat. nov.

ggph. Psilophyta * stat. nov. (1 gen.)

ggph. Neotelomophyta nom. nov., stat. nov.

mgph. Trimerophyta* stat. nov. (1 gen.)

mgph. Cenotelomophyta nom. nov., stat. nov.

grdph. Staurophyta* nom. nov., stat. nov. (3 gen.)

grdph. Anatelomophyta nom. nov., stat. nov. (large leaf with web of veins, midrib, and multiple traces, pseudopodial or monopodial and helical branching, recurvation of branch apices, trachieds with scalariform bordered pits, paired sporangiums grouped into terminal trusses, radially aligned xylem in larger axes, multiflagellate spermatozoids)

hprph. Lignophyta Doyle and Donaghue 1986 stat. nov. (secondary wood, phlem [bast], wood rays)

spph. Aneurophyta* (progymnosperms, 7 gen.) stat. nov.

spph. Metalignophyta nom. nov., stat. nov.

phyl. Archeophyta* (progymnosperms, 2 gen.)

phyl. Propityales* (progymnosperms, 1 gen.)

phyl. Spermatophyta Eichler 1883 (syn. Phanerogamae [auct.], Siphonogama Engler and Diels 1936) (80 ords. [12 others extinct], 480 fams., 14,000 gen., and about a quarter million sp.)(seeds, siphonogamy)

hprph. Psilotophyta (whisk ferns)(2 gen.) stat. nov.
hprph. Hypofilicophyta nom. nov., stat. nov. (horsetails, preferns)(1 extant gen., 3 extinct gen.; 3 extinct orders)
hprph. Filicophyta nom. nov., stat. nov. (ferns)(8 ords. [1 more extinct], 31 fams., 8,000 sp.)

Adanson, Michel. 1763. Famille des plantes.
Core, Earl. 1955. Plant Taxonomy. Prentice-Hall.
Doyle, J.A. and Donoghue, M.A.1992. Fossils and seed plant phylogeny reanalyzed. Brittonia 44: 89-106.
Doyle, James. 1998. Phylogeny of Vascular Plants. Annual Review of Ecology and Systematics 29: 567-599.
Kenrick, Paul and Crane, Peter. 1997. The Origin and Early Diversification of Land Plants: a cladistic study. Smithsonian Institute Press, Washington, D.C.
Lawrence, George. 1960. Taxonomy of Vascular Plants. Macmillan.
Tippo, Oswald. 1942. A modern classification of the plant kingdom. Chron. Bot. 7: 203-206.

Green Algae

In 1860 Stizenberger divided the green algae into 4 orders (families in parentheses):

Coccophyceae (Palmellaceae, Protococcaceae, Volvocaceae)
Siphonophyceae (Hydrogastreae (=Botryidiaceae), Vaucheriaceae)
Zygophyceae (Desmideae, Zygnemae)
Nematophyceae (Ulvaceae, Spheropleaceae, Confervaceae, Oedogoniacceae, Ulotrichaceae, Chroolepidiaceae (=Trentepohliaceae), Chetophoraceae)

This arrangement was followed by Rabenhorst and others, with later names usually being Protococcoideae, Conjugatae, Siphoneae, and Nematophyceae, respectively, a classification which remained in force till the turn of the last century. In more recent times, 4 classes have often been recognized, based largely on levels of organization:

Table 12-13. Traditional Classification for Green Algae.

Prasinophyceae (Pedinomonadales, Mamiellales, Pyramimonadales, Chlorodendrales (Tetraselmidales))
Volvocophyceae (Volvocales, Chlorococcales, Spheropleales, Schizogoniales, Chetophorales, Pleurastrales, and Oedogoniales)
Ulvophyceae (Ulotrichales, Ulvales, Cladophorales, Dasycladales, Caulerpales, Siphonocladales)
Charophyceae (Klebsormidiales, Zygnematales, Coleochetales, Charales)

Another arrangement has 4 different classes in 3 subphylums (from Roger Caratini's La vie des plantes (1971) and Encyclopédie Universalis (1995)):

Zygophytina

Euchlorophytina
Prasinophyceae
Euchlorophyceae
Monadophycidae (Volvocales and Tetrasporales)
Coccophycidae (Chlorococcales)
Septophycidae (2 subgroups):
isokontic
with an axial chloroplast (Prasiolales)
with a parietal chloroplast
with a pyrenoid surrounded by starch (Ulotrichales, Chetophorales, Ulvales)
with pyrenoids not surrounded by starch (Spheropleales, Acrosiphonales)
no pyrenoids and no starch (Trentepohliales)
stephanokontic (Oedogoniales)
Siphonophycidae (Siphonocladales, Dasycladales, and Siphonales).
Charophytina

A 3rd system, presented in the Botanique volume, edited by Fernand Moreau, in the "Encyclopédie de la Pléiade," from 1960, divides the green algae into 3 groups: Isokonta, Stephanokonta (Oedogoniales and Bryopsidales), and Aconta (Gamophyta).

The Cal. Acad. Scis., adapted largely from Fritsch from 1935, Smith from 1950, and Egerod from 1952, distinguished 2 phyla: Chlorophycophyta (11 orders)(including Zygnematales, and Coleochetaceae and Trentopohliaceae in Ulotrichales) and Charophycophyta (1 order).

The 23 orders, 80 families, 500 genera, and 9000 species are cladistically divided into 10 phyla as shown in Table 12-12. The arrangement is based on classical (Bremer, 1985; Mishler and Churchill, 1985; Mishler at al, 1994) and molecular (Lemieux et al 2007; Mishler at al, 1994; see also Lewis and McCourt, 2004, Graham et al, 2000, Bremer at al, 1987, and Leliaert et al, 2012) evidence.

There could be, however, 2 other phylums as *Mesostigma* might be an 11th (here it is placed in Charophyta because of similarities with it, a position shored up by DNA data (Not et al, 2004)), and *Chetospheridium* might be a 12th but probably belongs with *Coleochete* because of close similarities with it as found by Moestrup in 1974 (Mattox and Stewart, 1984). Volvocophyta contains Volvocopsida (chlamydophyceans) (misnamed as "chlorophytes"or "chlorophyceans" as the former name refers to all plants and the latter to all green algae) and Ulvopsida, thus the bulk of green algae, 14 orders.

Most of the evidence, both classical and molecular, supports the large 2-way split between most green algae on one hand and charophyceans sensu lato (Klebsormidiophyta, Chlorokybophyta, Coleochetophyta, *Mesostigma*, Charophyta) plus embryophytes on the other. The 26S analysis in Mishler at al (1994) seems to be the only exception. Sluiman (1982) had the same 2-way split but with a residual basal group with core prasinophytes as monophyletic. This dichotomy was apparently first proposed by Pickett-Heaps and Marchant in 1972 (Leliaert et al, 2012).

Prasinophyceae, which contain the most primitive green algae, was established by Christensen in 1962, separating the various genera from Volvocales, after Manton and Parke in 1960 discovered unusual features under the electron microscope, especially flagellar and body scales (Mattox and Stewart, 1984). *Prasinocladus* and *Platymonas* of the same order were subsequently combined with *Tetraselmis* by Mattox and Stewart.

The class includes 8 orders, perhaps 12 families, and 20 genera:

Pedinomonadales (*Pedinomonas, Resultor, Marsupiomonas*)
Pseudoscourfieldiales (Pycnococcaceae (*Pycnococcus, Pseudoscourfieldia*), Nephroselmidaceae

(*Nephroselmis*) (or as part of the previous order)

Prasinococcales (*Prasinococcus* (discovered in 1990 by Mishita et al, *Prasinoderma* (discovered by Fabien Jouenne in 2004) (sometimes placed in Pycnococcaceae; it is unclear who established the order but the ordinal name was coined before the 1960s probably by Smith or Fritsch; possibly belongs in Pycnococcaceae)

Pyramimonadales (Pyramimonadaceae (*Pyramimonas, Cymbomonas*) and *Halosphera*)

Pterospermatales (*Pterosperma* and *Pachysphera*) (with spider-web scales)(or as a family in the previous order or the next)

Mamiellales (Mamiellaceae (some with spider-web scales) (*Mamiella, Mantoniella, Dolichomastix, Crustomastix, Bathycoccus, Ostreococcus*), Micromonadaceae (*Micromonas*), Monomastigaceae (*Monomastix*)

Chlorodendrales (Tetraselmidales)(*Tetraselmis, Scherffelia*)

Mesostigmatales (*Mesostigma*) (formerly placed in Pyramimonadales)

Picocystis is of uncertain position and might be an independent lineage or belong in Pycnococcaceae. Chlorodendrales definitely belongs in Pleurastrophyceae as it shares several similarities with it, especially metacentric mitosis, and is placed there by Mattox and Stewart; *Mesostigma* belongs probably in Charales as already mentioned, and the others are a basal series. Pedinomonads show up as basal in most of the analyses in Mishler et al, but Mamiellales are considered the most primitive, so the latter might be better placed as basal. The macroscopic Palmophyllales, known since the 1980s and containing *Palmophyllum* and *Verdigellas*, might also be an early lineage like the prasinophyceans.

The internal arrangement for Ulvophyta is as follows based on O'Kelly and Floyd (1984) and van den Hoek et al (1995):

cl. Volvocopsida
- sbcl. Volvocidae (Volvocales, Spheropleales, Chetophorales, Oedogoniales)
- sbcl. Chlorellidae (Chorellales, Chlorococcales)

cl. Ulvopsida
- sbcl. Ulvidae stat. nov.
  - hpord. Ulvariae (Ulvales)
  - hpord. Siphonariae
    - spord. Cladophorarae (Cladophorales (Siphonocladales)-Dasycladales)
    - spord. Bryopsidarae (Caulerpales (Bryopsidales))
- sbcl. Pleurastridae stat. nov. (Pleurastophyceae Mattox and Stewart 1984) (Tetraselmidales, Pleurastrales, Prasiolales)
- sbcl. Trentepohlidae stat. nov. (Trentepohliales)

Ulvophyta have cruciate rootlets, Volvocopsida is distinguished by a collapsing interzonal telophase spindle in a phycoplast, and a theca, Volvocidae by a 1-7 o'clock cruciate arrangement and centripetal cleavage, Chlorellidae by a 12-6 arrangement and centrifugal cleavage, Ulvopsida by an 11-5 cruciate configuration.

Some orders, as often circumscribed, have both unicellular and multicellular, and sometimes also siphonous or semi-siphonous forms; these are Chlorococcales (coccoid, filamentous, and siphonous), Ulotrichales (coccoid, filamentous, filamentous-siphonous, and foliose), and Zygonematales (coccoid and filamentous). Cenobial (a gelatinous colonial condition found in some Volvocales and Chlorococcales), palmelloid (a tetrasporal colonial condition), sarcinoid (a cubical packet of cells), and ameboid (in Zygnematales) levels of organization also occur. Most of these 9 levels of organization occur in most algal groups. The ameboid level occurs also in Chrysophyceae, Xanthophyceae, Haptophyceae, Chlorarachnia, and Dinophyceae, and the siphonous level is present only in yellow, brown, and green algae. (Van Den Hoek, 1995).

The first published record of the name *Chara* as we know it today was apparently by herbalist Jacques Dalechamps in 1587 in volume 1 of his *Historia generalis plantarum,* which was the most complete botanical work of its time, and the first to describe the flora peculiar to the region around Lyons, who gave it as the popular name for an *Equisetum*-like acquatic plant used by the Lyonnais to scour plates and other utensils. Sébastien Vaillant formally established the genus in 1721. Many of the early botanists regarded it as species of horsetail (*Equisetum*) or mare's-tail (*Hippuris*). Adanson placed it in the aroids, De Candolle associated it with Naiadaceae, Richard and De Jussieu saw it as an angiosperm, but Linneus considered it algal, and many botanists of the 1800s, such as C. Agardh, who established *Nitella*, Wallroth, Endlicher, and Kützing had no hesitation in placing it among the algae. The reason for such disagreement was primarily misunderstanding concerning the nature of the reproductive organs. Braun, in addition to contributing much knowledge of the thallus and reproductive organs, in a long series of articles in the 1800s, laid the foundation for the later recognition of Charales as algal and as distinct from other algae. (Kessel, 1955).

It is in Ulvidae where we find the largest and loveliest green algae, including the sea lettuce or green laver (*Ulva* Linneus, 1753), which is edible and grows up to 2 feet long and about as wide, and maiden's hair or sea grass (*Enteromorpha* Link in Nees, 1820), in Ulvaceae Lamouroux ex Dumortier 1822 in Ulvales; green hair alga (*Chetomorpha* Kützing, 1845) and mermaid's hair and "boules de mousse" (called marimo in Japan where it is considered a national treasure)(*Cladophora* Kützing, 1843) in Cladophoraceae Wille, 1884 in Cladophorales; Neptune's shaving brush (*Penicillus* Lamarck, 1813) and *Halimeda* Lamouroux 1812 in Udoteaceae J. Agardh 1887, sea staghorn (*Codium* Stackhouse, 1797; a Mexican species, *C. magnum*, grows up to 26 feet long) in Codiaceae Kützing 1843, sea grape or feather algae (*Caulerpa* Lamouroux, 1809)(eaten in Japan, the Philippines, and Indonesia) in Caulerpaceae Kützing 1843, sea fern (*Bryopsis* Lamouroux, 1809), and *Derbesia* Solier 1846 in Bryopsidaceae Bory 1829, in Caulerpales (aka Bryopsidales, Codiales, and Siphonales); sponge weed (*Spongomorpha* Kützing, 1843) in Ulotrichales; mermaid's wineglass or cup (*Acetabularia* Lamouroux, 1812) in Dasycladaceae Kützing 1843 in Dasycladales. *Penicillus* and *Halimeda* are in Udoteaceae so are calcareous. *Cladophora*, *Chetomorpha*, and *Caulerpa* are used in aquariums as they are good nitrate absorbers. *Caulerpa* has been falsely labelled invasive by science and the "killer alga" by the press out of hysteria and possibly also as a ruse to get funding (see Caulerpa-Wikipedia).

The water net (*Hydrodictyon* A. Roth, 1797) in Hydrodictyaceae Dumortier 1829 in Chlorococcales is also notable and macroscopic, as is the water silk (also called mermaid's tressses or blanket weed)(*Spirogyra* Link, 1820) in Zygonemataceae Kützing 1843 in Zygophyta. In Charales we find the stonewort or muskgrass (*Chara* Linneus, 1753) and nitella (*Nitella* C. Agardh, 1824); the other 4 genera are the starry stonewort (*Nitellopsis* Hy, 1889), foxtail stonewort (*Lamprothamnus* Hiern, in D. Oliver, 1877), and *Lychnothamnus* (F.J. Ruprecht) A. Braun, 1857 in tribe Chareae, and the tassel stonewort (*Tolypella* A. Braun, 1857) in tribe Nitelleae.

Prasinophyceae are found as fossil phycomas (thick-walled resting stages) from the Precambrian. Calcified members of Dasycladales and Udoteaceae (Caulerpales) are found in Precambrian rocks. Some spherical, Precambrian microfossils have been placed in Chlorococcales and Chlorosarcinales, but their assignment there remains controversial. (Margulis et al, 1990). Peck in 1946 recognized 3 fossil families of Charales: Clavatoraceae Pia 1927, Trochiliscaceae Karpinski 1934, and Sycidiaceae Karpinski 1934 (Kessel, loc. cit.). Mädler in 1953 divided fossil charophytes into into 3 orders and 6 families: Sycidiales (1 fam.), Trochiliscales (1 fam.), and Charales, with Paleocharaceae Pia 1927, Clavatoraceae, Lagynophoraceae Stache 1880, and Characeae Richard 1815 ex C. Agardh 1924. Many of these fossils were marine.

Volvocopsida is predominantly freshwater, but are also found in soil; prasinophyceans are mostly marine but also brackish or freshwater and are an important part of the phytoplankton. Ulvopsida is mostly marine but also freshwater. Trentepohliales and Chlorokybophyta are terrestrial, Klebsormidiales is commonly terrestrial but may also be acquatic, Zygnematales is freshwater, Coleochetophyta and Charophyta are freshwater. Some green algae, such as *Chlamydomonas* and *Ankistrodesmus* in Volvopsida are snow algae found in the Alps and Arctic regions (Tiffany, 1958). *Raphidonema* in Klebsormidiales is also a snow alga (Hoek et al, 1995). Some pyramimonads (e.g., *Cymbomonas)* are mixotrophic (both phagotrophic and autotrophic) like many dinoflagellates and some Sphenomonadales in Euglenoida.

Only 1 phototroph or phototroph ally is pathogenic in humans: the green alga *Prototheca*, which causes protothecosis.

The largest families are Dasycladaceae Kützing 1843, the only family of Dasycladales, with 150 genera, but only 11 genera and about 50 species are extant (Parker, 1982); Characeae, with 60 genera, but only 6 extant; Desmidaceae Ralfs 1848, with 30 genera and about 4000 species (Parker, 1982) or 37 genera and some 10,000-12,000 species (Margulis et al, 1993), highly exaggerated figures (Pierre Bourelly in the Encyclopédie universalis gives 50 genera and 5000 species for all of Zygophyceae); and Ctenocladaceae (author unknown), with 32 genera but only 62 species (Parker, 1982).

The following are the higher level groups for green algae with citations for the botanists who established and named them and taken mostly from Kessel (1955) but also from Taxonomicon.Org. Synonyms are in parentheses, but the "or" refers to conflicting citations from different sources.

Table 12-14. Groups of Green Algae.

orders

Acrosiphonales Kornmann 1965 (part of Ulvales)
Bryopsidales Schaffner 1922
Caulerpales Setchell 1929 (Siphonales Wille in Warming 1884; Codiales Setchell 1929; Eusiphonales Feldmann 1946)
Charales Dumortier 1829 (Lindley 1836 as "alliance")
Chlorellales Bold And Wynne 1985
Chlorococcales Marchand orth. mut et emend. Pascher 1915
Chlorodendrales Fritsch in West 1927
Chlorokybales (author unknown)
Chlorosarcinales (established as Chlorospherales by Herndon in 1958; part of Volvocales)
Cladophorales Haeckel 1894 (or West 1904)
Coleochetales Chadefaud
Dasycladales Pascher 1931
Klebsormidiales Stewart and Mattox 1975
Mamiellales Moestrup 1984
Mesostigmales Cavalier-Smith 1998
Microthamniales Melkonian 1990 (in Trebouxiophyceae)
Monomastigales R.E. Norris ex Melkonian & Marin 2010
Microsporales Bohlin 1901(in Spheropleales)

Oedogoniales Blackman and Tansley ex West 1904 (or Heering 1914)
Prasiolaes Fritsch in West 1927 (Schizogoniales West 1904)
Pseudoscourfieldiales Melkonian 1991
Pyramidomonadales Chadefaud 1950
Siphonocladales (Blackman and Tansley) Oltsmann 1904 (Valionales Pascher 1931, nom. nud.)
Spheropleales Luerssen 1877 (or (Fritsch) Prescott 1951) (not usually recognized)
Tetraselmidales Mattox and Stewart 1984
Trentepohliales Chadefaud ex Thompson and Wujek 1997
Ulvales (Blackman and Tansley 1902) Oltsmann 1904 (Ulotrichales Borzi 1895 as Ulotrichiales, Chetophorales Wille 1884 (or 1909, Protococcales (Meneghini) Kirschner orth. mut. Engler 1892, Pleurococcales Chodat 1909, Chroopeloidales Chodat 1909, Microsporales Bohlin 1901 (part of volvicine line), Cylindrocapsales Prescott 1951 (sometimes considered as separate order))
Volvocales Oltmanns 1904 (Chlamydomonadales Fritsch in West 1927 (sometimes as suborder of Volvocales or as separate order), Tetrasporales Lemmermann 1915 (sometimes as separate order), Duniellales (usually considered as family in Volvocales, sometimes as separate order)
Zygnematales G.M Smith 1933 (Borge and Pascher as Zygnemales 1913, Conjugales (DeBary) Rabenhorst orth. mut. G.M. Smith 1920, Mesoteniales Fritsch in West 1927, Desmidiales Krieger 1933 nom. nud.)

classes

Brypsidophyceae Bessey 1907
Charophyceae G.M. Smith 1938 (or Rabenhorst 1863)
Chlorokybophyceae (author unknown)
Chlorophyceae Kutzing 1843 (or Wille in Warming 1884)(as syn. for phylum Chlorophycophyta) (Chlorospermae Harvey 1846, Pocillophycinées Chadefaud 1950, Prasinophycinées Chadefaud 1950)
Coleochetophyceae C. Jeffrey 1982
Conjugatophyceae Engler 1892
Dasycladophyceae van den Hoek, Mann, and Jahns 1995
Euchlorophycinées Chadefaud 1950
Klebsormidophyceae van den Hoek, Mann, and Jahns 1995
Loxophyceae Christensen 1962
Mamiellophyceae Marin & Melkonian 2010
Mesostigmatophyceae Marin & Melkonian, 1999
Pleurastrophyceae Mattox and Stewart 1984
Prasinophyceae Christensen 1962
Trebouxiophyceae Friedl 1995 (more or less synonymous with Pleurastrophyceae)
Zygnetmatophyceae Round 1971
Zygophyceae Rabenhorst 1868

phyla

Charophycophyta Papenfuss 1946
Charophyta Migula 1884
Chlorophycophyta Papenfuss 1946

Bremer K, Humphries, CJ, Mishler, BD, Churchill, SP. 1987. On cladistic relationships in green plants. Taxon 36: 339-349 (jstor).

Graham, Linda; Cook, Martha; and Busse, James. 2000. The origin of plants: Body plan changes contributing to a major evolutionary radiation. Proc Natl Acad Sci U S A. 97: 4535–4540 (pubmed).

Hoek, van den, C., Mann, G.D., Jahns, H.M. 1995. Algae: an introduction to phycology. Cambridge U. Press.

Kessel, Edward (ed.). 1955. A Century of Progress in the Natural Sciences, 1853-1953. California Academy of Sciences, San Francisco.

Leliaert, Frederik; Smith, David; Moreau, Hervé; Herron, Matthew; Verbruggen, Heroen; Delwiche, Charles; and De Clerck, Olivier. 2012. Phylogeny and Molecular Evolution of the Green Algae. Critical Reviews in Plant Sciences, 31:1–46 (phycology.ugent.be).

Lemieux, C., Otis, C. and Turmel, M. 2007. A clade uniting the green algae *Mesostigma viride* and *Chlorokybus atmophyticus* represents the deepest branch of the Streptophyta in chloroplast genome-based phylogenies. BMC Biology 5: 2 (biomedcentral.com).

Lewis, Louise, and McCourt, Richard. 2004. Green algae and the origin of land plants. Am. J. Bot. 91: 1535-1556 (amjbot.org).

Margulis, L. et al (eds.). 1993. Handbook of Protoctista.

Matttox, K.R., Stewart, K.D. 1984. Classification of the Green Algae (Chp. 2). In Systematics of the Green Algae, D.E.G. Irvine and D.M. John, eds. Academic Press.

Mishler, B. D., Churchill, S. P. 1985. Transition to a land flora: phylogenetic relationships of the green algae and bryophytes. Cladistics 1: 305-328.

Mishler, Brent; Lewis, Louise; Buchheim, Mark; Renzaglia, Karen; Garbary, David; Delwiche, Charles; Zechman, Frederick; Kantz, Thomas; and Chapman, Russell. 1994. Phylogenetic relationships of green algae and bryophytes. Ann. Msri. Bot. Gard. 81: 451-83 (jstor.org).

Not, Fabrice; Latasa, Mikel; Marie, Dominique; Cariou, Thierry; Vaulot, Daniel; and Simon, Nathalie. 2004. A Single Species, *Micromonas pusilla* (Prasinophyceae), Dominates the Eukaryotic Picoplankton in the Western English Channel. Appl. Environ. Microbiol. 70: 4064-4072 (asm.org).

O'Kelly, Floyd. 1984. Patterns and Features in Ulvophyceae (Chp. 4). In Systematics of the Green Algae, D.E.G. Irvine and D.M. John, eds. Academic Press.

Parker, Sybil (ed.). 1982. Synopsis and Classification of Living Organisms. McGraw-Hill, New York.

Picket-Heaps, J. D. and Marchant, H. J. 1972. The phylogeny of the green algae: a new proposal. Cytobios 6: 255–264.

Tiffany, Lewis, 1958. Algae: the grass of many waters, 2nd ed. Charles Thomas, Springfield, Ill.

Bryophytes

Liverworts

Traditionally considered a natural group, bryophytes are now generally regarded as paraphyletic as evidenced by differences in gametangial and sporophyte ontogeny, sporophyte structure, chromosome number, and rhizoid structure, with similarities attributed to parallelisms. Liverworts were first formally defined by de Jussieu in 1789 (Schuster, 1973). For these the traditonal classification is as follows (e.g., Parker, 1982):

Jungermannidae
  Calobryales (Takakiacea+Haplomitriacea)
  Jungermanniales
  Metzgeriales
Marchantidae
  Sphaerocarpales
  Monocleales
  Marchantiales

Some cladistic molecular classifications are much the same (e.g., Forrest et al, 2006):

Haplomitriopsida Stotler and Crandall-Stotler 1977, emend. 2000
  Haplomitriales Buch ex Schljakov 1972 (syn. Calobryales Campbell ex Hamlin 1972)
  Treubiales Schljakov 1972
Jungermanniopsida Stotler and Crandall-Stotler 1977, emend. 2000
  Metzgeriales Chalaud, 1930 (A)
  Jungermanniales Klinggräff, 1858 (leafy liverworts)
  Metzgeriales Chalaud, 1930 (B)
Marchantiopsida Stotler and Crandall-Stotler 1977, emend. 2000
  Blasiales (Schuster 1984) Stotler & Crandall-Stotler, 2000
  Sphaerocarpales Cavers, 1910 (bottle hepatics)
  Marchantiales Limpricht, 1877 (complex thalloids)

Jungermannidae has only 1 cell type (containing oil bodies and chloroplasts), a multi-layered capsule wall, cytokinesis with cleavage furrow, monomorphic rhizoids, a well-developed seta, is leafy and mostly isophyllous, normally with small wind-blown spores of short duration, commonly with asexual gametophytic reproduction by specialized structures, and without short-lived or ephemeral gametophytes.

Marchantidae has cellular dimorphism (typically small cells with 1 oil body and lacking chloroplasts, and large cells with chloroplasts but no oil bodies), single-layered capsule wall, cytokinesis with cell plate (except in Monocleales [1 family]), dimorphic rhizoids, a reduced seta, is mostly thalloid and anisophyllous, occasionally with spores of short duration, rarely with asexual gametophytic reproduction by specialized structures, and commonly with short-lived or ephemeral gametophytes.

Because of the Doctrine of Signatures (a plant that resembled an organ was used to treat ailments of that organ), hepatics were used in the Middle Ages to treat liver ailments. Their value is not medical but as pioneer plants colonizing burned-over soil. (Margulis and Schwartz, 1998)

Liverwort fossils existed as early as the Paleozoic, specifically in the Upper Devonian (c. 300 mya) (Bold et al, 1987); molecular clock estimates suggest the origin of liverworts was at around 450–475 mya, but the earliest whole-plant fossils were polysporangiates that date back to the Silurian (c. 425 mya)(Villarreal et al, 2010).

The largest liverwort families are Lajeuneaceae with 92 genera and 1500 species and Plagiochilaceae with 7 genera and 1600 species.

Hornworts

There was traditionally only one order, Anthocerotales, and one family, Anthocerotaceae, and only 5 genera (*Anthoceros, Dendroceros, Megaceros, Pheoceros*, and *Notothylas*), but now as many as 14 genera are recognized. Notable classifications were done by Schuster in 1988, Argentinian bryologist Gabriela Gustava Hässel de Menéndez in 1992, Hyvønen and Piipo in 1993, and Hasegawa in 1994 (Stotler and Crandall-Stotler, 2005). The last 3 did a cladistic analysis of the phenotype. The first of these is dubious according to Stotler and Crandall-Stotler and the last 2 agree with each other exactly except for a few genera included or excluded. The Hyvønen and Piipo taxonomy is as follows:

Notothylales Hyvønen and Piipo 1993
  Notothyladaceae (Milde) Müller ex Proskauer 1960 (*Notothylas* Sullivant ex Gray 1846)
Anthocerotales Limpricht in Cohn 1877
  Anthocerotaceae Dumortier 1829, corr. Trevis., emend. Hässel 1988
    sbfam. Anthocerotoideae
      tribe Anthoceroteae (*Anthoceros* Linneus 1753)
      tribe Spheroceroteae (*Sphaerosporoceros* Hässel de Menéndez 1988+
        (*Leiosporoceros* Hässel de Menéndez 1986+*Folioceros* Bhardwaj 1971))
    sbfam. Mesocerotoideae (*Mesoceros* Piippo 1993+ *Phaeoceros* Proskauer 1951)
  Dendrocerotaceae (Milde) Hassel de Menéndez 1988
    (*Dendroceros* Nees in Gottsche et al 1846+*Megaceros* Campbell1907)

The other genera are *Nothoceros* (Schuster 1987) Hasegawa 1994, *Hattorioceros* Hasegawa 1994, *Paraphymatoceros* Hässel de Menéndez 2006, *Phaeomegaceros* Duff et al 2007, and *Phymatoceros* Stotler et al, emend. Duff et al, 2007.

Molecular taxonomies are essentially the same, the only major difference being the position of *Leiosporoceros,* which is basal and sister group to all other hornworts. In molecularly-derived classifications there are 2 classes and 3-5 orders, mostly with only 1 family each (see Duff et al, 2007; Stotler and Crandall-Stotler, 2005; Villarreal et al, 2010). These, however, should be arranged as 2 orders and 5 families, especially where all the 5 "orders" have only 1 family each. Any additional levels can be classified as suborders or subfamilies as was done by Hyvønen and Piipo.

Anthoceros has the most species with about 80. *Leiosporoceros*, *Hattorioceros*, and *Phaeoceros* have only 1 each.

Hornworts are especially different from other bryophytes. In contrast to them they have bilateral spermatozoid symmetry with symmetrical flagellar insertion, continuous production of spores (non-synchronous sporogenic development), multicellular elaters (elaters are absent in mosses), stomas in both *n* and 2*n* generations, and a cyanobacterial association; there is an absence of protonemas, setas, oil bodies, leaves, calyptras, peristomes, and annulus; the chromosome count is low (at 5-6), and the sex organs are sunken in the thallus. In contrast to liverworts they have a columella (central column of sterile cells) and stomas and in contrast to mosses their rhizoids are unicellular. Their sporogenic development is unique among extant and extinct embryophyte lineages. Also, most hornworts have an algal-like chloroplast with a central pyrenoid that contains the enzyme RuBisCO and therefore exhibits a carbon concentration mechanism not seen in other land plants. Colonies of the cyanobacteria are internal. The only other plant gametophyte that harbours a nitrogen-fixing bacterium is that of the liverworts in Blasiales. In those, contrary to hornworts, the homoplastic development of the *Nostoc* colonies is external to the thallus. (Schuster, 1979; Bold et al, 1987; Villarreal et al, 2010).

Fossilzed hornworts date as far back as the Cretaceous (144-66 mya)(Margulis and Schwartz, 1998). Molecular clock estimates suggest hornworts diversified within a window of only about 30 million years, and the differentiation of all embryophyte lineages, except seed plants, is estimated to have occured over a period of only some 70 million years (Villarreal et al, 2010).

Hornworts live in shady, cool, and humid areas in temperate or tropical regions (Anthocérotales-Enc. Universalis, 1993).

Mosses

Some early classifications were based on forms and ecology. In 1869 Lorentz recognized acquatic and terrestrial mosses and a special group, the *compactae*, which represented saxicolous (rock-dwelling) species, snow species, etc. In 1910, Giegenhausen described dendroid forms, pendant forms, and forms in fronds. In 1918 and 1932, H. Gams distinguished floating (Natantia) and fixed forms (Adnata), the latter including saxicolous forms (Epipetria) or submerged like Fontinalaceae (fountain, brook, or water moss). In 1935 H. Meusel identified orthotropic (erect) and plagiotropic (prostrate) forms. (Bryophytes-Encyc. Universalis).

Many moss genera were named by Johann Hedwig (1730–1799)(Wikipedia; Britannica.Com; Encyclopedia.Com), who was a German botanist notable for his studies of mosses, doing more than any other scientist to advance knowledge of them, for which he is sometimes called the father of bryology. He was born in Transylvania, then under Hungarian rule, graduated from the University of Leipzig with an M.D. in 1759, and practiced medicine for the next 20 years, during which time he pursued botany as a hobby, eventually becoming a professor at Leipzig (first in medicine in 1786, then in botany in 1789), director of the Leipzig Botanical Garden, a Fellow of the Royal Society in 1788, and a foreign member of the Royal Swedish Academy of Sciences.

Skilled in both microscopy and biological illustration, he was the first identify moss antheridia and archegonia, the true organs of reproduction in mosses, and directly observed the germination of spores and the formation of protonemas. In 1782–83 he produced *Fundamentum Historiae Naturalis Muscorum Frondosorum*, in 2 volumes ("Elements of the Natural History of Leafy Mosses"), in which he covered the anatomy, fertilization, and reproduction of mosses and introduced a new method of classification based on the distribution of spores. He identified more mosses than any other botanist of his time and produced a series of well-illustrated, informative books on them. His chief work, *Species Muscorum Frondosorum,* describes nearly all the moss species then known, is the starting point for nomenclature of all mosses, except for the sphagnum group, and became the official basis for moss names. It was published posthumously in 1801. He is commemorated by the moss genus *Hedwigia* Palisot de Beauvois 1804, the family Hedwigiaceae Schimper 1856, and the magazines Hedwigia and Nova Hedwigia.

German painter and naturalist Max Fleischer (1861-1930)(not to be confused with the famous Polish-American animator to whom he bears no relation either) named nearly half of the moss orders and perhaps more, 7 moss families, and many moss genera. He studied art in Breslau, later in Munich and Paris, and became an acclaimed impressionist and expressionist painter. His interest in geology and paleontology led him to Zurich and a few years later to Italy, where he spent 5 years studying mosses and also painted. In 1898 he was given the task by the Dutch government of doing paintings of Java on the occasion of the World Exhibition in Paris. In 1903 he was appointed botany professor at the University of Berlin. He illustrated several works on mosses and developed a new classification system which was used in *Die Natürlichen Pflanzenfamilien* in 1924-1925 and became standard. (Max Fleischer (botaniker)-Danish Wikipedia).

Moss Orders with Author Names

Andreaeales Dixon 1932

Andreaeobryales B.M. Murray 1988
Archidiales Limpricht
Bryales Limpricht (syn. Eubryales)
Buxbaumiales Max Fleischer
Calomniales
Dawsoniales Max Fleischer
Dicranales H. Philib. ex Max Fleischer
Diphysciales Max Fleischer
Encalyptales Dixon 1932
Eubryales (syn. Bryales)
Fissidentales
Funiarales Max Fleischer
Grimmiales Max Fleischer
Hookeriales Max Fleischer
Hypnales (M. Fleischer) W.R. Buck & Vitt 1986
Hypnobryales
Isobryales
Oedopodiales Goffinet & W.R. Buck 2004
Orthotrichales Dixon 1932
Polytrichiales Max Fleischer
Pottiales Max Fleischer
Schistostegales
Seligeriales
Sphagnales Max Fleischer 1904
Splachnales (M. Fleischer) Ochyra
Syrrhodontales
Takakiales S. Hattori & Inoue 1958 ex Stech & W. Frey 2008
Tetraphidales Max Fleischer

Traditional classifications done for mosses are presented in below (taken from Miller, 1979).

Table 12-15. Traditional Taxonomies for Mosses.

Fleischer, 1904-23, 86 fams.

sbcl. Sphagnales (1 fam.)
sbcl. Andreaeales (1 fam.)
sbcl. Bryales
  spord. Eubryinales
    Fissidentales
    Dicranales
    Pottiales
    Grimmiales
    Funariales
    Schistostegales
    Tetraphidales
    Eubryales

Isobryales
Hookeriales
Hypnobryales
spord. Buxbauminales (2 monotypic orders: Buxbaumiales, Diphysciales)
spord. Polytrichinales (2 monotypic orders: Dawsoniales, Polytrichales)

Brotherus, 1924-25, same as above except 93 families, 11 orders in Eubryinales, and 1 order (but 2 fams.) in Buxbauminales.

Dixon, 1932, 77 fams.

sbcl. Sphagnales (1 fam.)
sbcl. Andreales (1 fam.)
sbcl. Bryales
clan Nematodonteae
Tetraphidales (Georgiaceae)
Calomniales (1 fam.)
Schistostegales (1 fam.)
Buxbaumiales (2 monotypic suborders: Buxbaumineae, Diphyscineae)
Polytrichales (2 monotypic suborders: Dawsonineae, Polytrichineae)
clan Arthrodonteae
sbclan Haplolepideae
Fissidentales
Grimmiales
Dicranales
Syrrhodontales
Pottiales
sbclan Heterolepideae
Encalyptales (1 fam.)
sbclan Diplolepideae
Orthotrichales
Funariales
Eubryales
Acrocarpi
Pleurocarpi
Isobryales
Hookeriales
Hypnobryales

Reimers, 1954, 84 fams.

sbcl. Sphagnidae (1 fam.)
sbcl. Andreidae (1 fam.)
sbcl. Bryidae
Archidiales
Dicranales
Fissidentales

Pottiales (3 sbords.)
Grimmiales
Funariales
Schistostegales (1 fam.)
Tetraphidales (Georgiaceae)
Orthotrichales
Eubryales
Isobryales
Hookeriales
Hypnobryales
sbcl. Buxbaumidae
sbcl. Polytrichidae

Robinson, '71, 82 living fams. and 2 fossil fams.

sbcl. Sphagnidae
Protosphagnales* (2 fams.)
Sphagnales (1 fam.)
sbcl. Bryidae
Andreales (1 fam.)
Polytrichales (2 fams.)
Dicranales (Haplolepidae)
Tetraphidales M. Fleischer (Georgiaceae)
Bryales (Diplolepidae)

Frey, '77, 6 superorders (did not list all fams.)

Polytrichanae (2 ords.)
Dicrananae
Dicranales (including Archidiales)
Fissidentales
Pottiales
Grimmiales
Funarianae (1 order)
Bartramianae (2 fams., presumably 1 order)
Eubryanae
Eubryales s.s. (3 fams.)
Hypnobryanae
Isobryales
Hookeriales
Hypnobryales

(groups of uncertain position: Rhizogoninae, Spiridentineae, Buxbaumidae, Tetraphidales, Schistostegales)

Crosby and Magill in 1977 accepted 87 families and D.H. Vitt in Parker (1982) recognized 88-96 families in 3 classes: Sphagnopsida (sphagnum or peat moss) (1 genus), Andreopsida (2 families, 1 genus each), and Bryopsida, with the last containing 3 subclasses: Polytrichidae (hair cap moss)(1 family), Tetraphidae (2 families), and Bryidae (2 superorders: Diplolepideae [12 orders] and Haplolepideae [2 orders]).

French bryologist, paleobotanist, and geologist Wilhelm Philippe Schimper, based in Strasbourg, Alsace, who proposed and named the Paleocene Era and co-wrote, with Philipp Bruch, the 6-volume *Bryologia Europaea*, 1836-55, named the most moss families, 31, mostly in 1856; Finnish bryologist Viktor Brotherus, who penned the Musci section in *Die Naturlichen Pflanzenfamilien* and had a personal herbarium of 120,000 moss specimens, named 20, in the early 1900s; and Kindberg named 10, in the late 1800s; so between them they named some 2/3 of the families (Bryophyte Family Names in Current Use-Marshall Crosby, Musci, 1991).

A cladistic classification based on 6 analyses (5 molecular and 1 using classical evidence)(Tree of LifeWeb.Org-Bryophyta) is as follows:

Sphagnopsida
Metamusci nom. nov.
    Takandreidae (1 or 2 ords.; 3 fams.)
    Neomusci nom. nov.
        Oedipodidae (1 fam.)
        Cenomusci nom. nov.
            Polytrichidae
            Altamusci nom. nov.
                Tetraphidae
                Bryidae

In some of these, classes Andreaeopsida, Tetraphidopsida, Polytrichopsida, and Bryopsida are recognized. The analyses of Karen Sue Renzaglia et al (2000) give different results as follows:

Protobryophyta nom. nov.
  Takakia
  Musci
    Sphagnales
    Metamusci nom. nov.
        Andreales
        Polytrichales
Hepaticae
    Haplomitriales
    Metahepaticae nom. nov.
        Marchantiales
        Jungermanniales

This is based on 3 data sets, which mostly agree with each other: one based on spermatogenesis, one based on developmental and morphological characteristics, and one on combined nuclear and mtSSU rDNA.

There is one extinct order in mosses which is Protosphagnales, which dates from the Permian (280 to 245 mya) in Russia, and unambiguous moss fossils date from the Carboniferous (360 to 280 mya) in the Late Paleozoic (Bold et al, 1987).

Mosses flourish in mostly moist habitats including fresh water, with some growing on seaside rocks, but none are marine, and occur in tropical and temperate regions, mostly the former (Margulis and Schwatz, 1998). They are pioneer plants on exposed or burned-over soil (Bold et al, 1987).

The largest moss family is Dicraniaceae (wind-blown moss) with 45 genera often having wind-swept (falcate-secund) "leaves". A large and diverse family, it may have been disassembled phylogenetically.

Major members include the fascinating elfin moss (also called goblin gold, luminous moss, and cave moss), which is luminous because of light-reflecting, globose cells in the branches of the persistent protonema. The generic name, *Schistostega* D. Mohr 1803, means "split lid," implying the operculum splits radially into tooth-like sections, which is erroneous (Bold et al, 1987). It is in Schistostegaceae Schimper 1856 in Dicranales and is found on sandstone rocks and in caves and in other fairly dark places such as under buildings. Other major members include peat moss (*Sphagnum* Linneus, 1753), granite moss (*Andreaea* Hedwig 1801, Andreaeaceae Dumorter 1829, Andreaeales; named for J.G.R. Andreae, Hanoverian chemist, geologist, court pharmacist, polymath, and alchemist of the 1700s), the already mentioned wind-blown moss, the hair-cap moss or pigeon wheat (*Polytrichum* Hedwig 1801, Polytrichaceae Schwaegrichen in Willdenow 1830, Polytrichales), cord moss (*Funaria* Hedwig 1801, Funariaceae Schwaegrichen in Willdenow 1830, Funariales), pygmy moss (3 genera of Ephemeraceae Schimper, 1856, Funariales), urn or top moss (*Physcomitrium* Bridel 1827, Funariaceae, Funariales), the umbrella or petticoat moss (*Splachnum* Hedwig 1801, Splachnaceae Greville & Arnott 1824, Splachnales), elf cap (*Buxbaumia* Hedwig 1801, Buxbaumiaceae Schwaegrichen in Willdenow 1830, Buxbaumiales; named for Johann Christian Buxbaum, German physician and botanist who discovered it in 1712 near the Volga), extinguisher moss (*Encalypta* Schreber ex Hedwig 1801, Encalyptaceae Schimper 1856, Encalyptales), carpet moss (*Hypnum* Hedwig 1801, Hypnaceae Schimper 1856, Hypnales), horn-tooth moss (*Ceratodon,* S.E. Bridel 1826, Ditrichaceae Limpricht 1887, Dicranales), turkish-slipper moss (*Diphyscium* Mohr 1803, Diphysciaceae M. Fleischer in Engler 1920, Diphysciales), stair-step moss, glittering woodmoss, feather moss, or mountain fern moss (*Hylocomium* Bruch and Schimper 1852, Hylocomiaceae (Brotherus) M. Fleischer 1914, Hypnales), tree moss (*Climacium* Weber and Mohr 1804, Climaciaceae Kindberg 1897, Hypnales), and silvery bryum (*Bryum* Hedwig 1801, Bryaceae Schwaegrichen in Willdenow 1830, Bryales).

Bold, Charles; Alexopoulos, John; Delevoryas, Theodore. 1987. Morphology of Plants and Fungi. 5th ed. Harper & Row.

Duff, R. Joel, Juan Carlos Villarreal, D. Christine Cargill, and Karen Renzaglia. 2007. Progress and challenges toward developing a phylogeny and classification of the hornworts. The Bryologist 110: 214-243 (Abstract-BioOne.Org).

Forrest, Laura; Davis, Christine; Long, David; Crandall-Stotler, Barbara; Clark, Alexandra; Hollingsworth, Michelle. 2006. Unraveling the evolutionary history of the liverworts (Marchantiophyta): multiple taxa, genomes, and analyses. The Bryologist 109: 303–334 (researchgate.net).

Margulis, Lynn, Schwartz, Karlene. 1998. Five Kingdoms, 3rd ed. Freeman.

Miller, H.A. 1979. Phylogeny and Distribution of Musci (Chp. 2), pp 11-49. In Bryophyte Systematics (Systematics Assc. Spl. Vol. 14), G.C.S. Clarke, J.G. Duckett, eds. Academic Press, London and New York.

Parker, Sybil (ed.). 1982. Synoptic Classification of Living Organisms. McGraw-Hill, New York.

Renzaglia, Karen Sue; Duff, RJT; Nickrent, D L; and Garbary, D J. 2000. Vegetative and reproductive innovations of early land plants: implications for a unified phylogeny. Philos. Trans. R. Soc. Lond. B Biol.Sci. 355: 769–793 (jstor.org; researchgate.net).

Schuster, R.M. 1979. The Phylogeny of the Hepaticae. In Bryophyte Systematics (Systematics Association Special Volume), G.C.S. Clarke and J.G. Duckett (eds.), Chp. 3, Academic Press.
Stotler, R., Crandall-Stotler, B. 2005. A Revised Classification of the Anthocerotophyta and a Checklist of the Hornworts of North America,North of Mexico. The Bryologist 108: 16-26 (jstor.org).
Villarreal, J. C., Cargill, D. C., Hagborg, A., Söderström, L., Renzaglia, K. S. 2010. A synthesis of hornwort diversity: patterns, causes and future work. Phytotaxa 9: 150–166 (mapress.com).

Pteridophytes

Psilophytes

Psilophyta were established by Kidston and Lang in 1917 and named Psilopsida by Tippo in '42 with the addition of whisk ferns. It was sometimes construed as 4 orders, all extinct:

Rhyniales (Rhyniaceae and Horneaceae; Silurian to Early Devonian)
Psilophytales (Trimerophyta)(1 family; Lower to Middle Devonian)
Zosterophyllales (Zosterophyllaceae, Sciadophytaceae; Upper Silurian to Middle Devonian)
Asteroxylales (1 family; Lower to Middle Devonian)

Banks, in 1968 and 1975, divided them into 3 groups, which proved pleophyletic and which he ranked as subphyla—rhyniophytes, zosterophylls, and trimerophytes. As originally conceived by Banks the rhyniophytes were based primarily on 3 Rhynie Chert genera: *Rhynia*, *Aglaophyton*, and *Horneophyton*. Rhyniales was cladistically separated out from the psilophytes as 2 distinct clades. Three trimerophyte genera, *Eophyllophyton* Hao and Beck 1993, *Psilophyton* Dawson 1859, and *Trimerophyton* Hopping 1956 have been cladistically separated out. A 4th, *Pertica,* is sister group of lignophytes in Kenrick and Crane, together called Radiatopses, but this is based on only 2 genera, *Tetraxylopteris*, being the other, which is in Aneurophytales, so it could be related directly to this genus instead of all lignophytes. Three other trimerophytes, *Dawsonites*, *Hostimella*, and *Psilodendron*, are of uncertain position (Crane et al, 2004). *Yunia*, a putative trimerophyte, is related to lycophytes. Asteroxylales and Zosterophyllales were transfered to Lycophyta.

Rhyniophytes are named for the celebrated Rhynie Chert (els.net; abdn.com; rhyniechert.com; chertnews.com), which is located in Aberdeenshire in N Scotland, and which represents the preservation of an Early Devonian (c. 400 mya) terrestrial and freshwater community of vascular plants, algae, invertebrates, funguses, oomycetes, and cyanobacteria, many plants of which are preserved in such remarkable detail that they can be described down to the cellular level and include the best kept land plants, some in 3D, with the earliest full sporophyte-gametophyte life histories, known from this time or earlier, and as such form the cornerstone of paleobotany. The site contains the oldest well preserved terrestrial ecosystem in the world. It was discovered by William Mackie while mapping the western margin of the Rhynie Basin in 1910–1913. Robert Kidston and William Lang were the two most important early workers at the site, describing some of the plant fossils between 1917 and 1921. It is located in Old Red Sandstone (see Chapter 19).

The area is designated a Site of Special Scientific Interest (SSSI) and is considered a Lagerstätte (Gm., "layer site," "bed place"), which is a sedimentary deposit that exhibits extraordinary fossils with exceptional preservation. It is part of an Early Devonian hot-spring system that contained minor amounts of gold and is the oldest hot-spring system known where surface features such as geyser vents are preserved. The waters contained dissolved silica, and when the erupted water cooled, amorphous silica was deposited, which is called sinter (silica deposited by hot springs). Some of the silica coated and trapped plants and animals on the land surface or within shallow ponds, and the organic structures were mineralized. In other words, it was formed when silica-rich water from volcanic springs rose rapidly and petrified the ecosystem. Burial of the sinters over millions of years resulted in deposition of more silica and the eventual conversion of sinter to chert, which is a cryptocrystalline silica, and varieties are flint, occuring primaily in the Upper Cretaceous and Tertiary, and Lydian stone (lyddite), which is used as a touchstone (for determining the purity of gold or silver). In our own time, hot springs depositing siliceous sinters occur in many areas, notably Yellowstone and near Rotorua in New Zealand.

The plant genera are *Aglaophyton* D.S. Edwards 1986, *Rhynia* Kidston and Lang 1917, *Horneophyton* Barghoorn & Darrah 1938, *Asteroxylon* Kidston and Lang 1920, *Nothia* A.G. Lyon 1964, *Trichopherophyton* A.G. Lyon & D.S. Edwards 1991, and *Ventarura* Powell et al 2000, the nematophytes *Nematothallus* Lang 1937, *Cosmochlaina* Edwards 1986, *Nematoplexus* Lyon 1962, and *Nematasketum* Burgess and Edwards 1988, and the green algae *Paleonitella* Kidston and Lang 1921, *Mackiella* Edwards and Lyons 1983, *Rhynchertia* Edwards and Lyons 1983, and *Cymotiosphera* O. Wetzel 1933. *Asteroxylon, Nothia, Trichopherophyton,* and *Ventarura* go to Zosteropsida, the nematophytes, a convenience group, are of unknown position, but might be vasculars, *Paleonitella* is in Charales, *Mackiella* and *Rhynchertia* are tentatively assigned to Ulotrichales.

Lycophytes

Lycopodiales has a single family, Lycopodiaceae Palisot de Beauvois ex Mirbel, in Lamarck & Mirbel, 1802, containing the genera *Lycopodium* Linneus 1753 (club moss or lycopod (of course, meaning "wolf foot" in Greek)), with some 200 species, *Lycopodiella* Holub 1964, *Phylloglossum* Kunze 1843 (1 species), and *Huperzia* Bernhardi 1801. The only genus in Selaginellaceae Willkomm 1854 is *Selaginella* Palisot de Beauvois 1805 (spike moss, little club moss, or selaginella), with some 800 species). Several extinct orders are recognized:

Archeolepidophytales (syn.Drepanophycales Pichi-Sermolli, 1958)(Drepanophycaceae; Lower to Middle Devonian)
Protolepidophytales (Eleuthrophyllaceae, Protolepidodendraceae, Archeosigilariaceae; Devonian and Carboniferous)
Lepidophytales Bessey 1907 (syn. Lepidodendrales Stewart 1983; Lepidodendraceae, Sigillariaceae, Bothrodendraceae, Sublepidodendracee, Leptophleaceae, Lepidosigillariaceae; Carboniferous and Devonian; arborescent)
Lepidospermales (Lepidocarpaceae, Miadesmiaceae; Carboniferous)(this order includes also the living Selaginellaceae)
Pleuromeiales (Chaloneriaceae, Pleuromeiaceae; Permian to Early Triassic; paraphyletic relatives of the Lepidodendrales)
Sawdoniales Kenrick and Crane 1997

An evolutionary arrangement for lycophytes based on morphology including fossils was done by Kenrick and Crane (1997). Their matrix had 35 taxons and 20 traits and resulted in 18,500 MPTs, 45 steps, and a CI of .74, and for this one they used a heuristic search. That many MPTs is a possible indication of too many missing data. I ran the matrix but I don't know how many MPTs there were in my results as the analysis stops at 100. Like myself they used PAUP and ACCTRAN but employed

Wagner. I used Fitch and there were 52 steps, and the CI was .67, and the RI was .82. The majority consensus cladogram was essentially the same as would be expected--there was a zosterophyll clade, a lycopsid clade, Sawdoniales, and Barinophytales, there were the same outlier genera, except most were on top, and *Zosterophyllum* was polyphyletic. There was a *Rhynia*-Lycophyta-*Psilophyton* trichotomy as opposed to a *Rhynia-Psilophyton*-Lycophyta trichotomy in Kenrick and Crane. A difference in my classification was Lycopsida as more advanced than Zosteropsida. There were only 3 clades that had confidence estimates above 50: Lycophyta with 75/70, Barinophytales with 71/71, and the core lycopsid clade with 70/54.

I also ran another one of their matrices which was 34x33, had a CI of .75 and nearly 3300 MPTs, was branch-and-bound, and the outgroups were *Haplomitrium* and *Spherocarpos*. There was no Bremer support done for it. The heuristic search compares very favourably with it as all the groups were the same (excepting *Polytrichum-Sporogonites,* which did not occur, and *Cooksonia* 2-*Cooksonia* 1, which did) except that they were in reverse order, with the exception of *Haplomitrium* and *Aglaophyton*. *Pertica-Tetraxylopteris* and the horsetail-Cladoxylopsida clade were below 50, but coscinophytes, core lycopsids, core zosterophylls, and a fern-lignophyte clade were moderately supported, while stomophytes, euphyllophytes, anatelomophytes, and Sawdoniales were strongly supported. Sawdoniales was also strongly supported in one of the Kenrick and Crane analyses where Bremer support was done, with a value of 3. Here is a simplified tabulation with the percentage of MPT frequencies, bootstrap estimates, and jackknife estimates, respectively.

Table 12-16. Reanalysis of Kenrick and Crane's Polysporangiate Classification from 1997.

Haplomitrium
Stomophyta 100/80/-
  Coscinophyta 100/67/60
    Aglaophyta
      SG 3 100/-/-
        SG 4 100/64/-
          Lycophyta 100/-/-
          Cooksonia 2-*Cooksonia* 1 62/-/-
          Euphyllophyta 100/75/62
            Eophyllophyta
            Metatelomophyta 100/75/-
              Neotelomophyta 51/-/-
                Anatelomophyta 100/60/70
                  Calamophyta 100/-/-
                  Pertica-Tetraxylopteris 100/-/-
                Psilophyta 2
              Psilophyta 1
        Horneophyta 100/53/60
      Rhyniophyta 100/85/70
    Polytrichum
    *Sporogonites**
  Anthoceros
Spherocarpos (bottle liverwort)

One of the 3 species of *Cooksonia* Lang 1937 included, *C. cambrensis*, was in Lycophyta. The other 2 are *C. pertoni* (*Cooksonia* 1) and *C. caledonica* (*Cooksonia* 2). These other 2 branch independently of Lycophyta also in Kenrick and Crane. The genus was included in Banks's rhyniophytes. It was named for Australian paleobotanist Isabel Cookson, a colleague of Lang, and occurred from the Late Silurian to the Early Devonian. There are 5 valid species in all, the other 2 being *C. hemispherica* and *C. paranensis* (Cooksonia-steurh.home.xs4all.nl).

Calamophytes

Whisk ferns comprise 2 genera: *Psilotum* Swartz 1801 and *Tmesipteris* Bernhardi 1801.

The single living genus of horsetails is *Equisetum* Linneus 1753. There are 4 extinct orders: Pseudoborniales (Devonian; large, little-known trees); Noeggerathiales (Carboniferous-Triassic; dominant in the coal swamps of the Carboniferous); Calamitales (Devonian-Permian), and Sphenophyllales (Upper Devonian-Permian; understory swamp plants and vines).

Pseudosporochnales and Hyeniales were transfered to Cladoxylales/Cladoxylopsida.

Cenopteridales were a polyphyletic group comprising what is now Stauropteridales Rhacophytales, and Zygopteridales.

The following evolutionary taxonomy for ferns is the consensus phylogeny by Smith et al (2006) based on both the phenotype and the genotype. Extinct orders were excluded. It is derived from, for example, Pryer et al (1995), Rothwell (1999), and Pryer et al (2004), which more or less agree with each other. Schneider et al (2009) is essentially the same except whisk ferns are with horsetails, Gleicheniales are below Hymenophyllales, and Hydropteridales are with Polypodales.

Table 12-17. Consensus Phylogeny for Calamophytes (all the groups above order are nom. nov. except for Leptosporangiata, and all the ranks above order are new).

sbph. Protocalamophytina
    Ophioglossales (adder's tongue)(1 fam., 3 gen., 76 sp.)
    Psilotales (whisk ferns or psilophytes 3)(1 fam., 2 gen.)
sbph. Metacalamophytina
        infph. Equisetophytina (syn. Sphenopsida)(horsetails)(1 gen, 15 sp.)
        infph. Neocalamophytina
            pvph. Marattiophytina (1 ord. 1 fam., 4 gen.)
            pvph. Leptosporangiata (syn. Polypodiopsida, Pteridopsida)
                mcph. Osmundarae (cinnamon fern)(1 ord., 1 fam., 3 gen.)
                mcph. Metaleptarae
                    mgord. Hymenophyllarae (filmy ferns)(1 ord., 1 fam., 10 gen.)
                    mgord. Neoleptarae
                        grdord. Gleicheniarae (1 ord.)
                        grdord. Cenoleptarae
                            hpord. Schizearae (curly grass)(1 ord., 1 fam., 4 gen.)
                            hpord. Analeptarae
                                spord. Hydrapteridales (water ferns)(1 ord., 2 fams, 5 gen.)
                                spord. Metanaleptarae
                                    ord. Cyathearae (tree ferns)(8 fams.)
                                    ord. Polypodales (polypods)(15 fams.,

8,000 sp.)

Schneider et al (2009), however, present a tree based on various DNA results in which spermatophytes are between lycophytes and ferns, and their own morphological tree is in agreement with this, and is in general agreement. The inclusion of fossils obtains the same result of seed plants below ferns (Rothwell, 1999; Rothwell and Nixon, 2006). This is congruent with the fossil record as lignophytes appear in the Middle Devonian, while the earliest true ferns appear in the Late Devonian or Early Mississippian. Doyle (1998) presents a synopsis of morphological and molecular phylogenies for calamophytes based primarily on Pryer et al from 1995 and Kenrick and Crane (loc. cit.)(morphology and physiology) (part of a wider stelophyte synopsis), which is as follows:

Sphenopsida
Cladoxylopsida
Ferns
  Rhacophyton
  Zygopteridales
  Unnamed
    Ophioglossales+Psilotaceae
    Marattiales
    Filicales

In Rothwell, the inclusion of fossils gives a significantly different result from the Smith et al consensus tree. Stauropteridales is basal, with Cladoxylales, Rhacophytales, and Zygopteridales forming a clade together as sister group to horsetails, with this clade in turn sister group to lignophytes. In Rothwell and Nixon, the same thing occurs. And in both there is also another extinct fern clade or one that excludes *Skaaropteris*, which I dub *Botryopteridales.* The basic phylogeny is as follows:

Stauropteridales*
Unnamed
  Psilophyta*
  Unnamed
    Psilotophyta (whisk ferns)
    Euphyllophyta
      Lignophyta+(Equisetales+Cladoxylopsida*)
      Filicophyta

In spite of their results, Rothwell and Nixon say that the Calamophyta clade is far more compatible with commonly proposed patterns of trait evolution.

Rothwell's matrix had 52 genera and species (half extinct) and 101 traits, which resulted in 12 MPTs, 429 steps, a CI of .35, and an RI of .71, and strict consensus was used. I ran this matrix and the cladogram is essentially similar to the 1999 classification. The majority consensus would be very similar to a strict consensus as most clades had a frequency of 100%. There were 9 MPTs, 454 steps, a CI of .34, and an RI of .70. It is simplified as follows:

Table 12-18.

- Aglaophyta*
- SG 1 100/-/-
  - SG 2 100/-/-
    - Psilophyta*
    - Anatelomophyta 100/64/60
      - SG 4 67/-/-
        - SG 5 100/-/-
          - Cladoxylopsida* 100/-/-
          - Equisetales 100/96/96
        - Lignophyta 100/82/82
      - Filicophyta 100/63/-
        - Marattiales 100/99/100
        - Metafilicae 100/-/-
          - Leptosporangiata 100/60/-
            - SG 9 100/-/-
              - SG 10 100/-/-
                - Botryopteridales* nom. nov. 100/-/-
                - SG 11 100/-/-
                  - *Skaaropteris**
                  - Polypodales 100/50/- (including *Cyathea*)
                  - Hydropteridales 100/92/88
              - *Schizea*
            - *Osmunda*+(Gleichenales-*Hymenophyllum*) 100/-/-
          - Ophioglossales 100/79/82
  - Psilotales 100/100/100
- Stauropteridales* 100/82/80

Table 12-19. Fern and Ally Classes and Orders with Author Names.

Pteridopsida Ritgen 1828
Cladoxylopsida Novak, 1930
Marattiopsida Doweld, 2013
Polypodiopsida Ritgen, 1828
Psilotopsida Heintze, 1927
Equisetopsida C. Agardh, 1825 (Sphenopsida)

Barinophytales
Blechnales
Cladoxylales* ?C.J. Cleal & B.A. Thomas, 1994
Cenopteridales*
Cyatheales A.B.Frank, 1877
Equisetales Dumortier, 1829
Gleicheniales A.B. Frank in Leunis, 1877
Hydropteridales= Salviniales
Hyeniales*
Hymenophyllales A.B. Frank in Leunis, 1877

Marattiales Prantl, 1874
Ophioglossales Newman, 1840
Osmundales Bromhead
Polypodiales Mett. ex A.B.Frank, 1877
Pseudosporochnales*
Psilotales Prantl
Rhacophytales*
Salviniales Bartl. in Mart. 1835 or Britton 1901
Schizeales Schimper
Stauropteridales*
Zygopteridales*

English botanist Frederick Bower, in the 1920s, recognized 10 fern families, pteridologist E.B. Copeland (1947), father of H.F. Copeland, recognized 14, Pichi-Sermolli (1958) distinguished 38 families in 18 orders, Wagner (1969), recognized 7, using a cladistic analysis consisting of his ground-plan divergence method (Wagner parsimony was named for him), Tryon and Tryon (1982) admitted 24 families, Kew (Brummitt, 1992) recognizes 36, and Swale gives about 40 (New World Encyclopedia on-line). Engler and Diels treated Eufilicales as 8 families, Christensen 15, Copeland 19, and Holltum 11. The families recovered in the cladistic analyses for leptosporangiates are in much the same sequence as those previously presented in traditional or semievolutionary classifications: Osmundaceae, Schizeaceae, Gleicheniaceae, Hymenophyllaceae, Cyatheaceae, plus the remaining families of Eufilicales and the water ferns on top. (Water ferns, however, also include *Ceratopteris* which goes in Polypodiales).

The largest families were Polypodiaceae, with 63 genera, Pteridaceae with 60, and Aspidaceae, with 57, but they were split up, the 1st having 3 families separated out, the 2nd 4, and the 3rd was split into 3. Now the largest families are Polypodiaceae Berchtold & J. Presl 1820 and Dryopteridaceae Ching 1965, both with 47 genera (but with Woodsiaceae included, the latter would have 67); Copeland's Aspidaceae are no longer recognized and Pteridaceae have only 6 genera.

The 2 genera of Ophioglossaceae (R. Brown, 1810) C. Agardh 1822 are *Ophioglossum* Linneus 1753 (adder's tongue; of course, the Greek name means "snake tongue") and *Botrychium* Swartz 1801 (grape fern or botrychium). For Marattiaceae Berchtold & J. Presl 1820 they are *Marattia, Angiopteris, Christensenia*, and *Danea*. Pichi-Semolli saw all 4 as families and Holttum saw *Danea* and *Angiopteris* as families, recognizing also a Kaulffussiaceae. The genera for Hydrapteridales (Salviniales), sometimes seen as 2 orders (Salviniales and Marsileales), are *Marsilea* Linneus 1753 (water clover or pepperwort), named for Italian botanist Giovanni Marsigli, with over 50 species; *Salvinia* Séguier 1754 (floating moss or salvinia), named for Italian botanist Antonio Salvini, with about a dozen species; and *Azolla* Lamarck 1783 (mosquito fern, so called because it is grown to control mosquitoes). *Azolla* is also sometimes seen as a family.

Bower, F. O. 1923-28. The Ferns, 3 vols. Cambridge U. Press.
Brummitt, R.K. (ed.). 1992. Vascular Plant Families and Genera. Kew Gardens.
Copeland, E.B. 1947. Genera Filicum.
Kenrick, P., Crane, P. 1997. The Origin and Early Diversification of Land Plants: a Cladistic Study. Smithsonian Institute Press.
Pichi-Sermolli, Rodolfo E.G. 1958. The higher taxa of Pteridophyta. Acta Univ. Upsaliensis 6: 70-90.
Pryer, Kathleen; Smith, Alan; Skog, Judith.1995. Phylogenetic Relationships of Extant Ferns Based on Evidence from Morphology and rbcL Sequences. American Fern Journal 85: 205-282.

Pryer, Kathleen, Schuettpelz, Eric, Wolf, Paul, Schneider, Harald, Smith, Alan, Cranfill, Raymond. 2004. Phylogeny and Evolution of Ferns (monilophytes) with a Focus on the Early Leptosporangiate Divergences. American Journal of Botany 91: 1582–1598 (amjbot.org).

Rothwell, Gar. 1999. Fossils and Ferns in the Resolution of Fern Phylogeny. Bot. Rev. 65: 819-212 (springer.com).

Schneider, Harald, Smith, Alan, Pryer, Kathleen. 2009. Is Morphology Really at Odds with Molecules in Estimating Fern Phylogeny? Systematic Botany 34: 455–475 (pryerlab.biology.duke.edu).

Smith, A.R., Kathleen Pryer, Eric Schuettpelz, Petra Korall, Harald Schneider, & Paul Wolf. 2006. A classification for extant ferns. Taxon 55: 705–731 (duke.edu).

Tryon, Rolla, Tryon, Alice. 1982. Ferns and Allied Plants.

Wagner, Warren H., Jr. 1969. The construction of a classification. In Systematic Biology, Publication 1692, pp. 67-90, National Academy of Sciences, Washington, D.C.

Spermatophyta

The following is a cladistic classification for seed plants (based on Doyle and Donoghue, 1986, 1992; Doyle, 1994, 1998; all the inclusive groups are stat. nov. and nom. nov. except where otherwise indicated).

Table12-20. Cladistic Classification for Spermatophytes.

sbph. Elkinsiophytina*
spph. Lagenostomae
  infph. Lyginophytina*
  infph. Metaspermae
    pvph. Medullosae*
    pvph. Neospermae
      mcph. Peltaspermae*
      mcph. Cenospermae
        nnph. Cycadophytina (1 fam., 9 gen., c. 100)
        nnph. Platyspermae
          pcph. Callistophytina*
          pcph. Eusiphonogama
            sbpcph. Pinophytina
              cl. Pinopsida
                sbcl. Gingyoidae (1 sp.)
                sbcl. Pinidae
                  spord. Cordaitarae*
                  sp.ord. Pinarae
                    ord. Pinales (Conifera)(7 fams., c. 50 gen., c. 500 sp.)
                    ord. Voltziales*
            sbpcph. Neosiphonogama
              infpcph. Corystophytina*
              infpcph. Glossophytina
                pvpcph. Glossoptera*
                pvpcph. Anthoidophytina (Anthophyta)
                  mcpcph. Caytonia*

mcpcph. Neanthoidophytina
nnpcph. Pentoxyla*-Bennettitales* (syn. Cycadoidea)
nnpcph. Anaspermae Loconte & Stevenson 1990
pcpcph. Gnetophytina
cl. Gnetopsida (syn. Chlamydospermae, Saccovulata) (3 orders, 3 fams. 3 genera, c. 50 sp.; joint fir, welwitschia, ephedra)
pcpcph. Anthophytina (Anthhophyta Bessey 1907, Angiospermae) (flowering plants)

Gymnosperms

In 1827 Robert Brown coined the term "gymnosperms". In 1898 Van Tieghem distinguished Astigmatae (Gymnospermae) and Stigmatae (Angiospsermae). In 1917 Coulter and Chamberlain recognized 9 ungrouped gymnosperm orders. Jeffrey in 1917 and Conard in 1919 organized them into Archigymnospermae and Metagymnospermae (the latter being conifers and gnetales). Van Tieghem and Costantin in 1918 organized them into 4 classes: Pteridospermae, Natrices (Cycadineae, Ginkyoineae), Vetrices (conifers), and Saccovuleae (Gnetales). Berry, in 1917, did not recognize them as monophyletic, substituting 3 phylums: Pteridospermophyta, Cycadophyta, and Coniferophyta. In 1920, Sahni distinguished Stachyospermae (Cordaitales, Coniferales, Ginkyoales, Taxales) and Phyllospermae (Pteridospermales, Cycadales, Bennettitales (Cycadeoidales)) based on the origin of the ovule (axillary or foliar, that is, borne on the axil or on the frond, respectively).

Chamberlain's 1935 classification is as follows:

Cycadophytes
Cycadofilicales (Pteridospermales)
Bennettitales (Cycadeoidales)
Cycadales
Coniferophytes
Cordaitales
Ginkyoales
Coniferales

The 2nd "g" in the misspelling of *Ginkyo* is an orthographic error originating with Kaempfer in 1712 and formalized by Linneus in 1771, as pointed out by several botanists notably Moule in 1937 and 1944 and Thommen in 1939, and some botanists have correspondingly rejected the false name and adopted the right one (e.g., Pulle in '46 and Widder in '48)(Lawrence, 1960, p. 358) and also in Caratini (1971).

In 1948, Arnold recognized Cycadophyta (Pteridospermales, Cycadales, Cycadeoidales), Coniferophyta (Cordaitales, Ginkyoales, Taxales, Coniferales), and Chlamydospermophyta (Ephedrales, Gnetales). Johnson's 1951 scheme had Pteridospermatophyta (1 order), Cycadophyta (4 orders, including Nilssoniales), Ginkyophyta (Ginkyoales and Cordaitales), Coniferophyta (Voltziales and Coniferales), and Ephedrophyta (1 order, Gnetales). 1957 saw Pant's taxonomy which delimited 3 phyla as follows:

Cycadophyta
cl. Pteridospermopsida (orders Lyginopteridales, Medullosales, Glossopteridales, Peltaspermales, Corystospermales, Caytoniales)

cl. Cycadopsida (1 order)
cl. Pentoxylopsida (1 order)
cl. Bennettitopsida (1 order)
Chlamydospermophyta (Gnetales, Welwitschiales)
Coniferophyta
cl. Coniferopsida (Cordaitales, Ginkyoales, Coniferales)
cl. Ephedropsida
cl. Czekanowskiopsida
cl. Taxopsida

Andrews' arrangement from 1961 had 6 phyla: Pteridospermophyta (1 order), Cycadophyta (Cycadales, Bennettitales), Cycadales (1 order), Ginkyophyta (1 order), Coniferophyta (Coniferales, Cordaitales), Gnetophyta (Ephedrales, Gnetales).

Bierhorst's (1971) classification of spermatophytes was based on Pilger's and Melchior's 1954 scheme and went like this:

Cycadopsida
- Pteridospermales
- Cycadales
- Cycadeoidales
- Caytoniales

Coniferopsida
- Cordaitales
- Coniferales
- Taxales
- Ginkyoales

Gnetopsida
- Ephedrales
- Gnetales
- Welwitschiales

Angiospermopsida

Sporne's 1974 scheme was similar to Bierhorst's, differing only in excluding Caytoniales, recognizinig Pentoxylales, seeing Cycadales as most advanced in Cycadopsida, and seeing Gnetopsida as having only 1 order, Gnetales. Taylor's 1981 scheme had Progymnospermatophyta, Pteridospermatophyta, Cycadophyta, Cycadeoidophyta, Ginkyophyta, and Coniferophyta. Stewart's 1983 taxonomy did not see gymnopserms as a group and distinguished Progymnospermopsida (3 orders), Gymnospermopsida (12 orders), and Gnetopsida (1 order). In 1984, Meyen arranged Pinophyta as Ginkyoopsida (9 orders, including Ephedrales), Cycadopsida (6 orders, including the other 2 gnetophyte orders), and Pinopsida (Cordaitales and Pinales). Gifford and Foster (1989) arranged gymnosperms as 7 phyla: Progymnospermatophyta, Pteridospermatophyta, Cycadophyta, Cycadeoidophyta, Ginkyophyta, Gnetophyta, and Coniferophyta. Bhatnagar and Moitra (1996) presented 4 classes in 1996: Progymnospermopsida (3 orders, 3 families, like Stewart), Cycadopsida (6 orders), Coniferopsida (5 orders), Gnetopsida (3 orders and 3 families).

The synapomorphies for Platyspermae are flat seed, saccate pollen, and sealing off of the micropyle after pollination; Diplophytina's are double fertilization, heteroxylous wood, columnar tracheid, perianth, short cambial initials, thin megaspore wall, and lignin syringal groups; Pinophytina has resin canals, little pith, and pycnoxylic wood; and that for Glossophytina is a bract-cupule complex. Metaspermae is distinguished by a nucellar beak, nucellar vasculature, and sarcotesta, and Neospermae by a sealed micropyle and axillary branching.

The progymnosperms were established by Beck in 1960 to accomodate Archeopteridales and Aneurophytales. The seed ferns include Pteridospermales (Lyginopteridaceae and Medullosaceae) of the Paleozoic and Caytoniales (Caytoniaceae, Corystospermaceae, Peltaspermaceae) of the Mesozoic.

The taxonomy here comprises the groups and positions most recovered in 18 cladistic analyses based on morphology and chemistry (Parenti, 1980 (extant only); Hill and Crane, 1982 (extant only); Crane, 1985 (2 analyses); Doyle and Donoghue, 1986 (2 analyses), 1992 (2 analyses); Loconte and Stevenson, 1990 (extant only); Nixon et al, 1994 (5 analyses); Rothwell and Serbet, 1994; Doyle, 1996, 2006; Hilton and Bateman, 2006). There is no dispute over the monophyly of Platysperma nor for the early divergence of the progymnosperms and the polyphyly of the seed ferns (pteridosperms=Cycadofilicales). The Anasperma clade is recovered in 17 of the studies and Gnetales is monophyletic in 13 of them and where it is a grade is only in Nixon. Six studies show a Ginkyo-Pinopsida clade while 10 show it as paraphyletic, but the former usually with fossils and the latter usually without, and 5 of these latter are in Nixon (3 of these are for only living taxons). And cycads are the most primitive living taxon in seed plants in 15 of the analyses, the exceptions being Hill and Crane (1982) and Doyle and Donoghue (1986), in which *Ginkyo* is the most primitive.

The gnetophyte-conifer-ginkyo grouping goes back to Chamberlain (1935) and the gnetophyte-angiosperm grouping goes back to Ernst Haeckel (1894) and Arber and Parkin (1907, 1908). Conifers and *Ginkyo* together had been supported by Bierhorst in 1971. Conifers and gnetophytes have linear leaves, reduced sporophylls, and circular bordered pits with toruses in the protoxylem, and together with *Ginkyo*, they uniquely share metaxylem that lacks scalariform pitting (Mathews, 2009).

There is a dispute only concernng the monophyly of gymnosperms and this is because of molecular analyses, which are not necessarily reliable, most having Gnetophytina grouping with typical gymnosperms instead of angiosperms, conflicting with morphological results. The contradictory results are due to low sampling and systematic errors (Sanderson et al, 2000; Rydin and Kallersjo, 2002; Zhong et al, 2011). And the genotypic results contradict the stratigraphic evidence (Doyle, 1998a,b). Morphology from fossil forms, which genotypic analyses can't include, continues to support the current taxonomy (Rothwell, Crepet, and Stockey, 2009). There has been a fossil gnetophyte found from the Permian which is claimed to provide evidence of a gnetophyte-gymnosperm clade (Zi-Qiang Wang, 2004), but the resemblance could be plesiomorphous. The synapomorphies for Anasperma (Diplophyta) are double fertilization, a tunica in the apical meristem, and Maule reaction (in lignin chemistry) (Doyle, 1998a).

The grouping together of *Caytonia*, Bennettitales, Gnetophytina, and Anthophytina (and sometimes also *Pentoxylon*) is called the anthophyte hypothesis and was introduced by Doyle and Donaghue (1986), but this is a misnomer as anthophyte refers to flowering plants so I have renamed the clade Anthoidophytina. The synapomorphies are granular, nonsaccate pollen, syndetocheilic stomas, simple leaves, and aggregated sporophylls.

The monophyly of gnetophytes in both morphological and molecular phylogenies is surprising, since it contains 3 very heterogenuous genera, but appears solid. There are 2 analyses based on classical evidence which support Gnetophytina paraphyly (Nixon et al, 1994; Hickey and Taylor, 1996).

Conifers were treated as 2 families and 11 subfamilies or tribes by Engler and Diels in 1924. In 1927, Pilger treated them as 7 families, 48 genera, and 520 species. Buchholz in 1934, distinguished 2 suborders and 10 families. There are 6 conifer families (16 gymnosperm families are often recognized, e.g., Kew) in suborders Cupressineae, containing superfamilies Cupressareae (Cupressaceae (cypress) and Taxaceae (yew)) and Popodocarpareae (Araucariaceae (kauri pine) and Podocarpaceae (Southern pine or podocarp) and Pinineae containing only 1 family, Pinaceae (Northern pine (pine, fir, spruce, cedar)). The largest extant families of gymnosperms are the cypress family (Cupressaceae), with 18 genera, the Southern pine family (Podocarpaceae), with 17, and the Northern pine family (Pinaceae), with 12. Taxodiaceae (bald cypress) has 10, Zamiaceae 8, Taxaceae 5, Araucariaceae 2, and the other 7 have 1 each.

A phylogenetic classification based on molecules from Wikipedia-Conifers (derived from research papers by Farjon Remaining Diversity of Conifers) and Quinn and Price (Phylogeny of the Southern Hemisphere Conifers) presented at the 4th Annual Conference of Acta Horticulturae (ISHS) in 2003, which basically agrees with morphology, is as follows:

sbord. Pinineae (1 fam.)
sbord. Metapinineae
    hpf. Podocarpariae (2 fams.)
    hpf. Neopinariae
        spf. Sciadopitydariae (1 fam.)
        spf. Cupressariae
            Cupressaceae
            Taxaceae

The synapomorphy for the Araucariaceae+Popodocarpareae clade recovered in molecular analyses is the uniovulate strobilus, although the 2 ovules in the Triassic podocarpaceans, *Rissikia* and *Mataia,* pose a problem. All analyses agree on the unity of Cupressales s. l. (i.e., Cupressaceae including Taxoidaceae as a grade within it plus Taxaceae-Cephalotaxaceae or Taxoideae-Cephalotaxoideae) (Doyle, 1998a).

Information on historical taxonomies was gleaned from Bhatnagar and Moitra (1996) and the CAS (1955).

Arber, E. A. N. and Parkin, J. 1907. On the origin of angiosperms. Botanical Journal of the Linnean Society 38: 29 – 80 (as cited in Doyle, 1998a).

Arber, E. A. N. and Parkin, J. 1908. Studies on the evolution of the angiosperms: the relationship of the angiosperms to Gnetales. Annals of Botany 22: 489 – 515 (as cited in Doyle, 1998a).

Bhatnagar, S.P. and Moitra, Alok. 1996. The Gymnosperms. New Age International, New Dehli.

Bierhorst, D. W. 1971. Morphology of Vascular Plants. Macmillan, New York (as cited in Doyle, 1998a).

Burleigh, J. Gordon and Mathews, Sarah. 2007. Assessing Systematic Error in the Inference of Seed Plant Phylogeny. Int. J. Pl. Sci. 168: 125-35 (jstor.org).

California Academy of Sciences. 1955. A Century of Progress in the Natural SciencesSciences, 1853-1953. San Francisco.

Crane, P. R. 1985. Phylogenetic analysis of seed plants and the origin of angiosperms. Ann. Mo. Bot. Gard. 72: 716–793.

Caratini, Roger (ed.). 1971. La vie des plantes (vol. 10 in Encyclopédie Bordas).

Chamberlain, C. J. 1935. Gymnosperms: Structure and Evolution. University of Chicago Press, Chicago, Illinois (as cited in Doyle, 1998a).

Crane, P. R. 1985. Phylogenetic analysis of seed plants and the origin of angiosperms. Ann. Mo. Bot. Gard. 72: 716–793 (as cited in Doyle, 1998a).

Doyle, J.A. 1996. Seed plant phylogeny and the relationships of Gnetales. Int. J. Plant Sci. 157 (6,

Suppl.): S3–S39.
Doyle, J.A. 1998a. Phylogeny of Vascular Plants. Annu. Rev. Ecol. Syst. 29: 567-99.
Doyle, J.A.1998b. Molecules, Morphology, Fossils, and the Relationship of Angiosperms and Gnetales. Mol. Phylgntcs. Evol. 9: 448-62 (botany.ubc.ca).
Doyle, J.A. and Donoghue, M. J. 1986. Seed plant phylogeny and the origin of angiosperms: an experimental cladistic approach. Bot. Rev. 52: 321–431.
Doyle, J.A. and Donoghue, M. J. 1992. Fossils and seed plant phylogeny reanalyzed. Brittonia 44: 89–106.
Gifford, Ernest and Foster, Adriance. 1989. Morphology and Evolution of Vascular Plants, 3rd ed. Freeman, New York.
Haeckel, E. 1894. Die systematische Phylogenie. Georg Reimer, Berlin (archive.org).
Hickey, L.J., and Taylor, D. W. 1996. Origin of the angiosperm flower. In Flowering Plant Origin, Evolution, and Phylogeny, D. W. Taylor and L. J. Hickey (eds.), pp. 176–231. Chapman and Hall, New York.
Hill, C.R., Crane, P.R. 1982. Evolutionary cladistics and the origin of angiosperms. In K.A. Joysey and Friday, A.E. (eds.) 1980. Problems of phylogenetic reconstruction, Proceedings of the systematics association symposium, Cambridge. Academic Press, New York, NY.
Hilton, J., Bateman, R. 2006. Pteridosperms are the Backbone of Seed-Plant Phylogeny. J. Torrey Bot. Soc. 13: 119-68.
Lawrence, G.H.M. 1960. Taxonomy of Vascular Plants. Macmillan, New York.
Loconte, H. and Stevenson, D.W. 1990. Cladistics of the Spermatophyta. Brittonia 42: 197–211 (jstor.org).
Mathews, Sarah. 2009. Phylogenetic Relationships Among Seed Plants: persistent questions and the limits of molecular data. Am. J. Bot. 96: 228-36 (jstor.org).
Nixon, K.C., Crepet, W.L., Stevenson, D., Friis, E.M. 1994. A re-evaluation of seed plant phylogeny. Ann. Mo. Bot. Gard. 81: 484-533 (jstor.org).
Parenti, Lynne. 1980. A phylogenetic analysis of the land plants. Biological Journal of the Linnean Society 13: 225–242 (as cited in Doyle, 1998a).
Rothwell, G. R. and Serbet, R. 1994. Lignophyte phylogeny and the evolution of spermatophytes: a numerical cladistic analysis. Syst. Bot. 19: 443–482 (as cited in Doyle, 1998a).
Rothwell, G.W., Crepet, W.L., Stockey, R.A. 2009. Is the anthophyte hypothesis alive and well? New evidence from the reproductive structures of Bennettitales. Am. J. Bot. 96: 296-322 (amjbot).
Rydin, C., Kallersjo, M. 2002. Taxon sampling and seed plant phylogeny. Cladistics 18: 485-51 (mendeley.com abstract).
Sanderson M.J., Wojciechowski M.F., Hu J.M., Khan T.S., Brady S.G. 2000. Error, bias, and long-branch attraction in data for two chloroplast photosystem genes in seed plants. Mol. Biol. Evol. 17: 782-97 (public.asu.edu).
Wang, Zi-Qiang. 2004. A New Gnetalean Permian Cone as Fossil Evidence for Supporting Current Molecular Phylogeny. Ann. Bot. 94: 281-88 (aob.oxfordjournals.org).
Zhong, Bojian et al. 2011. Systematic Error in Seed Plant Phylogenomics. Genome Biol. Evol. 3: 1340-1348 (oxfordjournals.org).

Flowering Plants

Classification

"Classification of Flowering Plants" (Vol. 1, Gymnosperms and Monocotyledons, 1904; Vol. 2,

Dicotyledons, 1925, as cited in Lawrence) by Cambridge-educated, English botanist A.B. Rendle (1865 -1938) was known for the clarity and completeness of the descriptions, and the system was similar to Engler's, and, like him, he did not claim it to be strictly phylogenetic, and he admitted the Monochlamydeae, Dialypetalae, and Sympetalae groups as grades as Engler did (Lawrence, 1960).

Rendle was Assistant in the Botany Dept. of the British Museum (Natural History) from 1888-1906, lecturer on botany at Birkbeck College from 1894-1906, Keeper of Botany at the Natural History Museum from 1906 to 1930, became F.R.S. in 1909, and was botany editor for the Encyclopedia Britannica's 11th edition of 1911, President of the Botanical Section of the British Association in 1916, President of the Linnean Society from 1923-27, and editor of the Journal of Botany from 1924-38. In 1905, he attended the International Botanical Congress in Vienna, where he was appointed to the editorial committee for the International Rules of Botanical Nomenclature (subsequently the International Code of Botanical Nomenclature), a role he continued in until 1935. (whoislog.info; Cambridge Alumni Database-cam.edu.uk)

Hallier (L'Origine et le système phylétique des angiospermes, Arch. Néerl. Sci. Exact. Nat., Ser. III B, 1: 146-234, 1912, as cited in Lawrence) recognized 36 orders and 210 families in his taxonomy. Ranales is the most primitive and Rubiales the most advanced in dicots. Heliobae contain the water plantain family. (Lawrence, 1960).

Here is his taxonomy:

Dicots
- Protérogènes (5)
- Anonophyles (5)
- Rhodophyles (7)
- Ochnigènes (12)

Monocots
- Liliflorae
- Artorhizae
- Ensatae
- Enenioblastae
- Spadiciflorae
- Cypereales
- Heliobae

Hans Hallier (1868-1932) was a German botanist who authored numerous articles on his classification, including 2 major revisions. The material presented in Lawrence was taken largely from the 1912 article which is much more detailed and very different from the commonly cited brief account of his classification in English in 1915. He worked in the herbariums of Bogor on the island of Java, in Hamburg, and in Leiden, where he was born. He considered the most primitive plants to be those with bisexual, insect-pollinated flowers, a developed perianth, and spirally-arranged floral parts. (freedictionary.com)

Bessey (1915) recognized 32 orders and 300 families arranged into higher categories as follows:

cl. Alternifoliae (monocots)
 sbcl. Strobiloideae (5)
 sbcl. Cotyloideae (3)
cl. Oppositifoliae (dicots)
 sbcl. Strobiloideae

spord. Apopetalae-Polypetalae (7)
spord. Sympetalae-Polypetalae (3)
spord. Sympetalae-Dicarpellatae (4)
sbcl. Cotyloideae
spord. Apopetalae (7)
spord. Sympetalae (madder, bellflower, and daisy orders)(3)

The Bessey family is of French extraction, the original name being Bessé. The tradition is that the early members of the family, who were Huguenots, were compelled on account of religious persecution to flee to England from the ancestral home near Strassburg in Alsace in the 17th century after which the y was substituted for the é. Some of the Besseys immigrated to America about the mid-18th century.

Charles Bessey (1845-1915) received his PhD from the U. of Iowa in 1879 and studied at Harvard under Asa Gray in 1872-73. He was botany professor at the Iowa Agricultural College, today known as Iowa State U., from 1870 to 1884. From 1884 onwards he was associated with the University of Nebraska, where he was botany professor, later botany department dean, and later chancellor. He became a member of the AAAS in 1872, a fellow in 1880, and served as president in 1911. He was a charter member of the Botanical Society of America and served as its president in 1895. He was also botanical editor of the American Naturalist from 1880-97 and also of Science from 1897 on. He authored several textbooks including "The Geography of Iowa" in 1878, "Botany for High Schools and Colleges" in 1880 (3 eds.), "The Essentials of Botany-Briefer Course" in 1884 (7 eds.), "Elementary Botanical Exercises" in 1892 (2 eds.), "Elementary Botany" in 1904, and "Outlines of Plant Phyla" in 1909, did a revision of McNab's Botany in 1881, and co-authored several other books, including "Essentials of College Botany" (with his son Ernst in 1914, who was professor of botany at Michigan State). In 1967, Iowa State U. built a plant industry building, which was named for him. Today the building is used by departments in the biological sciences. In 2009 he was inducted to the Nebraska Hall of Fame. (digitalcommons.unl.edu by Raymond Pool, Brief sketch of the life and work of Charles Edwin Bessy, R.J. Pool Am. J. Bot. 2: 505-18, 1915; Wikipedia).

British botanist John Hutchinson (1884-1972) presented a classification in "Families of Flowering Plants" (Vol. 1, 1924, Vol. 2, 1936); this 1st version distinguished monocots (Calyciferae, Corolliferae, and Glumiflorae) and dicots (Archichlamydeae and Metachlamydeae (Sympetalae)). "British Flowereing Plants" in 1948 distinguished between Lignosae (33 orders, with Magnoliales as basal) and Herbaceae (54 orders, with Ranales as basal). The '59 version had the monocots descended from Ranales.

Hutchinson was trained in horticulture and was appointed a student gardener at Kew in 1904. His taxonomic and drawing skills were soon noticed and resulted in his being appointed to the Herbarium in 1905. He moved from assistant in the Indian Section to assistant for Tropical Africa, returning to Indian botany from 1915-1919. He made 2 extended collecting trips to South Africa in 1928-29 and 1930, which were recounted in "A Botanist in Southern Africa." He was in charge of the African Section until 1936 when he was appointed Keeper of the Museums of Botany at Kew. He retired in 1948 but continued working on the phylogeny of flowering plants and published two parts of "The Genera of Flowering Plants." He was awarded the Victoria Medal (horticulture) in 1944 and Linnaean Darwin-Wallace Gold Medal in 1968 and was elected a member of the Royal Society in 1947 and the O.B.E. in 1971.

Armen Takhtajan, in "Systema magnoliophytorum," 1987 (in Russian)(previous versions came out in 1954, 1959, and 1981; a later (English) version came out in 1997, "Diversity and Classification of Flowering Plants," Columbia University Press, New York), formulated this arrangement (number of superorders in parentheses):

Magnoliopsida (dicots)
Magnolidae (5)
Ranuculidae (1)
Hamamelidae (3)
Caryophyllidae (4)
Dillenidae (9)
Rosidae (9)
Asteridae (6)
Liliopsida (monocots)
Alismatidae (2)
Triurididae (1)
Lilidae (7)
Arecidae (4)

Takhtajan (1910 – 2009) was director of the Botanical Institute of the Academy of Sciences of Armenia in 1944-48 and professor at St. Petersburg State U., and worked at the Komarov Botanical Institute in St. Petersburg, where he developed a classification scheme for flowering plants in 1940, which did not become known to botanists in the West until after 1950. In the late '50s he began a correspondence and collaboration with American botanist Arthur Cronquist. Takhtajan was also a member of the Russian Academy of Sciences, a foreign associate of the U.S. National Academy of Sciences, academician of the Academy of Sciences of Armenia, president of the Soviet All-Union Botanical Society in 1973 and of the International Association for Plant Taxonomy in 1975, and member of the Finnish Academy of Science and Literature, the German Academy of Naturalists "Leopoldina", and other scientific societies. His system remains influential; it is used, for example, by the Montreal Botanical Garden. He also devised a phytogeographic classification in 1969 and 1978 (more on this in Chapter 17).

Cronquist's taxonomy (1968) was largely influenced by his collaboration with Takhtajan and other botanists at Komarov but had no superorders.

Magnoliopsida
Magnolidae
Hamamelidae
Caryophyllidae
Dillenidae
Rosidae
Asteridae
Liliopsida
Alismatidae
Arecidae
Commelinidae
Zingiberidae
Lilidae

Arthur Cronquist (1919-92), avid member of the Boy Scouts of America, through which he gained an appreciation for the outdoors, received his doctorate in botany at the University of Minnesota in 1944. In 1943 he entered the employ of the New York Botanical Garden where he would spend the rest of his career.

Dahlgren presented his arrangement in General aspects of angiosperm evolution and macrosystematics, Nordic J. Bot. 3: 119-49, in 1983. In The Families of Monocotyledons, R.T.M. Dahlgren, H.T. Clifford, & P.F. Yeo, 1985, Springer-Verlag, Berlin, added were 2 superorders, Cyclanthiflorae and Pandaniflorae, and Liliflorae replaced Alismatiflorae as the basal taxon. Dicotyledoneae had 25 superorders and Monocotyledoneae had 8; there were no subclasses. (Brummitt, 1992).

Rolf Dahlgren (1932-1987) was a Swedish-Danish botanist, who, like many botanists in earlier times, was born to an apothecary. He graduated from Lund University in 1963, worked on South African plants during expeditions in 1956-57 and 1965–66, while affiliated with the Botanical Museum in Lund as docent, was botany professor at the University of Copenhagen from 1973 till his decease. The presentation of his system included instructive diagrams called Dahlgrenograms. He also worked on family circumscription in the monocots with H. T. Clifford and Peter Yeo. In 1986, he was elected member of the Royal Swedish Academy of Sciences.

Thorne also presented classifications for angiosperms and in earlier versions did not recognize subclasses but recognized superorders: 19 for dicots and 9 for monocots. In the '92 version there were 28 superorders, 71 orders, and 437 families, in the 2000 version there were 10 subclasses, 31 superorders, 73 orders, and 490 families, and the 2007 revision had 12 subclasses, 35 supeorders, 87 orders, and 472 families.

The publications are: R. P. Thorne, Proposed new realignments in the angiosperms, Nordic J. Bot. 3: 85-117, 1983 (earlier versions in '68, '76, and '81); later versions: Thorne, R.F. & J.L. Reveal, 1992, An updated phylogenetic classification of the flowering plants, Aliso 13: 365-389; Thorne, R.F. & J.L. Reveal, 1992, Classification and geography of the flowering plants, Bot. Rev. 58: 225-348; Thorne, R. F., 2000, The classification and geography of the monocotyledon subclasses Alismatidae, Liliidae and Commelinidae, pp. 75–124, in Plant Systematics for the 21st Century, B. Nordenstam, G. E-Ghazaly, and M. Kassas (eds.), Portland Press, London; Thorne, R. F., 2001, The classification and geography of the flowering plants: dicotyledons of the class Angiospermae (subclasses Magnoliidae, Ranunculidae, Caryophyllidae, Dilleniidae, Rosidae, Asteridae, and Lamidae), Botanical Revue 66: 441–647; Thorne, R.F. & J.L. Reveal, 2007, An updated classification of the class Magnoliophyta ("Angiospermae"), Bot. Rev. 73: 67-181.

Robert Thorne (1920-) is an American botanist and currently Taxonomist and Curator Emeritus at Rancho Santa Ana Botanical Garden and Professor Emeritus at Claremont Graduate University in California. After serving in the Air Force in WW 2, he earned his Ph.D. in economic botany at Cornell in 1949. He was President of the American Society of Plant Taxonomists and the Southern California Botanists, is a Fulbright Research Scholar, National Science Foundation Senior Postdoctoral Fellow, and Cramer Fellow. He received the American Botanical Society Merit Award in 1996, the Southern California Botanists Lifetime Achievement Award in 1999, American Society of Plant Taxonomists Asa Gray Award in 2001, Rancho Santa Ana Botanical Garden Glory Award in 2006, and the Botanical Society of America Centennial Award in 2006. And he is on the editorial boards of several national botanical periodicals.

Reddy et al (2004) delimit the Englerian school, including, of course, Engler and his collaborators, plus Eichler, Rendle, Wettstein, and Lawrence, and the ranalian school, including Bessey, Hallier, Hutchinson, Tippo, Takhtajan, Cronquist, and Thorne. The former category was only 3 lists of orders so was unsatisfactory and inadequate even as convenience. The latter school did bother to classify the orders into higher groups, but among those of more recent times Takhtajan's was superior, as it recognized both subclasses and superorders. This was done also by David Young, and, as mentioned earlier, by Thorne in later versions. George Stebbins, in 1974, had the same dicot subclasses as Cronquist and the same monocot subclasses as Takhtajan. All of these classifications, however, had only limited evolutionary merit, and Cronquist was a gradist and probably most were, although they contained some important insights and recognized several natural groups; a considerable number of the traditional families and orders have been confirmed by numerical evolutionary analysis.

True evolutionary classifications using classical evidence have been done by Dahlgren and Bremer (1985, including only monocots and some dicots), Donoghue and Doyle (1989), Hufford (1992), Doyle (1996), Tucker and Douglas (1996), Loconte and Stevenson (1991)(which agrees essentially with Qiu et al's 1993 molecular phylogeny, mainly differing in the position of Nympheales misplaced in Magnolidae), Stevenson and Loconte (1995), and Nandi, Chase, and Endress (1998). (Gonzales (1999) provides a good synopsis of most of them as well as molecular analyses.) These differ considerably but most agree on a paleoherb clade or something similar, and a monocot, eudicot (tricolpate), asterid, and rosid clade.

It was only those by Larry Hufford, professor and director of the WSU School of Biological Sciences, and Owi Nandi, a German botanist, linguist, and poet of Indian extraction, of the University of Zurich's Institute for Systematic Botany, American-born Mark Chase, Keeper of the Jodrell Lab at Kew and Darwin-Wallace Medal Winner, and Peter Endress, also of the University of Zurich's Institute for Systematic Botany, which considered most angiosperms, the latter being the most comprehensive, using 161 taxons and 252 characteristics (Table 12-21). A consensus tree was not done or presented by Nandi et al, only one of the 17 MPTs was presented. In that analysis based on phenotype there was a CI of only .09 and an RI of .41, and there were no strongly nor moderately supported clades of note. The rbcL analysis in Nandi et al had a CI of .22 and an RI of .41 and had 5000 MPTs, although after successive weighting there were only 9, and a CI of .62 and an RI of .65. The combined tree after successive weighting had only 1 MPT and a CI of .16 and an RI of .38. In Hufford there were 80 taxons and 60 characteristics, the CI was only .18, and no confidence measures were done, but the low homology indicates only a few groups would be well supported in resampling. Nonetheless, several new results came out from the Hufford analysis that were later supported by molecular studies, such as the polyphyly of Hamamelidae, the close kinship of Loasaceae and Hydrangeaceae, and the position of Sarracenaceae (pitcher-plant) in Asteridae (Endress, 2002).

The APG (Angiosperm Phylogeny Group) is an informal international group of botanical taxonomists who came together to try to establish a consensus but is based only or primarily on molecular evidence. There have been 3 versions of their classification system: APG I in 1998, APG II in 2003, and APG III in 2009. There are 8 authors who did 1 or more research papers for it and 32 contributors, with 4 of the authors doubling as contributors. The 8 are Birgitta and Kare Bremer of Uppsala, Michael Fay of Kew, Mark Chase of Kew, James Reveal of the U. of Maryland and Cornell, Douglas and Pamela Soltis of the U. of Florida's Museum of Natural History, and Peter Stevens of Harvard and the Missouri Botanical Garden.

Table 12-21. Phenotypic Cladistic Tree by Nandi et al (1998).

Ceratophyllaceae
Unnamed 1
 Chloranthus
 Unnamed 2
  Amborella
  Neoangiospermae
   Winteraceae

Cenangiospermae stat. nov.
Illiciales-*Austrobaileya*
Unnamed 3
Eupteleaceae
Unnamed 4
Paleoherbs
Magnolidae
Berberidopsis+Nelumbonaceae
Unnamed 5
Nympheaceae
Unnamed 6
Magnoliales+Annonales
Unnamed 7

Ranuc.+Laurales
Unnamed 8
Myristicaceae
Arstl.-Mncts
Lower Hamamelidae [Platanaceae
(*Myrothamnus* (Didymelaceae
(Buxaceae+Proteaceae))]
Unnamed 9
Trochodendrales
Unnamed 10
Peonaceae
Unnamed 11
Crossosomataceae
Unnamed 12
Cephalotaceae (Brunelliaceae (Coriariaceae (Vitaceae)))
Unnamed 13
Asteridae
Paracryphiaceae (Medusagynaceae + Eucryphiaceae)
Unnamed 14
Saxifragales s.s. + Crassulaceae
Unnamed 15
Sarraceniaceae (Loasaceae+Balsaminaceae)
Unnamed 16
Ericales (Ebenales + Theaceae)
Unnamed 17
Asterids 3
Unnamed 18
Cyrillaceae
Unnamed 19
Asterids 1 and 2
Unnamed 20
Polemoniaceae
Unnamed 21

- Santales, *Sabia*, Gunneraceae, Araliaceae, Eucommiaceae, etc.
- Unnamed 22
  - Aquifoliaceae
  - Unnamed 23

- Rosidae
  - Geraniales
  - Unnamed 24
    - (Myrtales+Faganae)(Cunoniales (Hamamelidales (Rosaceae+Dilleniaceae)))
    - Unnamed 25
      - Quiniaceae+Linales+Passiflorales+Caricaceae (=Papayaceae)
      - Unnamed 26
        - Celastrales s.s. + (Violaceae (Flacourtiaceae (Ochnaceae (Capparales + Cucurbitaceae))))
        - Unnamed 27
          - Oxalidaceae+Connaraceae
          - Unnamed 28
            - Zygophyllaceae
            - Unnamed 29
              - Polygalaceae (Chrysobalanaceae +Eleagnaceae)
              - Unnamed 30
                - (Rhamnaceae ((Rutales + Fabaceae) (Urticales (Euphorbiaceae + Sapindales)))(Malvales)
                - Unnamed 31
                  - Malpighiales s.s.
                  - Unnamed 32
                    - Moringaceae
                    - (Bretschneideria (Akariaceae)
                    - Unnamed 33
                      - Salicaceae
                      - Caryophyllidae

Table 12-22 shows a time tree and Table 12-23 gives the ages in mya of the various key bloom-bearing families based on the APW (Angiosperm Phylogeny Website) and TimeTree.Org. The APW usually gives molecular estimates only for the crown groups so I have excluded these. There are often very different estimates depending on different researchers in the molecular clock methods. For instance, molecular estimates for Crossosomaceae are 95-91 and 47-40 mya (Time Tree). However, according to Time Tree, there is generally congruence between the fossil record and molecular clocks.

Table 12-22. Time Tree Based on Soltis et al, 2000.

- Amborellaceae
- Unnamed 1

- Nympheales
- Unnamed 2
  - Austrobaileyales
  - Unnamed 3
    - Chloranthales+Magnolidae
    - Unnamed 4
      - Monocots
      - Unnamed 5
        - Ceratophyllaceae
        - Tricolpatae
          - Ranunculales
          - Unnamed 6
            - Proteaceae
            - Unnamed 7
              - Sabiaceae
              - Unnamed 8
                - Buxaceae+Didymelaceae
                - Unnamed 9
                  - Trochodendrales
                  - Unnamed 10
- *Gunnera+Myrothamnus*
- Unnamed 11
  - (Santalales (Dilleneaceae + Caryophyllales) + Asteridae
  - Unnamed 12
    - *Berberopsis+Aextoxicon*
    - Unnamed 13
      - Saxifragidae [Hamamelales (+Peonia+ (Saxifragales+Cercidiphyllum))]
      - Unnamed 14
        - Vitaceae
        - Eurosids
          - Eurosids 2
            - Geraniales
            - Unnamed 15
              - Crossosomales
              - Unnamed 16
                - Tapisciaceae
                - Unnamed 17
                  - Brassicales
                  - Unnamed 18
                    - Neuradaceae+ Malvales
                    - Sapindales
          - Eurosids 1
            - Myrtales
            - Unnamed 19
              - Zygophyllales
              - Unnamed 20

Nitroflorae
Unnamed 21
Celastrales
Unnamed 22
Oxalidales
Malpighiales

Table 12-23. Fossil Record and Molecular Clocks for Bloom-Bearers.

fam., ord., sbcl.	fossils APW	fossils TT	prd./epch.	molecular APW	molecular TT
Monocots					153-143
Magnolidae					152-140
Chloranthaceae	125		Early Cret.	150-133	
Austrobaileyaceae	125-113		Early Cret.		
Dilleniaceae			Early Cret.	115	
Buxaceae	113-98				
Platanaceae	113-98				
Sabiaceae	c.100			135-90	
Magnoliaceae	98-91				
Trochodendraceae				113-7	
Proteaceae	90				
Asteridae					128
Rosidae					117
Saxifragaceae	90?		Late Cretaceous		73-96
Crossosomaceae			Late Cret.- Paleocene	101-53	95-40
Rosaceae	90	44	Late Cretaceous		44
Fagaceae	90		Late Cretaceous	16-84	
Theaceae	90		Late Cretaceous		
Brassicales		90	Late Cretaceous		90-71
Myrtales		88	Late Cretaceous		c. 110
Hamamelaceae	84-72		Late Cretaceous		92-102
Malpighiales	80-66		Late Cretaceous		Late Cret.
Malvales			Late Cretaceous-Eocene		75-67
Zygophyllaceae					c.70
Polygalaceae				66	
Cucurbitaceae				65	
Fagales				61	
Sapindales					61
Geraniaceae	70-40		Late Cretaceous-Eocene		92
Violaceae				Paleocene	60
Celastrales		30	Eocene		62-58

In order to calculate the rate at which a stretch of DNA changes, biologists must rely on dates estimated from other dating techniques, such as radiometric dating, because the clocks must first be calibrated against independent evidence about dates, such as the fossil record. However, for viral phylogenetics and ancient DNA studies, two areas of evolutionary biology where it is possible to sample sequences over an evolutionary time scale, the dates of the samples themselves can be used to calibrate the molecular clock.

The molecular clock runs into particular problems at very short and very long time scales. In the former, many differences between samples do not represent fixation of different sequences in the different populations. Instead, they represent alternative alleles that were both present as part of a polymorphism in the common ancestor. The inclusion of differences that have not yet become fixed leads to a potentially dramatic inflation of the apparent rate of the molecular clock. In the latter, the problem is saturation. When enough time has passed, many sites have undergone more than one change, but it is impossible to detect more than one.

Molecular clock users have developed work-around solutions using a number of statistical techniques including maximum likelihood and later Bayesian inference (Bayesian methods can provide more appropriate estimates of divergence times, especially if large data matrices, such as those yielded by phylogenomics, are employed). In particular, models that take into account rate variation across lineages have been proposed in order to obtain better estimates of divergence times. These models are called relaxed molecular clocks because they represent an intermediate position between the strict molecular clock hypothesis and Felsenstein's many-rates model and are made possible through MCMC techniques that explore a weighted range of tree topologies and simultaneously estimate parameters of the chosen substitution model. It should be noted that divergence dates inferred using a molecular clock are based on statistical inference and not on direct evidence.

The cladistic classification below is based on Nandi et al, Hufford, molecular analyses (especially the reviews by Judd and Olmstead (2004), Soltis and Soltis (2004), and Cortina et al (2007)), historical opinion, fossil evidence, and molecular clocks. For monocots I have followed Wm. Hahn (Monocots-tol.org), which is based on both phenotypic and genotypic analyses.

Table 12-24. Cladistic Classification for Flowering Plants (71 orders).

spcl. Ceratophyllatae (hornwort)(1 gen., 3 sp.) stat. nov.
spcl. Metangiospermae stat. nov.
 cl. Amborella (amborella)(1 gen.) stat. nov.
 cl. Neoangiospermae stat. nov.
  sbcl. Winteridae (winter's bark)(1 fam.) stat. nov.
  sbcl. Cenangiospermae stat. nov.
   infcl. Chloranthidae stat. nov.
   infcl. Anangiospermae stat. nov.
    pvcl. Magnoloidia stat. nov.
     mccl. Magnolidae stat. nov.
      mgord. Eumagnolidae stat. nov.
       grdord. Illicianae (illicium)(1 ord.; Illiciaceae [illicium or star-anise], Austrobaileyaceae) stat. nov.
       grdord. Magnolanae stat. nov.
        hpord. Lauriflorae (laurel)(1 ord.) nom. nov., stat. nov.
        hpord. Metamagnolaria
         spord. Magnoliflorae (Chloranthales [chloranthus], Magnoliales [magnolia])
         spord. Annoniflorae (custard apple)(1 order)
      mgord. Lactoriflorae (Juan Fernandez pepper) (1 sp.) stat. nov.

mccl. Paleoherbacea nom. nov. ("paleoherbs" Doyle & Donaghue 1987) stat. nov.
nncl. Nympheidae (water lily) (1 ord.) nom. nov., stat. nov.
nncl. Paramonocotylidae stat. nov.
pccl. Aristomonocotylae stat. nov.
sbpccl. Aristolochidae (birthwort)(1 ord.) stat. nov.
sbpccl. Monocotylidae (14 ords.) stat. nov.
sptrord. Alismatiflorae stat. nov. (Alismales [water-plantain], Arales [arum])
sptrord. Neomonocotylae stat. nov.
trord. Liliflorae (1 ord.) stat. nov.
trord. Orchidiflorae stat. nov. (Orchidales, Iridales, Asparagales)
trord. Pandaniflorae stat. nov. (Cyclanthales [Panama-hat palm], Panadanales [screw-pine])
trord. Dioscoriflorae stat. nov. (yam, 1 ord.)
trord. Apoperianthae stat. nov.
ggord. Areciflorae stat. nov. (Palmales [Principes])
ggord. Commelariae stat. nov.
spord.Zingiberiflorae stat. nov. (Bromeliales [pineapple], Zingiberales [ginger])
spord.Commeliflorae stat. nov. (Commelales [spiderwort], Poales [grasses])
pccl. Piperidae (pepper, 1 ord.) stat. nov.
pvcl. Tricolpatae (Eudicotyledonae) stat. nov.
mccl. Ranuculidae Takhtajan 1958 (buttercup, poppy, barberry, 1 ord.) stat. nov.
mccl. Metatricolpatae nom. nov., stat. nov
nncl. Proteidae (protea or sugar bush, 1 ord.) stat. nov.
nncl. Neotricolpatae nom. nov., stat. nov.
pccl. Buxidae stat. nov. (box or boxwood, 1 ord.)
pccl. Cenotricolpatae nom. nov., stat. nov.
sbpccl. Trochodendridae (hoop tree, 1 ord.) stat. nov.
sbpccl. Anatricolpatae nom. nov., stat. nov.
infpccl. Gunneridae (giant rhubarb, 1 ord.) stat. nov.
infpccl. Pentapetalae D. Soltis, P. Soltis, et Judd, 2007 stat. nov.
hptrord. Berberidopsidae (coral plant, 1 ord.) stat. nov.
hprtrord. Caryophyllidae (Centrospermae) (Caryophyllales [carnation], Polygonales [knotweed or buckwheat], Plumbaginales [leadwort]) stat. nov.
hptrord. Dillenidae stat. nov. (1 fam.)(dillenia)
hptrord. Cenopentapetalae

ggord. Sympetalae (syn. Gamopetalae, Metachlamydeae, Asteridae Takhtajan 1967 emend.) stat. nov.

mgord. Corniflorae Thorne 1968 (dogwood)(1 ord.)

mgord. Araliflorae Dahlgren 1983 (1 order, carrot and ginseng)

mgord. Eusympetalae (euasterids, Gentianidae)(aster) nom. nov., stat. nov.

grdord. Theiflorae (tea, 1 ord.)

grdord. Ericiflorae (heath)(1 ord.) nom. nov., stat. nov.

grdord. Primuliflorae Dahlgren 1983 (2 orders, primrose and ebeny)

grdord. Lamiflorae Thorne 1968 (euasterids 1, Lamidae)(Gentianales [gentian, madder, dogbane], Lamiales [mint], Solanales [nightshade (*belle-de-nuit*)])

grdord. Asteriflorae Thorne 1968 (1 order, bellflowers and asters)

ggord. Santalidae Thorne 1968 stat. nov. (Santalales [sandalwood (*bois de santal)*])

ggord. Superrosidae stat. nov.

sptrord. Peonidae stat. nov. (peony [*pivoine*], 1 family)

sptrord. Rosidae Takhtajan 1967 (50)

trord. Vitiflorae (grape, 1 order)

trord. Metarosidae

ggord. Crossosomata (rockflower, 1 order)

ggord. Neorosidae

mgord. Geraniflorae stat. nov. (geranium, 1 order)

mgord. Myrtiflorae Thorne 1968 (Myrtales

[myrtle, 1 order]) stat. nov.
mgord. Anarosidae
grdord. Rosiflorae Thorne 1968 stat. nov. (Hamamelales [witch-hazel]?, Saxifragales?, Cunoniales [cunonia]; Fabales [legumes]-Zygophyllaceae [caltrop, caper bean], Rosales, Rhamnales, [buckthorn), Fagales, Urticales
grdord. Coviflorae nom. nov., stat. nov.
hprord. Celastriflorae (staff-tree, 1 fam.) stat. nov.
hprord. Oxaliflorae (oxalis or wood sorrel and zebra wood, 2 fams.)
hprord. Violiflorae Thorne 1968 stat. nov. (Linales [flax], Malpighiales [malpighia], Violales [violet], Passiflorales [passionflower], Salicales [willow], Euphorbiales [spurge]; Cucurbitales [cucumber], Brassicales [mustard])
mgord. Rutiflorae Thorne 1968 (Rutales [rue, citrus fruits], Leitnerales [corkwood], Sapindales [soapberry,

maple] stat.
nov.
mgord. Malviflorae
Thorne 1968
stat. nov.
(Malvales
[mallow])

Table 12-25 gives a list of authors and orders. Most of the author citations for ordinal names are completely erroneous and usually contradictory. Taxonomicon gives 35 author citations for orders by Dumortier 1829, but his classifciation has only 20 orders and none of the names even come close to any cited (Lindley, 1853; Dumortier, 1829). The APG cites him for a few of the orders but also often cites Jussieu ex Berchtold and Presl, but Jussieu didn't do any ordinal names and never used the -ales suffix (Jussieu, 1789). Link (1929) has also been credited with several but I checked his book and nowhere are there any -ales. Reichenbach (1828) is credited with Myrtales, but I checked his book, too, and, again, there is no -ales anywhere in it. Grisebach (1854) is credited with Hamamelales, but the name he used was in fact Hamamelinae and the only -ales he used was for Gruinales. James Reveal gives a list of the names by Berchtold and Presl and assigns them the -ales suffix retroactively (Reveal, 2011). Lamentably, this practice is allowed by the Code. Linales is attributed to Baskerville (1839), but, although he did use the -ales for most of his orders and followed Lindley close in his taxonomy and dedicated his book to him and coined Celastrales, he did not use the name Linales. In actual fact, it is Lindley who is the true originator of most of the major ordinal names. Edwin Bromhead is no. 2 with 9 (Lindley, loc. cit.). Bromhead's first version was in 1836 in the Edinburgh Journal, with subsequent revisions in 1837 in the Philosophy Magazine, in 1838 in the Edinburgh Philosophical Journal, again in 1838, but in the Magazine of Natural History, and in 1840 again in the Magazine of Natural History. The author citation is supposed to be for the actual and complete name of the taxon, not just the root, and not the taxon, which sometimes changes considerably anyways, or it should be specified the original name was of this or that rank. The author citations for family names is apparently unreliable, too, as Brummitt (1992) cites Jussieu for nearly 80 angiosperm family names but Jussieu never used the -aceae suffix.

Table 12-25. Authors and Orders (there are 73 orders; Lindley had the most with 36, he had the admirable habit of simplifying the names, and he used Acerales for Sapindales, Nymphales for Nympheales, and earlier used Saxales for Saxifragales).

Lindley 1833

Asterales
Alismales
Anonales
Aristolochiales
Commelales
Cornales
Cruciales (Brassicales, Capparales)
Cucurbitales
Ericales
Euphorbiales

Gentianales
Geraniales
Labiales (Lamiales)
Laurales
Liliales
Malvales
Myrtales
Palmales (Principes, Arecales)
Pandales (Pandanales)
Passionales (Passiflorales)
Piperales
Plumbales (Plumbaginales)
Polygonales
Primulales
Proteales
Ranales (Ranunculales)
Rhamnales
Rosales
Rutales
Santalales
Solanales
Theales
Violales
Umbellales (Apiales)
Urticales

Lindley 1846
Garryales

Bromhead 1838

Asperagales
Brassicales
Fabales
Lamiales
Magnoliales
Nympheales
Orchidales
Passiflorales
Zingiberales

Baskerville 1839
Celastrales

Senft ? 1856
Aquifoliales

Hooker in Le Maout et Decaisne 1873
Dioscoreales

Oliver 1895
Rafflesiales

Engler 1897

Fagales
Leitneriales

Small ? 1903
Poales

Bessey 1915
Caryophyllales

Hutchinson 1924
Dilleniales

Heintze 1927

Oxalidales
Peoniales

Nakai 1930
Apiales

Cronquist 1957
Canellales

H.H. Hu ex Cronquist 1981
Illicilaes

Takhtajan ex Reveal 1992

Austrobaileyales
Crossosomatales
Gunnerales
Lactoridales

Takhtajan ex Reveal 1996

Buxales
Vitales

A.C. Smith ex Reveal 1992

Winterales

Takhtajan 1997

Petrosaviales
Zygophyllales

Melikyan, A.V. Bobrov, & Zaytzeva in APG 2009
Amborellales

Dowell in Reveal 2011

Berberidopsidales?
Chloranthales
Cunoniales
Hamamelales
Linales
Malpighiales
Polygalales

Ceratophyllaceae Gray 1821 (1 genus) are a family of aquatic, submerged herbs of widespread distribution. It is rootless, monecious (stamens and pistils on the same plant), with dichotomously divided, whorled, sessile, estipulate leaves, 1 pistil, 1 locule, 1 carpel, 1 stigma, a solitary, pendulous, anatropous (reversed), tenuinucellate (no nucellar tissue separating the megasporocyte(s)/embryo sac from the epidermis; the megasporangium is called the nucellus) and unitegmic ovule (1 integument, the integument being the outer envelope), a superior ovary, parietal placentation, a 2-celled, longitudinally dehiscing anther (opening and shedding contents longitudinally), 10-20 spirally arranged stamens, an actinomorphic (regular) and unisexual flower; the fruit is a nut; and the seed has a large, straight embryo and no endosperm. All of these characteristics are primitive, except the aquatic habitat, herbaceous habit, rootless condition, dissected leaf, the nut fruit (the follicle is primitive), the tenuinucellate and unitegmic ovule, parietal placentation, and possibly the unisexual flowers. In molecular systems it is higher up, after monocots.

Amborellaceae Pichon 1948 has only 1 species (*Amborella trichopoda* Baillon) and is a sprawling shrub or tree with alternate or decussate (opposite leaves in 4 rows up and down the stem alternating in pairs at right angles), simple, estipulate, persistent, dimorphic leaves, small flowers, with an inflorescence that has been described as "determinate thyrses" (thyrse being a compact and more or less compound branching raceme [panicle], which has indeterminate influorescence [terminal flowers open last] with pedicelled or stalked flowers), with up to 3 orders of branching, each branch being terminated by a flower, which is subtended by bracts, and which gradually transitions into a perianth (floral enveloppe, i.e., the corolla and calyx together) of spiralling tepals (corolla [petals] and calyx [sepals] undifferentiated) of 6 to 15 in the male flowers and 7 or 8 tepals in female flowers. The species is diecious (stamens and pistils on different plants), has xylem without vessels (as in Winteraceae and Trochodendrales), has a spiral of 4 to 8 free carpels, a single, crassinucellate (one or more layers of cells [not epidermally derived] separating the megasporocyte(s)/embryo sac from the epidermis), bitegmic, semianatropous, anatropous, or orthotropous ovule (depending on the researcher), and 10 to 21 spirally-arranged stamens, with longitudinally dehiscing anthers, granulate exine (outer coat of pollen grain), and a small embryo. The fruit is ovoid, red, and drupaceous (drupe being a fleshy 1-

seeded indehiscent fruit with seed enclosed in a stony endocarp), containing red juice. All of these are primitive except the orthotropous ovule (which it may not have), drupe, and possibly the determinate thyrses. It may be the most primitive group instead of the hornwort, which it is in most molecular phylogenies. (See Tobe et al, 2000; Amborellaceae – Wikipedia).

Winteraceae R. Brown ex Lindley 1830 (winter's bark) are a family of tropical trees and shrubs containing 4 genera (*Drimys, Pseudowinterana, Takhtajana*, and *Zygogynum*). It is an independent clade in Doyle and Donaghue and Nandi et al but branches within Magnolidae in genotypic studies and has been recently placed with Canellaceae Martius 1832 (Winteraceae)(canella or wild cinnamon, 6 genera: *Canella, Capsicodendron, Cinnamodendron, Cinamosma, Pleodendron, Warburghia*), which are tropical trees, forming Canellales. The winter's bark family was positioned in the magnolia family in Ranales by Bentham and Hooker, in Annonales by Thorne, and Magnoliales by Dahlgren and Cronquist, and as Winterales in Magnolianae by Takhtajan. The winter's bark (*Drimys*) is cultivated as an ornamental. The canella family was placed in Parietales by Bentham and Hooker, Ranales (including also magnolialean families) by Bessey, Annonales by Hallier, Thorne, Dahlgren, and Takhtajan, and Magnoliales by Hutchinson and Cronquist. The leaves and bark of *Canella* are used in medicine and as a condiment and the tree is cultivated as an ornamental, prized for its purple flowers and black berries. Both families have aromatic bark.

Chloranthaceae Blume 1827 are a family of tropical or subtropical herbs, shrubs, or trees, with 2 Old World genera (*Chloranthus*, with about a dozen species, and *Ascarina*, with 3) and 1 New World genus (*Hedyosmum*, with about 25 species). It is monecious or diecious, with simple, opposite, minutely stipulate, usually aromatic leaves, bisexual or unisexual, actinomorphic flowers in spikes, cymes, panicles, or heads, anthers 2-celled or 1-celled and vertically dehiscent, 1 or 3 stamens, 1 pistil, 1 carpel, 1 locule, a solitary, pendulous, orthotropous ovule, inferior ovary, a gamosepalous, 3-toothed calyx, and seed with abundant, oily endosperm and a minute embryo. It has been assigned to Microembryeae by Bentham and Hooker, Piperales by Engler, Annonales by Hallier, the laurel suborder in Annonales by Thorne, as an order in Magnoliflorae by Dahlgren, as an order in Magnolianae by Takhtajan, and in Piperales in Magnolidae by Cronquist. It branches with Laurelales in Doyle and Donaghue, and in most molecular phylogenies it is an independent lineage, between monocots and magnolids, and is an independent lineage in Nandi et al also.

Illiciaceae (A.P. De Candolle) A.C. Smith 1947 (1 genus) was 1st defiintely established by Smith in 1947; most taxonomists had previously treated *Illicium* as a tribe of Magnoliaceae, but it had already been regarded as distinct from the magnolia family by numerous botanists. *Illicium* has actinomorphic, bisexual flowers, persistent, alternating, simple, entire, estipulate leaves, tepals, an anatropous ovule, 1 carpel, 1 locule, 7-21 whorled pistils, and 2-celled, longitudinally dehiscing anthers.

Austrobaileyaceae Croizat 1943, a woody vine that can grow up to 50 ft., is endemic to the jungles of Queensland and has only 1 genus and 1 or 2 species (*A. scandesns* and *maculat*a; it is not certain if these are distinct species or one and the same). It was described in 1933, and the name commemorates 2 men, F. M. Bailey, a noted Queensland botanist, and I. W. Bailey, longtime staff member at Harvard's Arnold Arboretum and world-renowned wood anatomist who was particularly interested in primitive flowering plants and who published a detailed account of the morphology of the genus. It has spirally arranged flowers, about 2 dozen tepals, 7-11 stamens, about a dozen carpels, about 6 ovules, is borne on 2 placentas, 1 on each side of the carpel, and is fly-pollinated so has an unpleasant odour. (Notes from the Arnold Arboretum-Austrobaileya (arnoldia.arboretum.harvard.edu); Fact Sheet-Austrobaileya (keys.trin.org.au); Austrobaileya-Wikipedia). Most botanists have treated this family as part of Magnoliales or Magnoliflorae/Magnolianae. It is treated as an independent lineage by the APG.

The paleoherbs of Doyle are really Magnolidae sensu Takhtajan (i.e., Magnoliales, Piperales,

Laurales, Illiciales, Nympheales, Aristolochiales) with monocots included, and Ceratophyllaceae and Rafflesiales excluded (Rafflesiaceae formed Aristolochiales with Aristolochiaceae, but probably go to Rosidae), perhaps along with Illiciales (with Winterceae and Austrobailyaceae separated out), *Chloranthus*, Laurales (with *Amborella* separated out and possibly the most primitive), and Eupteleaceae (which is placed either in Trochodendrales or Ranunculales). The clade, however, may not be monophyletic.

In monocots, Arecidae, the palm subclass, has not held up cladistically, nor has the lily family and order, which had already been known to be heterogeneous and of dubious monophyly. The other orders of Arecidae were Arales (arum), Pandanales (screw pine), and Cyclanthales (Panama-hat palm). Alismidae are considered by most to be the most primitive group in classical cladistic analyses, however, some, like Thorne, identified Araceae as most primitive, and they do show up as such in molecular trees.

Ranunculales (Ranales) were treated by Engler and Diels as having the water lily (2 families), hoop tree (Trochodendrales; 2 families), buttercup (4 families), and magnolia (11 families) suborders, but subsequently the magnolia and hoop tree suborders were grouped together as Magnoliales, and the water lily and buttercup suborders were united as Ranales. Lactoridaceae, with the single monotypic genus *Lactoris*, endemic to Robinson Crusoe Isl. in the SE Pacific, was in the magnolia suborder. Lauraceae (laurel), Annonaceae (custard apple), Eupotamiaceae, and Ceratophyllaceae in Nympheoideae, have been separated out in 1 or more analyses, be it classical or molecular. Piperales had 3 families: Piperaceae, Saururaceae (lizard's tail family), and Chloranthaceae, and now have only the first 2. The Ranunculales of modern circumscription are still mostly dimerous but have a few pentamerous members.

Proteales, with a single family of some 70 genera, mostly trees and shrubs, are native to the southern hemisphere and include the Macadamia or Queensland nut, popular in Hawaii where it is a delicacy. The type genus is the protea or sugar bush and is native to southern Africa. They were considered primitive by Engler and Hallier, Rendle noted it was difficult to associate them phyletically to any other orders, and Lawrence pointed out there was a divergence of opinion on them and concluded they could not yet be allied to any existing order. They were placed in their own superorder by Thorne, Dahlgren, and Takhtajan, but the last 2 in Rosidae, and as an order in Rosidae by Cronquist. The Oriental family Sabiaceae Blume 1851 (sabia, 1 genus) sometimes comes up as their sister group (Judd and Olmstead, loc. cit.). It was usually placed in Sapindales; Lawrence said phylogenists are agreed it belongs there.

Trochodendridae contain at least 2 families: Trochodendraceae Prantl 1888 (trochodendron or hoop tree) and Tetracentraceae A.C. Smith 1945 (tetracentron), Trochodendrales being the lower Hamamelidae, with the higher Hamamelidae being the wind-pollinated, catkin-bearing Amentiferae. Early botanists allied *Trochodendron* with the magnolia family. Prantl considered also *Euptelea* (euptelea) and *Cercidiphyllum* (katsura) as part of the family. In 1936 Engler and Diels excluded *Cercidiphyllum*, regarding it as a separate and more advanced family. In 1946 A.C. Smith viewed these genera along with *Eucommia* (gutta-percha or Chinese rubber tree) and *Tetracentron* as 5 families following the studies of van Tieghem in 1900. Bessey assigned Eucommiaceae to Hamamelales, where it is typically placed, but is in Corniflorae in Dahlgren and Asteridae in the APG and Nandi et al's phenotypic tree. Wettstein regarded 4 of the genera as families, retaining *Tetracentron* in the magnolia family. Each of the 5 families has only 1 genus and all are apetalous and Oriental, mostly Chinese and Japanese.

Euptelea was placed in the magnolia family in Ranales by Bentham and Hooker, in the Englerian system it was in Ranales or Magnoliales, Hallier and Hutchinson included it in the hoop tree family and the latter placed it in Magnoliales, Thorne positioned it in the hoop tree suborder in

Hamamelales, Dahlgren included it in Cercidiphyllales in Rosiflorae, Takhtajan regarded it as an order in the hoop tree superorder in Hamamelidae, and Cronquist included it in Hamamelales. In Hufford it is in Buxidae, is an independent clade in Nandi et al's phenotypic tree, and is in Ranunculales in the APG. It is characterized by a special wind-pollination syndrome (reduced perianth, long, hanging anthers, and relatively large stigma with long papillas), which is also known in some Hamamelaceae and related families and Papaveraceae and Ranunculaceae. Also, Eupteleaceae are sizable trees, whereas this trait is not known in Ranunculales. Its afffinities are unknown.

Cercidiphyllum was long included in the magnolia family, but in 1949 Swamy and Bailey rejected this concept, refuting the views of earlier workers, and concluding it was not of close affinity with any other group, but was afterwards generally included as part of Trochodendrales but also positioned in the recent "Saxifragales." Lawrence described it as unlike any other extant angiosperm. *Euptelea* likewise, as just noted, was placed in the magnolia family but afterwards usually treated as a member of Trochodendrales or Ranunculales and its position is uncertain.

Platanaceae Lestib. ex Dumortier 1822 (plane-tree, 1 genus) are a family of tropical trees whose evolutionary position was long considered to be in Urticales, but the consensus in the '50s was that it was of closer affinity to the rose and witch-hazel families than to other taxons (Lawrence, 1960). It was placed in Unisexuales in Monochlamydeae by Bentham and Hooker, Rosales by Bessey, the witch-hazel order or suborder in the Englerian system, and in Hutchinson, Thorne, Dahlgren, Takhtajan, and Cronquist, in Proteales by the APG, and branches with the witch-hazel family in Hufford and within Proteales in Nandi et al. So it is of uncertain kinship, but because it has the primitive form of tricolpate pollen (tectate-reticulate), like Ranuculales and Trochodendrales, its position at the base of tricolpates is appropriate (Qiu et al, 1993).

Buxaceae Dumortier 1822 (box or boxwood, 5 or 6 genera) are dimerous and show up as pre-pentapetalous in both surveys. The family contains evergreen herbs, shrubs, or trees indigenous to the tropics and subtropics especially in the Old World and was variously placed in Euphorbiales, Sapindales, Celastrales, Hamamelales, and Buxales (in Rosiflorae). The placement of Didymelaceae (1 gen.), evergreen trees of Madagascar and the Comoros, in Buxineae by Thorne and Buxales by Dahlgren concurs with its molecular position.

The monotypic Gunneraceae Meissner 1841 are considered in reviews by Judd and Olmstead (2004) and Cortina et al (2007) to be sister group to Pentapetalae. *Gunnera* (giant rhubarb) is a very old genus as evidenced by its pollen agreeing well with *Tricolpites,* known from the Lower Cretaceous (Dahlgren, 1983). It also has primitive features: P-type sieve-tube plastids, cellular endosperm formation, crassinucellate ovules, copious endosperm, a small embryo, wind-pollination, a dimerous condition, and bi- or tetrasporic megagametophyte development (Dahlgren, 1983; Judd and Olmstead, 2004). It uniquely possesses root symbiosis with blue bacteria, specifically *Nostoc* (Judd and Olmstead, 2004). It was usually placed in Haloragaceae (water milfoil) in Myrtiflorae but has little connection with that family (Dahlgren, 1983). Its leaves can grow up to several feet long and wide and it bears a resemblance to rhubarb (which is in the unrelated knotweed family), hence the name (Wikipedia). *Myrothamnus* (resurrection plant), aromatic-resinous shrubs of Africa and Madagascar, was usually assigned to the witch-hazel family, suborder, or order, but is placed with *Gunnera* by the APG, but the the APW notes they are too different to be confidently placed in the same clade.

Berberidopsales appear outside of Rosidae in both phenotypic and genotypic analyses. *Berberidopsis* (coral plant) has only 2 species, one is the montain tape vine of Australia and the other is the coral plant of Chile, a woody, evergreen vine with striking red flowers, locally known as voqui fuco, whose stems are used in traditional basketry by the southern Mapuche people (Wikipedia). It is usually placed in Flacourtiaceae, which are usually placed in Violales. *Aextoxicon*, a Chilean evergreen tree, usually placed in Euphorbiales, is positioned with *Berberidopsis* by the APG, but went to

Saxifragales in Qiu et al and with Santalales in Nandi et al. Its position is uncertain.

Plumbaginaceae Jussieu 1789 (leadwort, 25 genera), mostly herbaceous and of widespread distribution, were earlier placed in Primulales in Sympetalae, in Primulales in Theiflorae by Thorne, as a superorder by Dahlgren, and in Caryophyllidae by Hallier, Takhtajan, and Cronquist. The latter placement was corroborated by genotypic evidence. Likewise, Polygonaceae Jussieu 1789 (knotweed, 50 genera), are mostly herbaceous and of widespread distribution, and were placed in Theiflorae by Thorne and in Caryophyllidae by Hallier, Takhtajan, and Cronquist, with the latter placement corroborated by genotypic evidence.

Caryophyllales have recently been found to possibly be the sister group of Asteridae by Hilu et al in 2003 (Judd and Olmstead, loc. cit.) and Soltis et al (2007) and this was presaged by Bessey, Hutchinson, and others; as Lawrence relates: "The evidence is reasonably conclusive that the Primulaceae and the Caryophyllaceae have fundamentally the same type of gynecia, and as concluded by Douglas (1936)(and essentially Dickson, 1936) '...the vascular pattern and the presence of locules at the base of the ovary point to the fact that the present much reduced flower of the Primulaceae has descended from an ancestor which was characteirzed by a plurilocular ovary and axial placentation. This primitive flower might well be found in centrospermal stock as Wernham, Bessy, and Hutchinson have suggested.' "

Dillenidae, like Hamamelidae and the older Ranales, also turn out to be polyphyletic, a status presaged by Hickey and Wolfe in 1975, the ones with palmate leaf venation, Violiflorae and Malviflorae, going to rosids, and the ones with pinnate leaf venation, Theiflorae, Ebenales, Ericales, and Primulales, going to asterids. Dilleniaceae Salisbury 1870 (11 gen.), a family of tropical or subtropical trees, shrubs, and occasionally herbs, apparently has parallel venation, which is common in monocots, and has been variously placed in Ranales by Bentham and Hooker, the tea order, suborder, or superorder, or its own order in Dillenidae. Lawrence related that evidence was accumulating the family was much more primitive than allowed by Engler and that it is probable its alliance is with or near the ranalian taxons as recognized by Bessey, Hallier, and Hutchinson; there is some evidence it is allied to Caryophyllidae.

The type genus is named for J. J. Dillen (Latinized as Dillenius [Britannica; Wikipedia]), German botanist of the 1700s, who moved to England in 1721 at the behest of wealthy British consul Wm. Sherard, who studied botany in Paris under Tournefort, and in 1728 became the first botany professor at Oxford according to provisions in Sherard's will. He wrote several descriptive works on plants, notably Hortus elthamensis in 1732, a catalogue of the rare plants in Eltham, London, in the garden of Sherard's younger brother James, which contains descriptions and 417 drawings, done by Dillenius himself, and "Historia muscorum" in 1741, which contains descriptions and illustrations of more than 600 species of true mosses, liverworts, lycopods, algae, lichens, and other lower chlorophyll-bearing forms. Linnaeus visited him in 1735, staying for a month, and dedicated his own Critica botanica to him.

Peonia Linneus 1753 was traditionally one of 3 tribes in the buttercup family, the others being the hellebore and wind flower, but in 1908 Worsdell suggested, and in 1946 Corner emphasized, there was strong evidence for its segregation as a family, removing it from Ranales to a position near Dilleniaceae in Parietales (families often with parietal placentation), which was at the time basically the tea and violet order. It had already been established as a family in 1830 by F. Rudolphi. Thorne positioned it in Annoniflorae, Dahlgren in Theiflorae, Takhtajan in Ranunculanae, and Cronquist in Dilleniales. In both Hufford and Nandi et al it is the sister group of Rosidae. The genus is sometimes placed in Saxifragales, based on molecular data, but with no clear phenotypic synapomorphies for its inclusion, and, along with Cercidiphyllaceae, the witch-hazel family, and a couple of minor families, is isolated in it, and its relationship to other members of the order are unclear. This molecular circumscription, however, concurs with the traditional inclusion of Crassulaceae (orpine), Grossulariaceae, Iteaceae, Penthoraceae, Pterostomonaceae, and, of course, Saxifragaceae, and the rare inclusion of Hamamelidaceae and Tetracarpaceae. So that out of the 14 families in the molecluar definition, 8 agree with other taxonomies, so 6 can be considered as unambiguous members and 2 as possible members. Haloragacaeae and Peoniaceae are out of place in it.

Santalales appear as an early-branching independent clade in Pentapetalae in molecular taxonmies and are in Asteridae in Nandi et al but are not included in Hufford. They were usually regarded as the sister group of Celastrales, but Rendle viewed them as allied to Proteales, and Lawrence placed them between Proteales and Aristolochiales, following the Englerian system. They have a largely parasitic habit, a derived feature, which has led some authorities to regard them as advanced (Core, loc. cit.).

Balanophoraceae Richard 1822 (balanophore (Gk., "acorn bearer", 17 genera) are a family of root parasites assigned to Achlamydosporeae in Monochlamydeae by Bentham and Hooker, as a superorder by Dahlgren, in the monster flower superorder in Magnolidae by Takhtajan, and otherwise in Santalales including by the APG, which is where it probably belongs.

Sympetalae (Asteridae) have traditionally been considered as advanced over the other angiosperms, however, this is in contradiction to the fossil record, molecular clock estimates, the Nandi et al phenotypic tree, and some molecular analyses.

The tea family is placed in Asteridae in molecular taxonomies and in Nandi et al but not in Hufford where it branches with Dilleniaceae. Interestingly, Hutchinson had derived Ericales from Theales. Ericales, like Theaceae, have theoid leaf teeth. Like the monster flower order, *Euptelea, Cercidiphyllum*, *Eucommia, Myrothamnus,* and Podostemaceae, it is of uncertain affinity. Pittosporaceae (pittosporum (Gk., "pitch seed") or cheesewood, known as ho'awa, tarata, or kohohu in New Zealand, 9 genera) are distributed in the warmer regions of the Old World, Australia, and the Pacific. They have been variously placed in Polygalineae by Bentham and Hooker, Rosales by Bessy, Tubiflorae by Hallier, as an order in Archichlamydeae by Hutchinson, in Saxifragineae in Rosales in the Englerian system, as an order in Rosiflorae by Thorne, as an order in Araliflorae by Dahlgren, as an order in Cornanae by Young and Takhtajan, in Rosales by Cronquist, and Asteridae by the APG, and branch within Asteridae in Hufford and Nandi et al. The family's position is well established in Asteridae.

In Eusympetalae, genotypic taxonomies show 2 groupings, and Claudia Erbar and Peter Liens say Lamiflorae can be characterized by late sympetaly (petals appear as distinct primordia and the fused parts appear only later) and Asteriflorae by early sympetaly (the fused part appears first as a ring meristem on which the individual petals later appear), but the latter state is primitive, occuring also in Corniflorae and Araliflorae, which is why Asteriflorae, as genotypically circumscribed, including also Apiales (carrot and ginseng), Aquifoliales (holly), and Dipsacales (teasel), are probably artificial. The teasel order probably belongs in Corniflorae where it is usually placed, and the holly family probably belong in Corniflorae also where it is assigned by Dahlgren. Also, Garryales (garrya) are said to be in Lamiflorae but probably belong as a monotypic family in either Cornales or Apiales where they are otherwise assigned and they share petroselinic acid with Cornaceae, Apiaceae, and Araliaceae (Garryaceae-Wikipedia). The common name for Cornaceae Dumortier 1829 has nothing to do with dogs as it comes from "dagger wood," since the wood was used to make wooden daggers and is as hard as horn, hence the Latin name *Cornus* Linneus 1753 (Jones and Luchsinger, 1979). The French name is *cornouiller.*

Callitrichineae (water starwort) is a cosmopolitan, herbaceous, and mostly aquatic group and has only 1 genus. Lawrence stated its phyletic position was obscured by its extremely reduced parts (having no petals nor sepals) and vascular anatomy and was placed by him as a suborder in Geraniales.

It was assigned to Haloragaceae (in Rosales) by Bentham and Hooker but to Tubiflorae (in Sympetalae) in Melchior's Englerian system, to Lamiales by Thorne, Dahlgren, and Takhtajan, and was an order in Asteridae in Cronquist, and was placed in Plantaginaceae (plantain) in Lamiales by the APG, so its position is quite firmly established in Lamiales. It was not included in Hufford nor Nandi et al.

Both Takhtajan and Cronquist recognized a Rosidae-Asteridae branch and this has been confirmed by both phenotypic and genotypic studies. Rosidae are especially unresolved, with a large polytomy, an observation made also by Pamela and Douglas Soltis (loc. cit.) and Soltis et al (2011). Soltis and Soltis suggest it is because there was a series of rapid radiations. However, there is a Fabid clade recognized by the APG, which is corroborated by Nandi et al's phenotypic tree (for the most part) and Time Tree, as well as a Myrtales-Fabid clade in the last 2. The Coviflorae clade is supported by floral structure, especially between Celastrales and Malpighiales. Among Celastrales, Lepidobotryaceae especially share special similarities with Malpighiales, including a diplostemonous andrecium with 10 fertile stamens, and epitropous ovules with an obturator and strong vascularization around the chalaza (Matthews and Endress, 2005). The COM (Celastrales-Oxalidales-Malpighiales) group of the APG is corroborated by Nandi et al, although it also includes Malvales.

Crossosomaceae Engler 1897 appear as an independent early-branching clade in Pentapetalae only in Judd and Olmstead's survey; it often shows up as a typical rosid in molecular analyses. There are 3 genera, *Crossosoma, Apacheria*, and *Glossopetalon,* and they are found in the SW US and Mexico. The type genus is a perennial, desert shrub with beautiful, fragrant, white or purplish flowers blooming in late winter or spring, growing on rocky slopes in mountain foothills, hence the common name, rockflower, but it is also called the crossosoma. Lawrence relates that most botanists had accepted the view that the family belongs in Rosales, but, interestingly, that Hutchinson, in '26, placed them in Dilleniales, an order he considered intermediate between Magnoliales and Rosales, and that Lemesle, in '48, proposed its transfer to a position intermediate between the peony family and the buttercup family, the latter views being, of course, more in line with cladistic analyses of classical data. Bentham and Hooker and Melchior also placed it in Dilleniales. Molecular analyses, however, tend to position it within Rosidae (Cortina et al (loc. cit.); Soltis and Soltis; APW)

Doyle identifies 3 major molecular clades: one containing Sapindales, Brassicales (Capparales), Malvales (including some Theales), with the possible inclusion of Myrtales and Geraniales; and another containing Euphorbiales (spurge), Malpighiales, Linales (flax), many Violales, with the possible inclusion of Celastrales and Oxalidales. None of the 3 are defined phenotypically and all are contradicted by both Hufford and Nandi et al's phenotypic analyses, and these 2 latter phylogenies do not often concur with each other. Judd and Olmstead identify a polytomy for Rosidae comprising Vitales, Crossosomatales, Geraniales, Myrtales, fabids, and malvids. Cortina et al identify a polytomy for Rosidae comprising Fabidae, Myrtales, Crossosomatales, Geraniales, and Malvidae. The APG identifies a Malvidae-Fabidae dichotomy with Vitales and Saxifragales as outgroups.

The tropical, Oriental Sabiaceae Blume 1851 (sabia, 3 genera: *Sabia, Meliosma, Ophiocaryon*) in genotypic analyses are between Ranunculales and Proteales, but sometimes group with Proteaceae. In Nandi et al's phenotypic tree they are with Santalales and the giant rhubarb family in Asteridae. They are not included in Hufford. Lawrence stated phylogenists are agreed the family is closely allied to Sapindaceae, although later, Cronquist placed it in Ranunculales.

Vitaceae Jussieu 1789 (in Brummit; Lindley 1836 in Lawrence)(grape, 14 genera) (with Leeaceae as Vitales but *Leea* was included as a genus in Vitaceae), was assigned to Celastrales by Bentham and Hooker and is placed in Santaliflorae by Dahlgren, Cornales by Thorne, and Vitanae just below Cornanae in Rosidae by Takhtajan. Like Berberidopsales, the family appears outside of Rosidae in both phenotypic and genotypic analyses. It was usually assigned to Rhamnales based mainly on obhaplostemonous flowers (having twice as many stamens as petals, those of the outer set being opposite the petals), but this does not stand up to scrutiny (Dahlgren, 1983).

Molecular analyses recover a clade including Rosales, Fabales, Cucurbitales, Rhamnales, and Fagales (core Amentiferae), and Urticales, which all have symbioses with nitrogen-fixing bacteria (*Rhizobium* in Leguminosaae and *Frankia* in the others) in the root nodules. There are 10 families which have such a symbiosis: Fabaceae; Betulaceae, Casuarinaceae, and Myricaceae in Fagales; Coriariaceae and Datiscaceae in Cucurbitales; Rhamnaceae and Eleagnaceae in Rhamnales; Cannabaceae in Urticales; and Rosaceae. However, the trait has evolved multiple times and various morphological features and modes of infection support this idea (Time Tree, citing 2 articles and a book), so it is doubtful they all form a monophyletic unit.

Rhamnaceae Jussieu 1789 were considered as having affinities to Rosales by Wettstein, Hallier, and Rendle (Lawrence, 1960). Bessey placed them in Celastrales, Hutchinson considered Rhamnales as sister group of Sapindales, Thorne and Dahlgren assigned them to Malviflorae, and Takhtajan regards them as derivatives of Saxifragales (Dahlgren, 1983).

Cucurbitaceae Jussieu 1789 were assigned to Loasales but derived from Rosales by Bessey, but were usually placed in Violales and are in Violiflorae in Nandi et al. Cucurbitaceae, Begoniaceae, and Datiscaceae are certainly members, Coriariaceae+Corynocarpaceae and Tetramelaceae are probably members, and Anisophylleaceae and Apodanthaceae are probably not. Morphological synapomorphies for the molecular Cucurbitales can't be identified probably because it is an artificial taxon. For the core of the order (that is, excluding Apodanthaceae and Anisophylleaceae), Nandi et al list absence of mucilage cells and fasciculate or stellate hairs, and presence of libriform fibers, storied wood structure, and a slightly oblique vessel end wall angle. A morphological study of the flowers of 7 of the 8 families (that is, excepting Apodanthaceae) by Matthews & Endress in 2004 could not identify a synapomorphy for the order but could for its subclades. (Shaefer & Renner, 2011).

Floral structure in some Anisophylleaceae and Cunoniaceae is exceedingly similar. In contrast, in molecular trees, Anisophylleaceae and Cunoniaceae appear far apart, Anisophylleaceae in Cucurbitales and Cunoniaceae in Oxalidales. From preliminary results of a comparative morphological study, Anisophylleaceae would fit better with Oxalidales than with Cucurbitales. Floral fossils with structural affinities to both Anisophylleaceae and Cunoniaceae, recently described from the Late Cretaceous of Sweden by Schonenberger et al in 2001 further emphasize a potential close kinship between the 2 families. Currently, Anisophylleaceae are molecularly insufficiently known; only rbcL was studied in 2 species. (Endress, 2002)

Leguminosae Jussieu 1789 were placed in Rosales by Engler, in the order Aesculinae by Hallier, in Rosales by Bessey, as the order Leguminosae derived from Rosales by Hutchinson, in Rutiflorae by Thorne, Fabiflorae by Dahlgren, Fabanae in Rosidae by Takhtajan, and Rosidae by Cronquist. They branch with Sapindales and Rhamnales as sister group to Myrtales in Hufford and with Rutales in a larger group in Nandi et al's classical analysis.

Polygalaceae R. Brown 1814 (milkwort, 20 genera) were about half the time placed in Geraniales/Geranianae/Gerianiflorae and about half the time in Rutales/Rutanae/Rutiflorae. The family occurs with Eleagnaceae-Chrysobalanaceae as an independent clade in Nandi et al, and in Fabales in molecular analyses. It bears a resemblance to the flower of Fabaceae (Leguminosae), but which is said to be superficial as they differ in details, and Fabaceae and Polygalaceae have green embryos, which might be a key characteristic (Judd & Olmstead).

Zygophyllaceae R. Brown 1814 (caltrop, caper bean, 27 gen.) was usually placed in Geraniales but exceptionally in Rutales/Sapindales by Takhtajan and Cronquist. Hallier derived Leguminosae from it and Krameriaceae+Zygophyllaceae are sister group to Fabaceae in 2 molecular studies (Hilu et al, 2003, cited in Time Tree) and Nandi et al. Krameriaceae (rattany or krameria, 1 genus) are usually assigned to Polygalales but positioned with Zygophyllaceae by the APG. The possible connection between Polygalales and Fabaceae, Zygophyllaceae and Fabaceae, Zygophyllaceae and Krameriaceae, and Krameriaceae and Fabaceae (the genus was often considered a tribe in Cesalpinoideae of the legume family (Lawrence) and has papilionoid-looking flowers (APW)) probably indicate an evolutionary kinship between the 4. *Krameria* is a genus of tropical American, perennial shrubs which act as root parasites. The root is used medicinally and, especially in Portugal, to colour wines ruby red (Wikipedia).

Hamamelaceae (usually Hamamelidaceae R. Brown 1818)(witch-hazel [*noisetier des sorcières*], 30 genera) were positioned in Rosales by Bentham and Hooker, Engler, and Bessey; as an order (including also Coriariaceae) derived from Magnoliaceae by Hallier; as an order in a branch containing also Salicales, Myricales, Fagales, and Urticales by Hutchinson; in superorder Hamameliflorae by Thorne; as an order in Rosiflorae by Dahlgren; in subclass Hamamelidae by Takhtajan; as an order in Rosidae by Cronquist; and in Saxifragales by the APG. They branch with *Platanus* as an independent clade between Buxales s.l. and *Peonia* in Hufford and with Rosaceae + Dilleniaceae in Nandi et al.

All Southern Hemisphere (Australia, Madagascar, and Africa) genera in the family are characterized by a unique feature of the anthers. Although they have 4 pollen sacs, organized into 2 thecas, each theca opens by a single valve, which is hinged on the theca's ventral side. The combination of such a unique feature and the exclusive occurrence of these genera in the Southern Hemisphere indicate they are an evolutionary unit, which was separated from the northern part of the family sometime in the Cretaceous. This was later supported by molecular studies. (Endress, 2002)

The witch-hazel extract is prepared from *Hamamelis* Linneus 1753 bark and used as a liniment, and witch-hazel, sweet gum or storax (*Liquidambar*), and winter hazel (*Corylopsis*) are grown for ornament (Lawrence, loc. cit.). The common name apparently comes from a Middle English word meaning "pliant" or "bendable" instead of "sorceress" (Etymology On-Line; Webster's 9th New Collegiate Dictionary), because of its flexible branches used by Amerinds for making bows and by others to make divining rods for dowsing (L'hamamélis-pharmelia.com), the French, *noisetier des sorcières*, then, coming from mistaking the English for "sorceress." There is coincidentally a similarity between the French *sourcière* (female dowser) and *sorcière*. The "hazel", of course, comes from a resemblance to hazel (*Corylus*) of the birch family. The generic name apparently comes from the Greek for "with fruit" (L'hamamélis-pharmelia.com) or "to bring fruit" (indigobeauty.com).

Saxifragales appear as an early-branching independent clade, or between Asteridae and Rosidae (in Soltis and Soltis's summary tree), or as sister group to Rosidae in molecular analyses but otherwise are almost always placed in Rosales or, with minor families added, as Saxifragales in Rosanae or Rosiflorae, and Lawrence says it is difficult to separate them from related families especially Rosaceae. In Hufford, a part of them is sister group to Fagales and another to Rosaceae Jussieu 1789. In Nandi et al they are sister group to Asteridae, with Rosidae on top. Saxifrage, mock orange (*Philadelphus*), and coral bells (*Heuchera*) are important ornamentals in the family, and the golden saxifrage (*Chrysosplenium*) is edible. Grossulariaceae (currants and gooseberries) and Crassulariaceae (stonecrop) are closely related. *Escallonia*, *Hydrangea*, and *Deutzia* were removed to Asteridae.

The amentiferans (Lat. amentus, "catkin") were designated as a single order by Eichler in 1883, considered as several separate orders in Archichlamydeae by Engler, and accepted as a phyletically homogeneous but primitive unit by Rendle, as the same by Hallier but advanced (as family Amentaceae in Terebinthales, which was essentially Rutales without Sapindaceae), and Hutchinson placed them with Urticales as descendants of the witch-hazels. They were placed in Unisexuales by Bentham and Hooker, except for the willow family, which they assigned to Ordines anomales. They were restudied by Helmquist in 1948 who considered them a natural assemblage of several orders he called Amentiferae: Salicales (willow, 2 gen.), Garryales (garrya, 1 gen.), Myricales (sweet gale, 3 gen.),

Leitnerales (corkwood, 1 gen.), Batidaceae (batis, 1 gen.), Balanopsidales (balanops, 1 gen.), Juglandales (walnut, 8 gen.), and Fagales (beech and birch, 10 gen.). All had only 1 family except for Fagales, which had 2.

Lawrence adds, however, that evidence accumulated which indicated it was neither natural nor primitive. It is not natural at least in the large sense, but a core of it may be (Juglandaceae, Myricaceae, Fagaceae, Betulaceae) and none are primitive. Garryales has only a catkin-like inflorescence and has been variously placed in Umbelliferae and Cornales. Salicaceae has been placed in Caryophyllales by Bessey, the higher Dillenidae by Cronquist and Takhtajan (in the superorder Violanae), in Violales in superorder Violiflorae by Thorne and Dahlgren, and as a separate clade in Nandi and colleagues. Leitneraceae Bentham 1880 have been placed in Ranales by Bessey, as an order in Archichlamydeae by Hutchinson, in Rutiflorae by Thorne and Dahlgren, in Rutanae by Takhtajan, by Cronquist as an order in Hamamelidae, and in Sapindales in molecular analyses. It was not included in Hufford nor Nandi et al. Its position is firmly established in Sapindales-Rutales. *Balanops* was placed in Unisexulaes by Bentham and Hooker, as an order in Archichlamydeae in the Englerian system, in Buxineae in Pittosporales in Rosiflorae by Thorne, as an order in Rosiflorae by Dahlgren, as an order in Hamamelanae by Takhtajan, and in Fagales in Hamamelidae by Cronquist. Batidaceae have been assigned to Rutiflorae or Violiflorae/Violanae. At any rate, Amentiferae remains a good convenience category.

Many molecular analyses do corroborate the group in part (Doyle, 1998; Judd and Olmstead, 2004), as well as Hufford and Nandi and colleagues, but including Urticales (nettle, elm, and mulberry families), which has catkin-like inflorescences (in elms), and sometines also Casuarinales (cassowary, Australian pine, 4 gen.). Urticales were said by Core to be considered by taxonomists as related to Fagales. In Hutchinson they were in the same branch as Hamamelales, Salicales, Myricales, and Fagales, in Hallier they were a family in Terebinthinae, and Thorne and Dahlgren saw them as an order in Malviflorae, Takhtajan as an order in Urticanae in Dillenidae, and Cronquist regarded them as an order in Hamamelidae. Casuarinaceae were in Unisexuales in Bentham and Hooker, were an order in Melchior in Archichlamydeae, an order in Hamameliflorae, Hamamelanae, and Hamamelidae in Thorne, Takhtajan, and Cronquist, respectively, and in Rosiflorae in Dahlgren.

Wind pollination is a primtive trait and the conifers are all wind-pollinated; insect pollination is more efficient and is advanced. Aside from the amentiferans, the former occurs in ragweed (*Ambrosia*, Compositae), salad burnet (*Poterium sanguisorba* in Rosaceae); in many basal eudicots such as the box family, Eupteleacae, the moonseed family (Menispermaceae), the plane-tree or plantain family (Platanaceae), meadow rues (in Ranunculaceae), Papaveraceae, Tetracentraceae, Trochodendraceae, and Gunneraceae (Qiu et al); in Piperales in the paleoherb group; in the monocots there is the largest group, Poales (the grasses, sedges, and rushes), and also Flagellaraceae (whip vine or bush cane), and Typhaceae (cattail); and it occurs as well in Phytolaccaceae (inkberry, in Caryophyllales), Anacardaceae (cashew, in Sapindales), and Scrophularaceae (foxglove, in Asteridae)(Thorne, 1976). Wind-pollinated flowers have vast amounts of pollen on large anthers. Those pollinated by insects have much less on smaller anthers, and the flowers are showy and scented to attract the insects, while the wind-pollinated ones are unscented, lack nectar, and their corollas are reduced and green or absent. Beetle pollination is believed to be the primitive form in insect pollination--cycads are both wind and beetle-pollinated. Most of the pollinators start with a "b": bees, beetles, butterflies, bats, and birds.

Night bloomers are another special group occuring in various related and unrelated families. There are about 40 genera in 21 families, especially in the cactus familiy (8 gen.) and Asteridae (12 gen.). Particularly well-known are the moonflower (*Ipomea*) in the morning glory family, evening primrose (*Oenothera*)(probably containing the most such species with about 20) in the evening primrose family, and night-blooming jasmine (*Cestrum*) in the nightshade family. They bloom at night to attract night-flying pollinators, especially moths, most of which are nocturnal, tend to be fragrant as the insects must depend on smell, and make for fascinating night gardens. The nightshade family (Solanaceae Jussieu 1789) contains such flowers in 4 genera, but the common name may not refer to nocturnal flowering; the origin is unknown but might refer to the black berries (Wordnik.Com) or the deadly effects of many of its fruits (On-Line Etymological Dictionary). The origin of the generic name *Solanum* is also unknown but may come from a perceived resemblance of certain solanaceous flowers to the sun and its rays--at least one species of *Solanum* is known as sunberry--but, alternatively, it may come from the Latin verb solari, meaning "to soothe", presumably in reference to the soothing pharmacological properties of some of the psychoactive species of the family.

Rafflesiaceae Dumortier 1829 (monster flower, so named because it can be as wide as a yard and can weigh up to 20 pounds; it is the biggest flower known; 9 genera), primarily tropical root and stem parasites, are variously placed in Myrtales by Bessey, Aristochiales by Hallier, their own superorder by Thorne, in superorder Magnoliflorae by Dahlgren, as a superorder in Magnolidae by Takhtajan, in Rosidae by Cronquist, close to Rutales in Nandi et al, and in Malpighiales as sister group to Euphorbiaceae in molecular analyses. The affinities for this family are undetermined.

Malpighiales is a large, heterogeneous order of about 30-40 families, and, although it has strong statistical confidence, it has no phenotypic synapomorphies. There are apparently 3 strongly supported clades that make up the order (APW). Several families assigned to it are poorly known (APW) so it might be that a special similarity is waiting to be discovered, but it could also be an artifact. In any case, some (about a quarter) of it correlates with Hallier's Passionales, which contained the passionflower, willow, and spurge families, and Achariaceae, Flacourtiaceae, Malesherbiaceae, Turneraceae, and Papayaceae (Caricaceae), all of which also occur in the molecular Malpighiales (except for Papayaceae), and his Polygalineae, which contained 10 families, 4 of which, the malpighia and violet family, and Dichapetalaceae, and Trigoniaceae, also occur in the molecular Malpighiales, with both groups considered derived from Linaceae by Hallier, and also partly correlate with Hutchinson's Malpighiales-Euphorbiales (in part) branch. So the most advanced suborder or superfamily is basically the traditional Violales, as 7 (Achariaceae, Violaceae, Lacistemataceae, Scyphostegiaceae, Turneraceae, Malesherbiaceae, and Passifloraceae) of the order's 10 or 11 families, along with Salicaceae, which has usually been assigned as a related order or suborder, are in this malpighian suborder, so that 8 of the 10 families of this suborder are Violales. The Flacourtiaceae family has proven to be polyphyletic as the cyanogenic members have been placed in Achariaceae and the ones with salicoid teeth were transfered to Salicaceae (Judd and Olmstead, 2004).

As circumscribed by Lawrence, the traditional Geraniales order consisted of several suborders, namely Geranineae (notably including Linaceae, Oxalidaceae, Rutaceae, Zygophyllaceae, besides Geraniaceae), Malpighineae, Polygalineae, and Tricocceae (Euphorbiaceae and Daphniphyllaceae, the latter family molecularly placed in Saxifragales), Dichapetalineae, and Callitrichineae. Ctenolophonaceae, Humiriaceae, Ixonanthaceae, and Limnanthaceae were sometimes separated from Linaceae. Balsaminaceae and Tropeolaceae were also included by some. Lawrence relates that it was considered several orders by several authors, the accumulation of evidence from all fields of botany indicates it is not a natural taxon, and that it may yet be accepted as comprising 4 or 5 more or less disjunctive orders--indeed, there are 8 unrelated lines: geranian (1 family), rutacean (1 family), polygaliacean, zygophyllacean, callitrichacean, malpighian (containing the rest), balsamine, and limnanthian.

There are 10 families of the old Geraniales that appear in the new Malpighiales: Linaceae, Erythroxylaceae, Achariaceae, Malpighiaceae, Trigonaceae, Dichapetalaceae, Euphorbiaceae, Ctenolophonaceae, Humiriaceae, and Ixonanthaceae, the last 3 being included in Linaceae by earlier workers but separated out by later ones. About half the families in the molecular Malpighiales were in the traditional Geraniales or Violales. The modern Geraniales (1 or 2 families) is an independent clade in phenotypic and some molecular phylogenies but in some of the latter is sister group to Myrtales. Melianthaceae were always placed in Rutales/Sapindales but the APG has placed them in Geraniales.

The other most polyphyletic major order of traditional taxonomies is Parietales, which had 10 suborders and from which many of the Malpighiales families come, namely Medusagynaceae, Ochnaceae, Hypericaceae (Guttiferae, Clusiceae), Caryocaraceae, Elatinaceae, and the Flacourtineae suborder, which was really Violales. The 1st 3 families make up what are called the clusioids.

Euphorbiaceae Jussieu 1789 have been placed in Unisexuales in Monochlamydeae by Bentham and Hooker, in Geraniales in the Englerian system and by Bessy, in Passionales by Hallier, as an order derived from Malpighiales by Hutchinson, as an order in Malviflorae by Thorne and Dahlgren, as a superorder in Dillenidae by Takhtajan, as an order in Rosidae by Cronquist, and in Malpighiales by the APG, with the phyllanthoid subfamily placed as a separate suborder. This large family probably belongs in the malpighian line.

The Chrysobalanaceae R. Brown 1818 (coco plum) family was usually assigned to Rosales or as an order in Rosiflorae, but exceptionally as an order in Myrtiflorae by Dahlgren, who stated it was out of place in Rosales, deviating from them in the zygomorphic (irregular) flowers, erect ovules, paracytic (relating to cells in a location where they are not usually present) stomas, the presence of silica and foliar sclereids (a sclereid is type of sclerenchyma, which is the supporting tissue in plants; the other type is fiber), the syncarpous pistil (with united carpels, which are foliar units of a compound pistil or ovary) with a common, gynobasic style (one that is at the base of the ovary), and by differences in pollen grains and in certain anatomical features. He also stated it was of uncertain affinities. It branches with Eleagnaceae (oleaster) as sister group to Polygalaceae forming with them an independent clade in Nandi and colleagues and is in Malpighiales in molecular and combined analyses and in other molecular analyses, but the group it is in in that order has no unique synapomorphies so its shared derived features occur elsewhere and are homoplasious, so its position is uncertain.

Podostemaceae Richard ex C. Agardh 1822 (riverweed), 50 genera of tropical lithophytes, were usually positioned in Rosales or as an order in Rosanae/Rosiflorae but assigned to Multiovulate aquaticae by Bentham and Hooker, to Caryophyllales by Bessey, to Ranales by Hallier, placed as a separate superorder by Dahlgren, in Malpighiales in molecular schemes, where they are placed in the clusioid suborder, Podostemonaceae and Hypericaceae, 2 of the 5 families in it, having secretory canals and xanthone pigments (Judd and Olmstead, 2004), and showed up in Caryophyllales in Nandi and colleagues, and were not included in Hufford. The Clusioid group (Guttiferae) was usually placed in Theales. It is incertae sedis.

Balanopsaceae Bentham 1880 (balanops) are trees that occur in Queensland, New Caledonia, and Fiji. They have been variously placed in Unisexuales in Monochlamydeae by Bentham and Hooker, as an order in Archichlamydeae in the Englerian system, in Buxineae (in Pittosporales in Rosiflorae) by Thorne, as an order in Rosiflorae by Dahlgren, as an order in the witch-hazel superorder by Takhtajan, in Fagales in Rosidae by Cronquist, and in Malpighiales in the APG. They were not included as a family by Bessey and Hallier and were not included in Hufford nor Nandi et al. Recent investigations have discovered zeylanol uniquely in both *Balanops* and *Dichapetalum* (Darbah et al, 2012), which are in the balanops clade in Malpighiales. But they are incertae sedis and the monophyly of the balanops clade as a whole has not been confirmed.

Rhizophoraceae R. Brown 1814 (mangrove) are a tropical family of 14 genera some of which, including the type genus *Rhizophora* (=*Mangium*), are famous for mangove swamps in the Florida everglades. Their extensive root systems protect the Gulf of Mexico coast from erosion and storm damage. The family was usually placed in Myrtales and otherwise in Cornales by Thorne and as an order in Rosidae by Cronquist. In genotypic analyses it is in Malpighiales. The affinities of this family have not been ascertained.

Balsaminaceae A. Rich. 1822 (balsam, 2 genera) contain the impatiens, garden balsam, and Himalayan balsam, which are grown as ornamentals. Not to be confused with the Canada balsam (*Abies*) in the pine family nor the balsam apple (*Momordica*) in the cucumber family. The family is traditionally placed either in Geraniales or Rutales/Sapindales and in Ericales by the APG. Its position is uncertain.

Limnanthaceae R. Brown 1833 (meadow foam, 2 gen.) was often placed in Geraniales, but in Sapindales by Melchior, with Tropeolaceae (nasturtium, 3 gen.) as Tropeolales in Rutiflorae by Dahlgren, as an order in Rutanae by Takhtajan, and in Brassicales by the APG. Tropeolaceae have an oil similar to that of *Nasturtium officinale* (watercress) (which is in Cruciferae in Brassicales), hence the name.

Oxalidaceae R. Brown 1818 (oxalis or wood sorrel, 6 gen.) have always been traditionally placed in the geranium order and considered as closely related to the geranium family. Connaraceae (zebra wood, 12 gen.) was included in Rosales by Bentham and Hooker, Engler (Melchior in Leguminosineae), Bessy, and Cronquist; Fabineae (in Rutales) by Thorne; Sapindales by Hutchinson and (in Rutiflorae) by Dahlgren; and as an order in Rutanae (in Rosidae) by Takhtajan. It branches with Geraniales in Hufford and with Oxalidaceae in Nandi et al. It is strongly supported in both morphology and molecules as sister group to Oxalaceae (APW). Along with several other families, the 2 together are usually an independent clade or weakly supported as sister group to another clade in genotypic trees. Their kinship to the other families in the molecular Oxalidales is weak (APW). Those other families are Cunoniaceae, Brunelliaceae, Cephalotaceae (in Saxifragales), Eleocarpaceae (in Malvales), and Huaceae (a relatively new family established in '47 usually assigned to Malvales). Cunoniales are generally traditionally circumscribed as containing Cunoniaceae, Brunelliaceae, Davidsoniaceae, Eucryphiaceae, and Baueraceae. The last 3 are all placed in the 1st by the APG, so the phenotype and the genotype mostly concur on the unity of Cunoniaceae/Cunoniales but not on the placement; Cunoniales do not belong with Oxalidales, they belong in Rosiflorae, Eleocarpaceae and Huaceae belong in Malvales, and the Western Australian insectivorous, monotypic pitcher herb family Cephalotaceae (cephalotus) probably belong in Saxifragales.

Huerteales is a newly-concocted order comprising 6 genera in 4 families taken from Celastrales, Sapindales, Malvales, and Violales that, like the molecularly defined Malpighiales, Saxifragales, and Celastrales, appears to be artificial, being heterogeneous and having no phenotypic traits in common. It is placed between Sapindales and Malvales or just below them.

Sapindales and Rutales were traditionally placed together most of the time and show up as such also in molecular treatments. Rhamnales, Fabaceae, and Sapindales form the sister group of Myrtales in Hufford, with Rutales not included, and Sapindales are in a large group containing Rutales in Nandi et al.

The following 16 tricolpate taxons are of uncertain position:

Aextoxicon
Balanopaceae
Caryophyllidae
Cercidiphyllum
Chrysobalanaceae
Clusineae
Daphniphyllaceae
Euptelea
Hamamelaceae

Haloragaceae
Rafflesiales
Platanus
Podostemonaceae
Rhizophoraceae
Sabiaceae
Saxifragales

The phylogeny for angiosperms, like for any large group, is complex and largely unresolved, so it is important to use also a convenience classification, which has 42 orders, and which is as follows:

Table 12-26. Convenience Classification for Angiosperms.

- Apetela (11 orders, Amentiferae, Santalales, Trochodendrales, etc.)
- Tripetala
  - Magnolidae (1 or more orders)
  - Monocots (9 orders)
    - Calyciflorae
    - Corolliflorae
    - Glumiflorae
- Dipetala
  - Proteales
  - Trochodendrales
  - Buxales
  - Ranunculales
  - Gunnerales
- Pentapetala (20 ords.)
  - Dialypetalae
    - Caryophylliflorae (1 order)
    - Malviflorae (Dilleniales, Theales, Violales, Malvales)
    - Disciflorae (Geraniales, Celastrales, Rutales (Sapindales))
    - Calyciflorae (Rosales, Myrtales, Cornales, Umbelliflorae)
  - Sympetalae
    - Heteromae (Ericales, Primulales, Ebenales)
    - Bicarpelatae (Gentianales, Lamiales)
    - Inferae (Rubiales, Campanulales, Asterales)

Table 12-27 gives statistics for the largest families—the ones with 150 or more genera and 3000 or more species, the average being about 20 species per genus, since there are about a quarter million species and 13,500 genera (according to Brummitt, but there may be more like 12,000), making a Top 17. If we use the limit of at least 100 genera or at least 2000 species, 14 more families can be added, which are, in descending order, Rosaceae, Ericaceae (heath, *bruyère*), Boraginaceae (borage, *bourrache*), Gesneraceae (gesneria or African violet, *violette africaine*), Sapindaceae (soapberry, *saponière*), Cactaceae (cactus, *cactus*), Annonaceae (custard apple, *anone*), Rutaceae (rue, *rue*), Araceae (arum, *arum*), Amaranthaceae (amaranth, *amaranthe*), Bignonaceae (trumpet flower, *fleur en trompette*), Bromeliadaceae (pineapple, *ananas*), Cucurbitaceae (cucumber, *concombre*), and Verbenaceae (verbena or vervain, *verveine*). As can be seen there are discrepant estimates, some varying considerably.

Asclepiadaceae R. Brown 1810 (milkweed, *laiteron*) have c. 100 genera and c. 2000 species but are placed within Apocynaceae Jussieu 1789 (dogbane, *apocyn, laiteron*), which have milky sap like milkweed, as the 2 families are most closely related, and Hallier placed them with/in Apocynaceae (Lawrence), and this is corroborarted by the genotype and the former is a subfamily in the latter (APW). The milkweed family had 75-315 genera and 1700-2000 species, and the dogbane family s. s. had 180-300 genera and 1300-1500 species. The totals for Apocynaceae in the table by Core, Caratini, Jones & Luchsinger, Kew, and the U. of Hawaii's Botany Dept. are combinations, as these families in the wide sense were not recognized by them.

Malvaceae Jussieu 1789 were always considered closely related to Sterculiaceae (sterculia or karaya tree, *arbre de karaya*), Bombacaceae (bombax or cotton tree, *bombax, arbre de coton*), and Tiliaceae (linden tree, *tilleul*), and this is confirmed by cladistic analysis of the phenotype (Judd & Manchester, 1997), but the latter 3 are found to be paraphyletic; the genotype also confirms this (APW). The former (s.s.) mallow family had about 80-120 genera and 1500 species. The totals for Malvaceae in the table by Core, Caratini, and Kew are combinations as these families in the wide sense were not recognized by them.

As usually traditionally circumscribed, Liliaceae Jussieu 1789 (lily, *fleur-de-lis*) had 240 genera and 4000 species (Core), but the family was polyphyletic. Of the 12 subfamilies and 35 tribes recognized by Krause in 1930, the usuallly accepted circumscription, Hutchinson, in 1934, segregated out 4 subfamilies, 2 tribes, and 2 subtribes as 8 separate families, and 3 tribes he removed to Amaryllidaceae Jussieu 1789. It was further fissioned by Dahlgren and Takhatajan, who separated some 25-30 families, and Brummitt recognizes only 11 genera for Liliaceae s.s.. Molecular analysis has corroborated the polyphyly but with somewhat different families. Most of the genera go to Asparagaceae; the APG recognizes 7 subfamilies, all of which were previously regarded as families.

Scrophulariaceae Jussieu 1789 (foxglove, *gant-de-notre-dame*) are polyphyletic also, as it was through the absence of traits characteristic of related families that plants were generally assigned to the foxglove family, and the genotype supports this view (Olmstead et al, 2001). The traditional Scrophulariaceae had c. 200-300 genera and c. 3000 species.

For the rose family, the generic total ranges from 95-122, with the average at 104. Lawrence states that the generic limits in the family cannot be sharply drawn, and that some authors regard *Malus, Pyrus, Sorbus*, and *Aronia* as one genus, *Pyrus* Linneus 1753. The total is probably around 60 as many of the genera are likely more properly placed in other genera, which is done by many authors. The number of species is about 2500-3400.

The cactus family estimates are especially discrepant, 25-300 genera (but Schumann acknowledged 21 in 1899; 26 were acknowledged by Parish in 1936 [Lawrence]), the latter in the Penguin, with the average at 125.5 (counting the figures in Lawrence it is 112). The number of species is estimated at 1000-2000.

For Aizoaceae F. Rudolphi 1830 (mesembryanthemum or ice plant, *mésembryanthème*) the range is 23-134 and 600-1900, with an average estimate of 98 genera.

The pepper family (Piperaceae C. Agardh 1824) estimates are 5-15 and 1300-3600, with the average for the latter at 1946.

At the opposite extreme there are 61 families with only 1 species (compiled from Caratini), though some of them are considered to have 2 or 3 species by some botanists.

Table 12-27. The Largest Bloom-Bearing Families (blanks mean the source did not include the information).

	Core	J & L	Caratini	Penguin Dict.	SCLO
	Brummitt	Britannica.Com	Botany.Hawaii.Edu	Plant List	
Asteraceae	aster or daisy	marguerite	Asterales		
	950; 20,000	1000; 20,000	900; 22,000	1100; 25,000	1100; 20,000
	1509	1620; 23,600	1100; 20,000	1765;27,773	
Orchidaceae	orchid	orchidée	Orchidales		
	450; 15,000	500; 20,000	735; 17,000	750; 18,000	1000;15-20,000
	835		1000; 15-20,000	925; 27,135	
Leguminosae	legume	fève	Fabales		
	550; 13,000	500; 14,000	590; 13,200	700; 17,000	600; 14,200
	678	700; 20,000	400; 10,000	917; 23,535	
Gramineae	wheat or grass	blé	Poales		
	500; 4500	450; 6000	620; 10,000	620; 9000	500; 8000
	657	800; 10,000	500; 8000	777; 11,461	
Rubiaceae	madder	garance	Rubiales		
	400; 5000	500; 6000	500; 6000	500; 7000	450; 6500
	606	660; 11,000	450; 6500	617; 13,548	
Euphorbiaceae	spurge	euphorbe			
	238; 7300	300; 8-10,000	290; 7500	300; 5000	300; 7500
	331	275; 7500	300; 7500	229; 6511	
Labiatae	mint	menthe	Lamiales		
	200; 3200	200; 4000	180; 3500	200; 3000	
	212	236; 7000	200; 3200	250; 7852	
Apocynaceae s.l.	dogbane	apocyn	Asteridae		
	400; 3000	500; 3300	310; 3500		200; 2000
	483	415; 4600	450; 4000	402; 5031	
Melastmomaceae	melastoma	mélastome	Myrtales		
	150; 4000	150; 4000			200; 4500
	194	182; 4500	200; 4000		
Cyperceae	sedge	carex	Poales		
	85; 3000	75; 4000	90; 4000	90; 4000	70; 4000
	102	102-123; 4-5000	70; 4000		
Cruciferae	mustard	moutarde	Brassicales		
	350; 2500	350; 4000	375; 3200	375; 3000	350; 3000
	381		338; 3710	350; 3000	
Malvaceae s.l.	mallow	mauve	Malvales		
	162; 2515		192; 2150		220; 3150
	269	243; 4225	210; 3000	236; 3704	
Acanthaceae	acanthus	acanthe	Asteridae		
	240; 2200	250; 2500			250; 2500
	228	220; 4000	250; 2500	225; 2894	
Umbelliferae	carrot	carotte	Apiales		
	125; 2900	150; 3000	275; 2850	300; 2500-3000	
	428	3-400	300; 3000	347; 2786	
Palmae	palm	palmier	Palmales		
	200; 1500	200; 4000	215; 2500	210; 2780	200; 3000

202			200; 3000	187; 2466	
Myrtaceae	myrtle	myrte	Myrtales		
80; 3000	100; 3500				140; 3000
127	150; 3300		140; 3000	144; 5774	
Asparagaceae	asparagus	asperge	Asparagales		
				143; 3632	

Economic Importance

Economically, the most important families are listed below and are the Top 16 based on the number of genera having the most economically important species (only the major ones of these)(1st column), the number of types of products or uses (2nd column), and the number of food genera (3rd column); the 4th column is the score.

Table 12-28. Top 16 Most Economically Important Families.

1. wheat	blé	30	6	8	44
2. legume	fève	24	6	11	41
3. mallow	mauve	28	6	2	36
4. rose	rose	22	4	8	34
5. aster	marguerite	21	3	5	29
6. mint	menthe	25	3	0	28
7. nightshade	belle-de-nuit	19	3	4	26
8. palm	palmier	16	4	5	25
9. carrot	carotte	16	4	4	24
10. asparagus	asperge	16	4	1	21
11. arum	arum	16	2	3	21
12. myrtle	myrte	13	4	4	21
13. mustard	moutarde	12	3	3	18
14. cashew	acajou	9	4	5	18
15. cucumber	concombre	12	2	3	17
16. madder	garance	14	3	0	17

The number of food genera is factored in as the food category is probably most important. A more definitive assessment would require comparative data on production for each use in each family, which sometimes varies from source to source and/or year to year, and possibly also the number of economically important species instead of genera. However, the criteria used here certainly give a good indication. And certainly no one or few would dispute the wheat, legume, and rose families as the most important economically, and the mallow family s.l. is new. All of the 16 are included in Bennett's Top 25 except for aster and asparagus. He also includes the yam, morning glory (for sweet potato), and banana families as these are top 10 crops but also the grape family, since wine is an important world product, as well as the rue family because of the importance of citrus fruits. But these families have only 1 or a few economically important genera, except for the rue family, which has a score of 15, tied with the pineapple and heath families just below madder and just above the orchid, sapodilla, mulberry (which he also includes), and olive families tied at 14. Bennett includes the pine family as well, as his list is for seed plants, which is apparently not based on any statistics.

The French names are also given, and it should be noted concerning the pronunciation that there is the bizarre practice of falsely representing sounds for the short "a" and short "i" in phonetic transcriptions for French (and Portuguese). The "a" at the beginning or in the middle of a word, when without a circumflex and unnasalized and not followed by another vowel, as a rule, as much in Europe as in Canada and the rest of the world, is pronounced like the short a in the English "pan", which is represented in the IPA system by the ash symbol (æ). This applies to all the a's above, that is, in Table 12-25. And Cannes, Galles, and "panne," for example, are pronounced like the English "can", "gal", and "pan" in the North American accent, the short "a" in other words.

Foliar Theory

The foliar theory (plantcell.org-Morphogenesis of Flowers-Our Evolving View; botany.wisc.edu) was first published in 1790 by the German poet and philosopher Johann Wolfgang von Goethe in an influential, extended essay titled “Versuch die Metamorphose der Pflanzen zu erklären” (An attempt to explain the metamorphosis of plants). The idea of metamorphosis had been proposed much earlier by Cesalpino and further developed by Linneus. In Linneus' words, “the flower [can be regarded] as the interior portions of the plant which emerge from the bursting rind; the calyx as a thicker portion of the shoot; the corolla as an inner and thinner rind; the stamens as the interior fibers of the wood, and the pistil as the pith of the plant”. Based on his observations of variations in normal plant growth and abnormalities that sometimes occur, Goethe provided an alternative view of metamorphosis: “The same organ which on the stem expands itself as a leaf, and assumes a great variety of forms, then contracts in a calyx—expands again in the corolla—contracts in the reproductive organs—and for the last time expands in the fruit.” Goethe named the generalized organ “Blatt” (leaf) and thought of it as a generalized plant organ rather than a leaf, leaves being just one of the forms it could adopt. This simple and attractive proposal has influenced the thinking of plant scientists until the present day and forms the basis for the current consensus that all parts of the flower are modified leaves: sepals, petals, stamens, pistils are believed to have all evolved from the same basic plan.

Floral Evolution

There are 2 competing theories on floral evolution. The pseudanthial theory states that flowers are the result of integration of unisexual structures similar to *Ephedra* (herba.msu.ru-Systematic Botany-Lecture 33-Minot State U.) and the pseudanthium of *Euphorbia*--the transformation of male and female "flowers" of Gnetales along with the subtending leaves into the floral organ (Cornelia Löhne, A Case Study of Nympheales (dn-b.info)). It is described by Qiang (Early Cretaceous *Archefuctus*- frostburg.edu) as having a bisexual axis with bracts and attached secondary reproductive axes. It is said to have a branched axis by Sarah Mathews and Elena Kramer (Evolution of reproductive structures in seed plants-dash.harvard.edu). It emphasized angiosperms as derived from gymnosperms. Proposed by Eichler in 1876 and Engler in 1897 (Steussy, 2004) and Wettstein in 1924, and Zimmerman in 1930 (Lohne, loc. cit.).

Euanthial theory (Lohne, loc. cit.)(also called the strobilar or foliar theory) holds to an origin from more complicated bisexual generative shoots, flowers arose de novo, and are derived from a large, compound bisexual strobiloid, flower-like organ with enrolled megasporophylls, and stamens and perianth, organs similar to cycadeoids. It is decribed by Qiang as having a simple bisexual strobilus with bracts and fertile sporophylls. It is said to have an unbranched axis by Mathews and Kramer. It supports cycadeoid origins and woody magnolids as earliest angiosperms. Proposed by Delpino in 1890, Bessy in 1893, and Arber and Parkin in 1907 (Steussy, 2004).

Angiosperm Origins

Darwin's "abominable mystery"(paleo.gly.bris.ac.uk-The Origin and Evolution of Angiosperms; Berendse and Marten, 2009) and "perplexing phenomenon", which he stated in a letter to Hooker, is the massive boom in diversity, complexity, and abundance of angiosperms, comparable to the Cambrian Explosion, which occurred in the Early Cretaceous in the Upper Mesozoic, about 130 mya when flowering plants 1st appeared in the fossil record, which shows that angiosperms replaced their precursors, the gymnosperms, as the dominant plants within 40 million years--a wink of an eye in evolutionary terms. By the early part of the end of the Cretaceous many of the lineages we recognize today were represented. Originally, this was noted in deposits containing leaf and pollen remains. The flowering plant radiation is one of the evolutionary events that most puzzled Darwin. The seemingly sudden appearance of so many angiosperm species in the Chalk Age (Cretaceous) conflicted strongly with his gradualist perspective on evolutionary change.

The oldest definite angiosperms date back to around 123 million years (in the early Aptian in the Late Cretaceous), and the earliest fossils unequivocally identified as angiosperms are pollen grains of the Hauterivian (around 130 mya) from Late Cretaceous sediments). In slightly later sediments from the late Barremian-early Aptian several major lineages, including Nympheales, Chloranthales, magnolids, monocots, and tricolpates are reliably documented (Time Tree).

However, some biologists do not consider them as the first true angiosperms and believe they maybe emerged much earlier, but have either not been preserved or discovered. Margulis and Schwartz (1996, p. 416), for instance, state flowering plants "...undoubtedly evolved earlier in the Permian." And some estimates place the time of angiosperm origin as far back as the Carboniferous or Permian in the Upper Paleozoic about 300-350 mya (Strickberger, p. 296, 1996, citing Savard et al, 1994), and later molecular dating gives ages of 175 mya (in the Middle Jurassic), 170 mya (in the Middle Jurassic), and between 198 (in the Early Jurassic) and 140 mya (in the Early Cretaceous)(Time Tree). Susana Magallon of Time Tree claims it was 175 mya and that there was a rapid diversification which gave rise to Chloranthales, Magnolidae, monocots, Ceratophyllaceae, and tricolpates.

Though there have been many reports of the "earliest angiosperm" from the Triassic (245-208 mya), no definite contenders have been discovered yet, although the enigmatic *Sanmiguelia* could provide an insight. All such fossils except *Archefructus* Ge Sun 2002 remain controversial in that none preserved enough detail for them to be assigned without doubt to the water lily family; *Archefructus* itself cannot be classified unequivocally either as a common ancestor or as part of a derived lineage.

Archefructus was discovered in 1998 in the Yixian Formation of the western Liao Ning Province of northeastern China. The formation is a layer of volcanic ash some 2000-2500 m. (6562 -8202 ft.) thick between 2 sedimentary layers containing many plant and animal remains consisting of reproductive axes bearing fruit in a spiral shape, where they ripen from the outside inwards. The organs are clearly those of angiosperms because they enclose the seeds completely. No pollen or pollen-producing organs have been found, but a structure resembling a rudimentary flower has been discovered, with "crumpled leaf-structures" that may have served the same purpose as petals to attract pollinators. It has been hypothesized that both wind and insect pollination played a part in *Archefructus* reproduction. When it was found, it was excitedly hailed as the definite Jurassic angiosperm that had been so long awaited. It was later learned that the rock was Cretaceous, nixing the theory that angiosperms had originated in Asia.

Whatever the case, the Cretaceous radiation is widely accepted, and today, angiosperms are among the most successful organisms in the world in terms of species diversity, ecological dominance, and modes of reproduction and methods of propagation.

While the reasons for this presumed sudden take-over are still unclear, 2 major hypotheses have been suggested: the competitive displacement theory and the non-competitive model. The 1st states that flowering plants were able to outcompete their rivals for almost every resource, shouldering them out of the way as we see at a lower, interspecific level today. For example, angiosperms have a shorter life-cycle than gymnosperms, which enables them to increase their numbers more rapidly. The 2nd states that since the mass extinction at the end of the Triassic left many niches open when a large proportion of gymnosperm and pteridophyte lineages died out, there was no real competition, and angiosperms could settle in unimpeded.

A 3rd is Bob Bakker's suggestion that angiosperms and dinosaurs may have co-evolved, with dinosaur feeding habits influencing the evolution of angiosperms. A change in communities of herbivores occurred just before the evolution of angiosperms, and he argued that the new dinosaur groups were instrumental in stimulating the explosive radiation of the angiosperms. The fossil remains from around 160 mya (Late Jurassic) tell us that around 95% of the herbivore biomass was made up of high-browsing stegosaurs and sauropods, like *Ultrasaurus*, with cranial and dental adaptations for feeding on conifers. This browsing pattern put pressure on the mature trees but still permitted the gymnosperm seedlings to flourish. So there was a shift from high-browsing sauropod dinosaurs in the Late Jurassic to the low-browsing ornithischian dinosaurs, which strongly increased the mortality of gymnosperm saplings and created disturbed environments from which gymnosperms were excluded due to their low tolerance to herbivory (Bakker, 1978). Although this sounds plausible, there is no convincing evidence in the fossil record for correlations between major events in the evolution of herbivorous dinosaurs and the origin of angiosperms (Barrett & Willis, 2001).

A 4th theory (Berendse and Scheffer, 2009) proposes that the massive scale of the shift in the composition of vegetation may be explained as the result of a positive feedback between angiosperms and their environment. Although factors limiting plant growth probably varied strongly among the many different habitats (as they do nowadays) they assume that soil nutrients, especially nitrogen and phosphorus and water have been the major growth-limiting factors. During the Cretaceous, CO_2 levels in the atmosphere were several times higher than today resulting in increased stomatal closure and reduced water stress, which probably reinforced the important role of soil nutrients. Proponents contend that angiosperms have higher growth rates and need higher nutrient levels than the gymnosperms that dominated before, whereas at the same time angiosperms promote soil nutrient levels by producing litter that is more readily decomposed. In this view, the prolonged dominance by gymnosperms could reflect a classical example of hysteresis ending in a rapid shift to an alternative stable state.

Hysteresis (thefreedictionary.com; chemeurope.com; wordiq.com; infoplease; Wikipedia) (coined by James Ewing, from the Greek for "lagging behind") is the dependence of a system on both its current and past environment, i.e., the state of a body at a given instant is determined by external conditions not only at that instant but also at preceding instants, so it can also be defined as the dependence of the state of a system on the history of its state, a phenomenon in which the response of a physical system to an external influence depends not only on the present magnitude of that influence but also on the previous state of the system, a property in which the order of previous events can influence the order of subsequent events, a phenomenon whereby 2 (or more) physical quantities bear a relationship that depends on prior history, the delay between action and reaction, or the lagging of an effect behind its cause. It is a path-dependent (the state is dependent on the path taken to achieve it) and rate-independent (it displays irreversible behaviour whose rate is practically independent of the driving force rate) property, and this behaviour is the key feature that distinguishes it from other dynamic processes in many systems. The plot of response vs. input gives a closed curve called the hysteresis loop. The response lags the input. The family of hysteresis loops, from the results of different applied varying voltages or forces, forms a closed space in 3D called the hysteroid. Mathematically, the response to the external influence (input) is expressed as a double-valued function and is bistable (one value applies when the input is increasing and the other when it is decreasing).

The phenomenon occurs primarily in physics, e.g., the magnetization of a material such as iron depends not only on the magnetic field it is exposed to but on previous exposures to magnetic fields. This "memory" of previous exposure to magnetism is the working principle in audio-tape and hard-disk devices. The phenomenon is also designed into thermostats. Deformations in the shape of substances that last after the deforming force has been removed, as well as phenomena such as supercooling, are also examples. It occurs also in respiration, cell division, neurology, and economics.

The conditions under which the co-existence of competing species (or other groups) is unstable are well known from classical competition theory. The general rule illustrated by Lotka-Volterra models is that competition can lead to unstable co-existence (implying alternative stable states) if it is more favourable to have conspecifics around than individuals of the other species. The authors maintain this may well characterize the competition between gymnosperms and angiosperms. They hypothesize that gymnosperms do relatively well under low nutrient conditions and also maintain low nutrient levels in the soil due to the nature of their litter. Angiosperms do not grow well under such conditions, but once they are present in sufficient densities they enhance soil fertility through their litter implying a positive feedback that might produce a runaway process once angiosperms have reached a certain critical abundance.

Another idea is that peaking CO_2 levels in this period may have played a role (Barrett & Willis, 2001). Elevated CO_2 would have allowed higher water-use efficiency and enabled plants to colonize drier areas.

Yet another has to do with leaves (de Boer et al, 2012)(and like in Barrett and Willis, involves CO_2), where it is shown that the period of rapid angiosperm evolution initiated after the leaf interior (postvenous) transport path length for water was reduced beyond the leaf interior transport path length for CO_2 at a critical leaf vein density (number of veins). Their data and modelling approaches indicate that surpassing this critical vein density was a pivotal moment in leaf evolution that enabled evolving angiosperms to profit from developing leaves with more and smaller stomas in terms of higher carbon returns from equal water loss and therefore facilitated evolving angiosperms to develop leaves with higher gas exchange capacities required to adapt to the Cretaceous CO_2 decline and outcompete previously dominant coniferous species in the upper canopy.

The authors state evidence is now emerging that ancestral angiosperms existed in low evaporative niches during the Early Cretaceous before the period of their rapid diversification in the Middle Cretaceous. During the Late Cretaceous, evolving angiosperms spread poleward and gained ecological dominance in most of the world's ecosystems by replacing needle-leaved (gymnosperm) conifer tree species in the evaporatively more demanding upper canopy. Recent insights suggest that the evolution towards more reticulated leaf venation was linked to the escalation of angiosperm leaf gas exchange capacity in relation to falling atmospheric CO_2 concentrations (Ca). The resulting rise in productivity likely enabled evolving angiosperms to outcompete conifers in the upper canopy. However, the underlying mechanisms involved remain unknown.

Falling atmospheric CO_2 concentrations before and during the angiosperm radiation probably put evolutionary pressure on terrestrial C3 plants because carbon uptake for photosynthesis is intrinsically linked to transpirative water loss through the stomatal pores on their leaf surfaces. (C3 plants produce, as a 1st step in photosynthesis, phosphogylceric acid, which has 3 carbon atoms; C4 plants produce, as a 1st step in photosynthesis, oxaloacetic acid, which has 4 carbon atoms.) Natural selection may therefore have favoured those species with the most plastic stomatal traits capable of optimizing photosynthesis with minimal transpiration. Consequently, falling CO_2 drove leaf evolution towards higher maximal stomatal conductance (*g* s max).

The exchange of water vapour for CO_2 through stomas is principally determined by diffusion according to Fick's law and Stefan's law, where the former relates the rate of steady-state diffusion to the concentration gradient across the stomatal pore, and the latter describes the "end correction" component of the pore resistance. As smaller stomas reduce the diffusion distance across the stomatal pore and limit the resistance of the end correction, evolution towards higher *g* s max was necessarily linked to a reduction in stomatal size combined with an increase in stomatal density. Contemporary with this rise in g s max, the geological record reveals a 2-phased rise in angiosperm leaf vein density (Dv) during the Cretaceous. The first phase in rising angiosperm Dv occurred around 100 mya during the Late Albian when, for the first time, the non-angiosperm maximum Dv was surpassed. The second phase occurred in the latest Cretaceous and earliest Tertiary (around 60 mya), during which the first modern values of angiosperm Dv are observed and angiosperm leaf gas exchange capacity escalated.

Although the evolution towards more reticulated leaf venation appears to have occurred on the backdrop of the Cretaceous CO_2 decline, relating rising angiosperm leaf venation to falling CO_2 has not been convincingly demonstrated. Plant physiological modelling does suggest that the combined effect of higher Dv and higher gsmax may have enabled evolving angiosperms to increase leaf gas exchange rates to offset the adverse effects of falling CO_2 for productivity, but to benefit from such highly conductive leaves in terms of additional growth and reproduction, the required additional carbon investments in water transport tissue should have been paid off by a larger increase in carbon gain.

On the basis of a combination of data and modelling approaches, they propose a mechanism to explain why evolving angiosperms could suddenly expand their leaf gas exchange capacity during the Cretaceous, and why conifers could not. This proposed mechanism is based on the consequences of morphological differences between planar-shaped hypostomatous (angiosperm) broad leaves and tubular-shaped (conifer) needle leaves for the leaf interior transport paths of CO_2 and water. They note that non-angiosperm species also evolved planar-shaped leaves with a variety of venation structures, but that these morphologies did not evolve the high Dv values common to modern upper canopy angiosperms. In their schematization of the broad-leaf morphology they differentiate between the Early Cretaceous with low Dv and low Ds and the modern with high Dv and high Ds.

The authors explored the role of both mechanisms and found that the period of rapid angiosperm evolution initiated after the leaf interior (postvenous) transport path length for water became shorter than that for CO_2, a transition which may have enabled the evolving angiosperms to develop leaves with higher gas exchange capacities required to adapt to the Cretaceous CO_2 decline and outcompete previously dominant coniferous species in the upper canopy.

Common for all leaf morphologies is that CO_2 moves from the leaf boundary layer, through the stomatal pore and substomatal cavity into the intercellular airspaces, and from there, CO_2 is absorbed in the mesophyll cells where photosynthesis occurs in the chloroplasts. Water moves from the soil through the plant's water transport system up to the leaf veins before it evaporates and passes through the stomata.

The exact mechanisms responsible for leaf interior water transport beyond the vein endings up to the stomata remain debated, with 2 being proposed: a liquid-phase flow through the apoplast across individual cells and through plasmodesmata with evaporation occuring close to the stomata in the substomatal cavitiy or a vapour-phase flow where evaporation occuring deeper inside the leaf throughout the mesophyll or close to the leaf vein endings, with water being transported as a vapour across the intercellular air spaces up to the stomata, but both may occur in parallel or involve a mixed phase.

There are 3 principal reasons why angiosperms are among the most diversified life forms on the planet, along with holometabolous insects and funguses: enclosure of ovules in carpels, nonsexual propagation (which allows for swift colonization of a sparsely populated area, ensuring dominance), and rapid growth (which allows them to take over and spread fast when in competition with other groups), innovations their rivals, principally the gymnosperms, do not have.

Proposals of seed ferns as the specific gymnosperm ancestor were put forward by Thomas in 1936, Gaussen in 1946, Stebbins in 1974, and Doyle in 1978 and 1996 (Steussy, 2004).

The original flowering plant was either herbaceous or woody. A woody origin is based on the assumption that Magnoliales is most primitive, so that the first flowers were numerous, large, and complex, spirally arranged on a cone-like axis, borne on trees or woody shrubs with thick branches and large leaves, and living below the forest canopy.

Several theories propose that modern flowering plants evolved from gymnosperms having a bisexual (hermaphroditic) strobilus (T. N. Taylor et al., 2009, p. 878, as cited in gigantoperoid.org). Arber and Parkin's strobilus theory of the angiosperm fructification (Arber and Parkin, 1907, loc. cit.), Delpino's ranalian theory, Alexander von Braun's strobilus theory of the 1800s, Takhtajan's hypothesis on a "neotenous origin of flowering plants" from 1976, and the woody magnolid hypothesis may be grouped together, as evolution of the flower from hermaphroditic gymnosperm strobili is a common thread of these proposals. "From this one may conclude that the angiosperm flower arose most probably from a bisexual entomophilous strobilus, especially as among the earliest gymnosperm examples of bisexuality associated with entomophily are known within the Bennettitales" (Takhtajan, 1969).

Woody magnoliid and ranalian theories tie-in with Becker and Theissen's out-of-male and out-of-female hypotheses (Becker and Theissen 2003), Baum and Hileman's model of cone and floral development from hermaphroditic strobili (2006), and modifications of the latter 2 ideas on bisexual strobilus development suggested by Theissen and Melzer (2007) and Melzer et al (2010).

The paleoherb theory relies on the assumption that angiosperms evolved in response to disturbed soils, e.g., sunny stream-sides and flood plains. This implies that the first angiosperms were rapid colonizers that tolerated disturbance, thanks to an herbaceous nature which does not require so much maintenance, and by having leaves capable of high photosynthetic rates and small simple flowers to reduce the cost of producing them.

Evidence from fossils fails to pin down one of the hypotheses as the more probable. Promoters of the woody theory maintain that most features of Magnoliales, such as the long floral axis, the woody stem, the simple, alternating, oval-shaped leaves, and the simplicity of the seed coat and pollen grains, are all primitive features, and that compound leaves and complex structures are advanced. Herbs are said to be more advanced in structure compared to woody taxons, and, in fact, no gymnosperm is herbaceous, so evolving herbs from the gymnosperms involves changes that later had to be reversed to produce trees. The flowers of Magnoliales are also often compared to, and associated with, the reproductive structures of cycadians. The 1st land plants, however, were herbs. The herbaceous or woody habit may not apply to algae as these do not have stems, an herb being defined as lacking a persistent stem--algae do not have stems at all. At any rate, herbs are primitive in plants in general, woody fibers being an innovation.

There is also the matter of an aquatic or terrestrial origin. Ceratophyllaceae as most basal supports an acquatic origin for angiosperms, while *Amborella* as most basal supports a terrestrial one. From observation of the basal terrestrial lineages, a terrestrial origin would involve a woody, tropical angiosperm with an unusual seedling development in a humid, low-lit, forest understory where the soil is often disturbed. In fact, *Amborella* grows in 2 phases: it first establishes itself as a small, ground-hugging seedling with a creeping root system and many shoots from the base, changing to a shrub-like growth once it is well established. Multiple, low-lying shoots allow it to better resist possible breakage resulting from ecological adversities due to shifting sediments, while a developed root system enables better anchorage in unstable substrates such as the understory stream margins and upland-forest washout zones where terrestrial stem groups are currently found. Although a phylogenetic reconstruction based on a terrestrial origin is the most parsimonious, there is no direct fossil evidence for primitive angiosperms occupying such habitats. However, in the case of upland slopes, this may be the result of a bias in the fossil record since these areas are rarely preserved; stream margins and channels are common depositional settings.

An aquatic origin implies an herbaceous plant with extensive root systems and emerged leaves and flowers at the surface of the water. Seedlings establish at the bottom of the water column, either in low light conditions in stable sediments, later extending leaves to the surface to better photosynthesize or in clear streams where the sediment is constantly disturbed. Seasonal, ephemeral pools, where most living water lilies are established today, could explain the evolution of the angiosperm characteristics of rapid growth and accelerated reproduction which are thought to be the basis for their success.

Demands on organisms differ greatly under water and in the air. Buoyancy provided by water supports the cells, exposing the organisms to forces such as currents and waves rather than gravity. It is hard to imagine how flowering plants migrated from water to land and were able to evolve quickly enough to cope with gravity. In addition, aquatic angiosperms show strong convergence between different lineages in their mechanisms to deal with currents and the low diffusivity of gases in water impacting on their photosynthetic abilities. This suggests that a limited number of solutions are available, thus restricting the scope for widespread radiation and innovations which would lead to migration on to land. Fewer than 2 percent of angiosperm species are aquatic.

So angiosperms probably had a terrestrial origin, but tested the aquatic habitat early in their evolution. Ideas about why this happened include escaping competition and overcrowding by gymnosperms on land; evolving in areas with environmental conditions similar to those under water such as low-light levels under a tree canopy; and living in areas where there were frequent, ephemeral, seasonal pools, leading to an aquatic lifestyle.

Evidence supporting the aquatic hypothesis occurs in both living and fossil forms. *Amborella*, supposedly the most basal angiosperm, has vestigial gas exchange canals, useful in submerged stems and roots, and similar to those found in water lilies. A fossilized lily flower found in 2001 establishes the lineage as already existing in the Early Cretaceous, around 125 mya. Another fossil, the earlier mentioned *Archefructus*, from Chinese lake deposits, dated at 124 million years ago, further proved that angiosperms had already colonized the water by the Early Cretaceous, as fossil fishes were found among the leaves.

Theories as to topographical centers of origin and radiation of the first flowering plants (gigantopterid.org - Origin of Angiosperms) include an uplands hypothesis by Daniel Axelrod in 1952 and a coastal hypothesis by Retallack and Dilcher in 1981. In a later review of early angiosperm history based on new evidence concerning the antiquity, center of origin, and evolution of flowering plants, Axelrod in 1970 suggested angiosperms originated in Triassic-Jurassic tropical uplands "... Spreading out into a new adaptive zone, presumably in equable, warm upland areas ..." and proposed they and their insect pollinators dispersed and radiated into semiarid forest openings and into arid lowlands from fold belts during the Early Cretaceous. Stebbins in 1974 and 1984 proposed alpine biomes of northern latitudes as possibly having been the center of the early radiation of angiosperms. Peter Raven, in 1977 (Strickberger, 1996), pointed out that tropical lowland conditions with their large insect populations are more favourable to entomophilous plants and such climates expanded significantly during the Cretaceous allowing for early angiosperms to disperse and become dominant. A similar idea, the East-Asian-centers hypothesis, was put forward by Sun et al. in 2001. Based on the recovery and study of fossil pollen casings (palynomorphs) recovered from deep-sea drill holes, Hochuli and Feist-Burkhardt in 2004 suggested that early flowering plants might have evolved in a boreal cradle.

The chloranthoid hypothesis suggests that flowering plants evolved from chloranthoid ancestors not unlike the modern Chloranthaceae (Leroy, 1983). Arguments presented by Stuessy (2004) in favour of Leroy's proposal include reports of data from molecular taxonomy that place *Hedyosmum* (Chloranthaceae) in a clade basal to other extant angiosperms (Qiu et al. 1999).

The major developmental and genetic hypotheses on the origin of angiosperms and the flower are Asama's growth-retardation theory, from 1982 and 1985, Sergei Meyen's gamoheterotrophic hypothesis, from 1986 and 1988, the mostly-male theory by Frohlich and Parker from 2000, Becker and Theissen's out-of-male and out-of-female hypotheses from 2003, Baum and Hileman's model of 2006 on cone and floral development from hermaphroditic strobili, and modifications of the latter 2 ideas on bisexual strobilus development suggested by Theissen and Melzer in 2007 and Melzer et al in 2010.

There is also Stuessy's multi-disciplinary transitional-combinational theory from 2004, which proposes that carpels evolved first, followed by double fertilization, and then flowers, "slowly over many millions of years... perhaps more than 100 million ...".

The so-called anthophyte hypothesis (Doyle and Donoghue 1986, 1987, Donoghue and Doyle 2000), which is misnamed as anthophytes include all flowering plants and not cycadeoids, "... proposes that the outer seed integument and carpel are derived from fertile structures that already were aggregated into a flower-like reproductive organ (i.e., a strobilus or cone) ..." Rothwell et al (2009) contend Doyle's arguments in favour of Thomas's classical 1925 caytonian seed fern hypothesis, specifically the glaring absence of transitional fossils and any definable heterochronic lineage linking *Caytonia* with basal angiosperms and impossibly complicated evo-devo (evolution-development) from a deeply conserved cone and floral tool kit perspective, renders his proposal moot, and that fossilized material of *Caytonia* is fragmentary and poorly preserved, leaving interpretation of developmental reproductive structures conjectural.

According to Rothwell et al, the mostly-male theory and gamoheterotrophic hypotheses fall into a category that proposes that ovules developed in positions on leaves or stems occupied by pollen-bearing organs.

Gene duplications may consist of (small) WGDs (polyploidy), which are stated by the authors as important in the evolution of flowering plants. The term "insect-mediated" is the same as entomophily or may refer to the importance of pollination by insects in the origin of angiosperms.
Studies of wood pedomorphosis may offer new clues on a possible Mesozoic origin for angiosperms (Carlquist, 2009), but studies of potentially neotenous gymnosperm secondary xylem development in deep (Paleozoic) time are lacking.

Melville develops his earlier ideas on a gonophyll theory of 1969 in a review published in 1983 that proposes a Permian origin of angiosperms from glossopterids. Retallack and Dilcher in 1981 presented an in-depth discussion of Melville's ideas including a reanalysis of glossopterid fructifications.

Mary White in 1986 proposed glossopterid Microfructi as basal to several parallel but sometimes branching and reticulate lines of evolution leading to Caytoniales, angiosperms, *Cycas*, Podocarpaceae, Araucariaceae, and certain amentifers including the cassowary family. As well, White proposed that the glossopterid Megafructi were a second basal group on which ranunculans, monocots, including *Pandanus* (Pandanaceae, Pandanales, Arecidae), *Williamsonia* (a cycadeoid), cycads, and certain other angiosperms evolved.

Barrett, Paul & Willis, Katherine. 2001. Did dinosaurs invent flowers? angiosperm coevolution revisited. Biol. Revs. 76: 411-47 (abstract).
Baskerville, Thomas. 1839. Affinities of Plants, with some observations on progressive development.

Taylor and Walton.
Berendse, Frank and Scheffer, Marten. 2009. The angiosperm radiation revisited, an ecological explanation for Darwin's 'abominable mystery'. Ecol. Lett. 9: 865–872 (ncbi.nlm.nih.gov).
Bessey, Charles. 1915. The phyogenetic taxonomy of flowering plants. Ann. Mo. Bot. Gard. 2: 109-64.
de Boer, Hugo Jan; Eppinga, Maarten; Wassen, Martin; & Dekker, Stefan. 2012. A critical transition in leaf evolution facilitated the Cretaceous angiosperm revolution-uu.nl Nature Communications.
Brummitt, R.K. 1992. Vascular Plant Families and Genera. Kew Gardens.
Caratini, Roger. 1971. La Vie des plantes (Encyclopédie Bordas, vol. 10). Encyclopédie Bordas.
Core, Earl. 1955. Plant Taxonomy. Prentice-Hall.
Cantino, Philip et al. 2007 Towards a phylogenetic nomenclature of Tracheophyta. Taxon 56: 822–846 (washington.edu).
Cronquist, A. 1968. The Evolution and Classification of Flowering Plants, 1st ed. (2nd ed.: An Integrated System of Classification of Flowering Plants, 1988)
Dahlgren, R.T.M. & Bremer, K. 1985. Major clades of angiosperms. Cladistics 1: 349- 368.
Darbah, V.F., Oppong. E.K., Eminah, J.K. 2012. Chemical investigation of the stem bark of *Dichapetalum madagascariense* Poir. Intl. J. Appl. Chem. 8: 199-207 (euw.edu.gh; freedictionary.com).
Doyle, J. A. 1996. Seed plant phylogeny and the relationships of Gnetales. Int. J. Plant Sci. 157(Suppl.): S3–S39.
Donoghue, M.J. & Doyle, J.A. 1989. Phylogenetic analysis of angiosperms and the relalionships of Hamamelidae. In P.R. Crane & S. Blackmore (eds.), Evolution, Systematics, and Fossil History of Hamamelidae, Vol. 1, pp. 17-45. Clarendon Press, Oxford.
Endress, Peter. 2002. Morphology and angiosperm systematics in the molecular era. Botanical Review (highbeam.com).
Gonzales, Favio. 1999. Monocotiledóneas y dicotiledóneas: Un sistema de clasificación que acaba con el siglo. Rev. Acad. Colomb. Cienc. 23: 195-204 (accefyn.org.co).
Grisebach, Rudolf. 1854. Grundriss der systematischen Botanik (books.google).
Hufford, Larry. 1992. Rosidae and its relationships to other nonmagnolid dicotyledons: a phylogenetic analysis using morphological and chemical data. Ann. Missouri Bot. Gard. 79: 218-48.
Jones, Samuel & Luchsinger, Arlene. 1979. Plant Systematics. McGraw-Hill.
Judd, W.S., Manchester, S.R. 1997. Circumscription of Malvaceae (Malvales) as determined by a preliminary cladistic analysis of morphological, anatomical, palynological, and chemical characters. Brittonia 49: 384-405
Judd, W.S. & Olmstead, R.G. 2004. A survey of tricolpate (eudicot) phylogenetic relationships. Amer. J. Bot. 91: 1627–1644 (amjbot.org).
de Jussieu, Antoine. 1789. Genera plantarum (1791 ed.)(gallica.bnf.fr).
Lawrence, G.H.M. 1960. Taxonomy of Vascular Plants. Macmillan, NY.
Leroy, J.F. 1983. The origin of angiosperms: an unrecognized ancestral dicotyledon, *Hedyosmum* (Choranthales) with a strobiloid flower is living today. Taxon 32: 169–175.
Lindley, John. 1853. The Vegetable Kingdom, 3rd ed. (books.google).
Loconte, H. & Stevenson, D.W.. 1991. Cladistics of Magnolidae. Cladistics 7: 267-296.
Matthews, Merran and Endress, Peter. 2005. Comparative floral structure and systematics in Celastrales (Celastraceae, Parnassiaceae, Lepidobotryaceae). Botanical Journal of the Linnean Society, 149: 129–194.
Nandi, O.L., Chase, M.W., & Endress, P.K. 1998. A combined cladistic analysis of angiosperms using rbcL and non-molecular data sets. Ann. Missouri Bol. Gard. 85: 137-212 (docstoc.com).

Olmstead, R. G., de Pamphilis, C. W., Wolfe, A. D., Young, N. D., Elisons, W. J., & Reeves P. A. (2001). Disintegration of the Scrophulariaceae. American Journal of Botany 88: 348–361.

Perleb, C.J. 1838. Clavis classium, ordinum, et familiarum atque index generum regni vegetabilis (biodiversitylibrary.org).

Qiu, Y. L., Chase, M.W., Les, D.H., & Parks, C.R. 1993. Molecular phylogenetics of Magnolidae: Cladistic analyses of nucleotide sequenees of the plastid gene rbcL. Ann. Missouri Bol. Gard. 80: 587-606.

Reddy, S.M., Rao, M.M., Reddy, A.S., Reddy, M.M., and Chary, S.J. 2004. University Botany, vol. 3. New Age International (books.google).

Reichenbach,. Ludwig. 1828. Conspectus regni vegetabilis per gradus naturales evoluti (books.google).

Reveal, J.L. 2011. New ordinal names established by changes to the botanical code. Phytotaxa 30: 42-44 (mapress.com).

Rothwell, G. W., Crepet, W. L., Stockey, R. A.. 2009. Is the anthophyte hypothesis alive and well? New evidence from the reproductive structures of Bennettitales. American Journal of Botany. 96: 296-322.

Schaefer, Hanno & Renner, Susanne. 2011. Phylogenetic relationships in the order Cucurbitales and a new classification of the gourd family (Cucurbitaceae). Taxon 60:122–138 (umsl.edu).

Soltis, Pamela; Soltis, Douglas. 2004. The origin and diversification of angiosperms. American Journal of Botany 91: 1614–1626.

Stuessy, Tod. 2004. A transitional-combinational theory for the origin of angiosperms. Taxon 53: 3-16.

Takhtajan, Armen. 1969. Flowering Plants: Origin and Dispersal (translated by C. Jeffrey), p. 35, Oliver and Boyd, Edinburgh.

Tobe, Hiroshi; Jaffré, Tanguy; Raven, Peter. 2000. Embryology of *Amborella* (Amborellaceae): Descriptions and Polarity of Character States. J. Plant Res. 113: 271-280 (horizon.documentation.ird.fr).

Tucker, Shirley & Douglas, A.W. 1996. Floral structure, development, and relationships of paleoherbs: *Saruma, Cabomba. Lactoris*, and selected Piperales. In Flowering plant origin, evolution, and phylogeny, pp. 141-175, D.W. Taylor & L. J. Hickey (eds.), Chapman & Hall.

Chapter 13 Chromophytes

The air was cool and fresh and smelled of the kelp and salt that streamed in off the bay at the full of the tide. - Anne Rivers-Siddons, American novelist-topfamousquotes.com.

Chromistan Algae

Chlorarachnians, found in cultures of tropical or subtropical siphonous green algae, have an ameboid-plasmodial stage with reticulopds (forming a web, hence the name), a walled coccoid stage, and a uniflagellate stage, and can form zoospores. They have a 4-membrane chloroplast envelope, a chloroplast in the RER (rough ER), a nucleomorph, chlorophyll a and b, paramylon as the storage product, and tubular cristas. There are 5 genera and 6 species (*Chlorarachnion* Geitler 1930, *Cryptochlora* Calderon-Saenz et Schnetter 1987, *Gymnochlora* Ishida et Y. Hara 1996, *Lotharella* Ishida et Y. Hara 1996, and *Bigelowiella* Moestrup 2001); 4 genera are monotypic and *Lotharella* has 2 species.

Cryptophyceae, specifically *Cryptomonas* and *Chilomonas,* were first described by Ehrenberg in 1832, who, 6 years later, established for it the family Cryptomonadina. In 1841, Dujardin placed them in his Infusoires, dubbed Flagellaten by Cohn in 1853. Klebs, in 1892, placed them as 1 of 2 families in Chromomonadina, the other being Chrysomonadina. Although various early authors regarded them as algae, cryptomonads retained their position in Flagellata for a long time, with general acceptance as a group of plants s. l. begining only in 1900, when Senn gave a treatment for them in Engler and Prantl's "Die Natürlichen Pflanzenfamilien." Klebs said Chromomonadina could be refered to as chrysophytes, a designation formally adopted by Pascher in 1914 as the name for the chrysomonads, heterokonts, and diatoms. Pascher placed Cryptophyceae with chrysomonads in Pheochrysidales in 1911 and 3 years later relocated them in his phylum Pyrrhophyta, which he established the same year, along with Desmophycidae and Dinophycidae, and recognized only 2 orders: Cryptomonadales and Cryptococcales, for motile and nonmotile forms, respectively. Families recognized in the 1930s and 1940s (as arranged by Pascher, Pringsheim, and Skuja) were: Cryptochysidaceae, Cryptomonadaceae, Cyathomonadaceae, Kathablepharidaceae, and Senniaceae. (Margulis et al, loc. cit.; Kessel, 1955).

The group is now divided into 3-5 orders: Cryptomonadales, Thecomonadales, Cyathomonadales, Katablepharidales, and Goniomonadales. Butcher (1967) recognized 3 families, 13 gen., and 60 sp.; Bourelly (1970) listed 8 families and 23 gen.; Van Den Hoek (1995) reports 12 gen. and 200 sp., 100 freshwater (mostly lakes) and 100 marine (nanoplankton in mostly tidal pools); 15 gen. are covered in Lee et al (1997) but are not classified.

Haptophyceae, specifically the most familiar examples, the coccolith-bearing species, were discovered by Ehrenberg in 1835, the coccoliths (name coined by Huxley in 1858) being calcified plates or scales, often invested with elaborate ornamentation. The distinctive organelle, the haptonema, was described as a "3rd flagellum" by Scherffel in 1900 but in the 1950s, in a series of articles, Parke et al demonstrated that the organelle was really a thin filamentous appendage, differently structured from a flagellum, which they called a haptonema, hence the descriptive name for the group.

The great majority of haptophyceans are unicellular and flagellate (monadoid) but some species also have ameboid, coccoid, palmelloid, or filamentous stages. The great majority, also, are marine and planktonic (nanoplanktonic [2-20 microns] or picoplanktonic [.2-2 microns]). Some also have an alternation of generations.

Some 50-75 genera and about 500 species are usually recognized. The traditonal convenience classification has 4 orders: Pavlovales, Prymnesiales, Isochrysidales, containing those species with isokontic flagella, and Coccolithophorales, containing those species with coccoliths. A more phylogenetic arrangement, recognized by some (e.g., Christensen in 1980, Green and Jordan in 1994, and Jordan and Green also in 1994), has only 2 orders, subsuming the other 2--Pavlovales, without plate-scales and with anisokontic flagella, and Prymnesiales, with plate-scales, and with isokontic flagella.

Heterokonts were first recognized as a kingdom by Leedale in 1974. The informal synonym, stramenopiles (from the Latin for "straw hairs," refering to the tubular mastigonemes of the heterokont condition), was introduced by Patterson in 1989, and first formalized as Stramenopila by Alexopoulos et al in 1996 in their book "Introductory Mycology" and with kingdom status, and as Kingdom Straminipila by Michael Dick in 2001 in his book "Straminipilous Fungi", using the adjectival form, which has an i instead of an e, and using the middle i, which is appropriate for Latin, instead of an o, which is appropriate for Greek, but using a instead of the plural i, and the name Heterokontae was used by Luther in only for yellow algae (Blackwell, 2009). (Neomura is also properly spelled with an i in the plural.)

Pheophyceae (brown algae) are traditionally divided into subclasses Pheosporeae, where the zooids are either unilocular, unicellular, and forming several reproductive elements, or plurilocular, multicellular, and forming only 1 reproductive element, where there are both spores and gametes (alternation of generations), and where isogamy, anisogamy, or oogamy, variously occur; and Cyclosporeae, which have only gametes and oogamy, and are usually interpreted as having no alternation of generations, only a diploid, sporophyte generation, but some believe the gametophyte is reduced and included in the sporophyte (a case comparable to/analogous with the female prothallus gametophyte [embryo sac] of angiosperms). Cyclosporeae contain only 1 order, Fucales Kylin 1917, but often Acroseirales and Durvilleales are separated out, while Pheosporeae are divided into the following:

Table 13-1. Pheosporeae Taxonomy.

Isogeneratae (sporophyte and gametophyte generations are morphologically identical [isomorphic])
- haplostichous
  - Ectocarpales Setchell et Gardner 1925
- polystichous
  - Sphacelariales Oltmanns 1922
  - Dictyotales Kjellman 1893
  - Cutleriales Oltmanns 1922 (both iso- and heteromorphic)

Heterogeneratae (generations dissimilar, the gametophyte being a microscopic, more or less branching prothallus)
- haplostichous
  - Chordariales Setchell et Gardner 1925
  - Sporochnales Sauvageau 1926
  - Desmarestiales Setchell et Gardner 1925
- polystichous
  - Dictyosiphonales Setchell et Gardner 1925
  - Scytosiphonales
  - Laminariales Kylin 1917

A phylogenetic arrangement is presented by Van Den Hoek (1995) where Ectocarpales forms 1 clade, Sphacelariales and Dictyotales form a 2nd, Scytosiphonales and Cutleriales form a 3rd, and Heterogeneratae and Cyclosporeae form a 4th. Heterogeneratae has 2 groups: Dictyosiphonales+Chordariales and Sporochnales+(Desmarestiales+Laminariales), but the latter is sometimes grouped with Cyclosporeae. An evolutionary link by others also is postulated between Chordaria, Dictyosiphonales, and Cyclosporeae as they share cryptostomas, conceptacles, and apical growth, and there is a striking similarity between the mastigotes of the unilocular structure in the conceptacles. Sporochnales, Desmarestiales, and Laminariales share a heteromorphic life history, anisogamy, female pheromones eliciting 2 responses, reduction of the male plurilocular gametangia, and filamentous gametophytes. The 3rd clade should be united with the 4th on the basis of heteromorphic generations and trichothallic meristems.

The recently discovered (in 1999) genus (with 2 species) *Bolidomonas*, probably goes to brown algae because of pigmentation and laterally inserted flagella, instead of with diatoms based only on genotypic data, and comprises a family and order included in the Pheophyceae totals.

The most primitive order is Ectocarpales and the most advanced is Fucales. The largest families are Ectocarpaceae C.Agardh 1828 and Chordariaceae Greville 1830, both with about 30 genera. The largest families in Chromista are probably Fragillariaceae Greville 1833 and Bacillariaceae Ehrenberg 1831 (in Diatomae).

Brown algae are almost entirely marine and benthic and inhabit the intertidal and subtidal zones, especially rocky shores. In these algae we find the kelps (mainly *Laminaria* Lamouroux, 1813 and *Macrocystis* C. Agardh, 1820, both in Laminariaceae Bory de St. Vincent, 1827) (the blade and giant or great kelp can grow to 50 yards long! and make enchanting underwater kelp forests and a productive and dynamic ecosystem), sea lace, bootlace weed, or mermaid's tresses (*Chorda* Stackhouse, 1797) in Chordariaceae Greville 1830, winged kelp (*Alaria* Greville, 1830), and sea colander (*Agarum* Bory de St. Vincent, 1826) in Costariaceae (author unknown) in Laminariales; rockweed or wrack (*Fucus* Linneus, 1753), knotted wrack, yellow tang or tangle, or sea whistle (*Ascophyllum* Stackhouse, 1809) in Fucaceae Adanson, 1763, and sargassum weed (*Sargassum* C. Agardh, 1820) (found in the Sargasso Sea near Bermuda) in Sargassaceae Kützing 1843 in Fucales; and mermaid's hair and landlady's wig (*Desmarestia* Lamouroux, 1813) in Desmarestiaceae (Thuret) Kjellman 1880 in Desmarestiales; among other large algae (Lee, 1977/1986; Lamoureux, 1985).

The phylogeny for Heterokonta, based on data by Williams (loc. cit.), Cavalier-Smith and Chao (loc. cit.), and Saunders et al (loc. cit.), is given in Table 13-2. There are some 70 orders.

Table 13-2. Heterokonta.

(all are stat. nov.; the totals are from the same sources as for the Tally Table and the ones for genera and species are approximations)

sbphyl. Bicophycotina nom. nov. (Bicosecales)
sbphyl. Gyrista Cavalier-Smith 1998
    infphyl. Hydromyxa (Labyrhinthulata)(slime nets)
    infphyl. Opalinata (aka paraflagellates, protociliates)
    infphyl. Pseudofungi (water molds)(Oomycetes, Hyphochytridiomycetes)
    infphyl. Ochrophycotina nom. nov.
        grdcl. Pedinellophyceae (Pedinellales)
        grdcl. Euochrophyceae nom. nov.
            hpcl. Pelagophyceae (Pelagococcales)

hpcl. *Rhizochromulina*
hpcl. Limnioraphidi (Chloromonada)
hpcl. Cellulosemurus nom. nov.
    cl. Sarcinophyceae (Sarcinochrysidales)
    cl. Pheophyceae
hpcl. Silicophyceae nom. nov.
    spcl. Silicoflagellata (Dictyochae)
    spcl. Silicamurus nom. nov.
        cl. Diatomeae
        cl. Vaucheriae (Xanthophycidae)
        cl. Chrysariae nom. nov.
            spord. (Chrysomeridales) + (Chrysophyceae, Synurophyceae)
            spord. Eustigmata -Pelagoraphidi (Chloromonada)

Bicosecida Grassé, 1926, Grassé and Deflandre 1952 (syn. Bicosecales Bourrelly, 1968, 1981)(4 fams., 8 gen., 16 sp.)
Hydromyxa (syn. Labyrhinthulata Cienkowski 1867; Labyrinthulomycota Whittaker, 1969; Labyrinthomorpha Page in Levine et al., 1980)(slime nets)(1 ord., 4 fams., 8 gen., 40 sp.)
Opalinata Wenyon 1926, emend. Cavalier-Smith 1993, 1997 (syn. Slopalinida Patterson, 1989)(aka paraflagellates, protociliates)(1 ord., 2 fams., 15 gen., 400 sp.)
Pseudomycota Barr 1992 (syn. Pseudofungi Cavalier-Smith 1986)(water molds) (2 cls., 7 ords., 14 fams., 70 gen., 800 sp.)
Oomycetes Winter in Rabenhorst 1879
Hyphochytridiomyctes Sparrow 1959
Pedinellales Kristiansen 1990 (as Pedinellea)(10 gen.)
Pelagophyceae Andersen and Saunders 1993 (1 or 2 gen.)
Rhizochromulina
Limnioraphidi (Chloromonada) (1 ord., 1 fam., 6 gen., 12 sp.)
Sarcinochrysidales (1 gen.)
Pheophyceae De Bary 1881 (syn. Pheophyta Wettstein 1901)(12 ords, 41 fams., 260 gen., 2000 sp.)
Silicoflagellata Borgert 1891 (syn. Dictyochae Haeckel 1894, Silicophyceae Rothmaler 1951) (1 gen.)
Diatomeae Dumortier 1821 (syn. Bacillariophyceae Haeckel 1878)(40 ords., 81 fams., 200 gen., 10,000 sp.)
Xanthophyceae Allorge in Fritsch 1935 (6 ords., 20 fams., 100 gen., 600 sp.)
Chrysomeridales (1 gen.)
Chrysophyceae Pascher 1914 (7 or 9 ords., 24 fams., 123 gen., 1000 sp.)
Eustigmata (1 ord., 1 fam., 6 gen., 12 sp.)
Pelagoraphidi (Chloromonada) (1 ord., 1 fam., 6 gen., 12 sp.)

Here is Williams' classification from 1991 (the high level names are my own):

Table 13-3. Chromista by Williams.

sbph. Protoheterokonta
    Pseudofungi
    Hydromyxia

Proteromonada
sbph. Metaheterokonta
infph. Bicosecida
infph. Photoheterokonta
pvph. Pedinellales
pvph. Euphotoheterokonta
mcph Pelagophyceae
mcph. Neoheterokonta
mgcl. *Rhizochromulina*
mgcl. Silicophyceae
grdcl. Xanthophyceae
grdcl. Metasilicophyceae
hpcl. Limnioraphidi
hpcl. Neosilicophyceae
spcl. Sarcinophyceae
spcl. Diatomae+(Pheophyceae+Haptophyceae)
spcl. Limnistia
cl.(Eustigmata+Cryptophyceae)
+Mariniraphidi
cl. Chrysophyceae

The CI was .42, 27 taxons were included, there were 79 characteristics and 27 MPTs, and strict consensus was applied. Hennig86 was used so there were no confidence measures. Silicoflagellates and chlorarachnians were not included. Chrysophyceae includes Synurales and Chrysomeridales. Eustigmata-Metaraphidia agrees with Saunders et al (1997) and Cavalier-Smith and Chao (1996). The separation of Raphidophyceae agrees with the latter. The freshwater genera are *Goniostomum*, *Meritricha*, and *Vacuolaria*, the 1st and 3rd occuring in bog pools. The marine genera are *Chattonella, Chlorinimonas, Fibrocapsa, Haramonas, Heterosigma*, and *Seriola,* along with possibly a few more.

Haptophyceae were also nested within Heterokonta in Saunders, Potter, and Andersen (1997), based on phenotypic, molecular, and combined data, but excluding Cryptophyceae. However, the latter the combined data show Haptophyceae as sister group to Heterokonta, with the strict consensus for the phenotypic results not well resolved, and the molecular results grouping Haptophyceae with Alveolata+Heterokonta. The bootstrap support is very strong, at 100 for Haptophyceae+Alveolata-Heterokonta, 100 for Alveolata, and 96 for Alveolata-Heterokonta. The CI was .52 and the RI .64. Chlorarachnia were excluded here as well. In Yoon et al (2002), the ME bootstrap values for Euchromista are 93 and 96 and for Neochromista are 84 and 87.

A new class of photosynthetic heterokonts has been described by Kawachi et al (2002), Pinguiophyceae, which contains 5 monotypic genera: *Glossomastix, Phaeomonas, Pinguiochrysis* (type genus), *Pinguiococcus*, and *Polypodochrysis,* all unicellular microalgae including some picoplanktonic genera. It possesses an unusually high percentage of polyunsaturated fatty acids, especially EPA (eicosapentaenoic acid) and is corroborated as monophyletic by analyses of nuclear-encoded 18S rRNA and chloroplast-encoded rbcL gene sequence data.

Another new class of photosynthetic heterokonts, Synchromophyceae, for the genus *Synchroma,* was described by Horn et al (2007).

The fossil record of pheophyceans is rather sparse, as these organisms do not produce hard parts, such as calcified red and green algae, nor resistant spores. Also, in the absence of pigments, fossil brown algae may be almost impossible to distinguish from these other algae, since there are many morphologically convergent forms among the 3 groups. These difficulties are compounded by the lack of trained paleobiologists who specialize in algae and phycologists who examine fossils.

Possible fossil pheophyceans have been found in North American and Asian rocks as old as the Vendian, but some experts question their identity. More likely fossil pheophyceans have been found in Ordovician and later rocks, but even these algae are placed in Pheophyceae with some doubt. The oldest fossils which definitely belong to Pheophyceae are kelps from the Tertiary. Miocene deposits have yielded taxons assignable to the extant orders Dictyotales, Laminariales, and Fucales. The lack of recognizable brown algal fossils from the latter Paleozoic or the Mesozoic has been a problem for many biologists who feel that the divergence must have occurred in the Precambrian. Others have proposed that the larger, recognizable forms, such as the kelps, may not have radiated until the Miocene, the time when animals associated with kelp-dominated communities first appear in the fossil record. In any case, the relatives of brown algae all first appear as fossils in the Jurassic or Cretaceous, suggesting that pheophytes may have originated at that time as well. (UCMP.Berkeley.Edu-Pheophyta: Fossil Record).

Dinoflagellates

There are about 50 families, 130 genera, and 2000 living species. Much like euglenoids, some 50% are colourless. Most are microscopic, but some are larger, the largest is *Noctiluca* at 2 mm. so is visible to the naked eye and was the first discovered; this was in 1753 by Henry Baker. Stein, in 1883, regarded them as animalcules and placed them as a suborder, arthrodele Flagellaten, in his order Flagellaten, arranging them into 5 families: Prorocentrinen, Noctiluciden, Peridiniden, Dinophysiden, and Cladopyxiden. Pascher, starting in the 1920s, partitioned them into 3 groups, Desmokontae, Cryptophyceae, and Dinophyceae. Fritsch and Graham separated out Cryptophyceae, and the former regarded the 2 remaining groups as forming class Dinophyceae. The classification presented by the California Academy of Sciences is a synthesis of the systems of Pascher from 1914, 1927, and 1931, Lindemann from 1928 (a volume in Engler and Prantl's "Natüralichen Pflanzenfamilien"), Schiller from the '30s (in 2 volumes), Fritsch from 1938, and Graham from 1951 (Table 13-4), in which there are 10 orders and 41 families. (Kessel, 1955).

Table 13-4. Cal. Acad. Scis. Classification for Dinophyceae Fritsch 1935 from 1955 (in parentheses are the family totals)

- Desmophycidae (Pascher) Graham orth. mud. Papenfuss (syn. Desmokontae (Pascher 1914) Graham 1951, subdiv. Adiniferae (Bergh) Lindeman 1928, ord. Adiniferidea (Bergh) Kofoid and Swezy 1921))
  - Desmomonadales Pascher 1914 (syn. Desmocapsales Pascher 1914, Athecatales Lindemann 1928) (2)
  - Thecatales Lindemann 1928 (syn. Tribe Thecatoidae Kofoid and Swezy 1921 nom nud., Prorocentrales Pascher ex Graham 1951)(1)
- Dinophycidae (Fritsch) Graham orth. mud. Papenfuss (syn. Dinokontae Fritsch 1935 elevated to subclass by Graham in 1951), Dinophyceae Pascher 1914)
  - Gymnodiniales (Poche) Lindemann 1928 (syn. Amphilothales (Kofoid and Swezy) Lindemann 1928)(6)
  - Blastodiniales Schiller 1935 (5)
  - Dinophysiales (Kofoid) Lindemann 1928 (4)
  - Peridiniales Schütt 1896 (15)

Rhizodiniales Pascher 1931 (1)
Dinocapsales Pascher 1914 (1)
Dinococcales Pascher 1914 (4)
Dinotrichales Pascher 1914 (2)

A phylogeny for dinoflagellates was proposed by Fensome and colleagues (Lee et al, 2000) and is as follows:

Table 13-5.

Prorocentrales
Dinophysiales
Oxytoxaceae
Unnamed 1
    Peridiniales
    Unnamed 2
        Cladopyxinae
    Unnamed 3
        Gonyaulacales
        Unnamed 4 (8 orders)

A phylogeny for dinoflagellates, after Van Den Hoek et al (1995) is presented in Table 13-6. Ebrids and ellobiophyceans are not mentioned, but ebrids have a dinokaryon and share an internal siliceous skeleton with some genera and an inverted Y-shape in 1 (*Dicroerisma*), and ellobiophyceans are 1 of 4 subclasses recognized by Loeblich (1982), the others being Dinophycidae, Ebriophycidae, and Syndiniophycidae (1 order, no dinokaryon).

Table 13-6. Dinoflagellata.

sbph./cl. Protodinophycidae (Oxyrrhinales)
sbph. Metadinophycidae
  cl. Gymnodiniophycidae (11 orders) (gymnodinioid thecal organization)
  cl. Syndinophycidae (fusion of thecal plates)
    spord. Peridiniarae (Gonyaulacales, Peridiniales) (gonyaulacoid-peridinioid thecal type)
    spord. Desmokonta (Desmocapsales, Desmomonadales, Prorocentrales) (prorocentroid thecal type and desmokonty)

Dinoflagellates might be considered a basically protozoan group as phagotrophy is common even among photosynthetic members, that is, they are mixotrophic, and there appears to have been several endosymbioses in this taxon, as some species harbour a cryptophyte or chrysophycean endosymbiote suggesting the taxon was originally heterotrophic and that chloroplasts were acquired from a variety of algal sources. Prorocentrales, Dinotrichales, Phytodinales, Thoracospherales, and Gonyaulacales are phototrophic; Blastodiniales, Oxyrrhinales, Noctilucales, Syndiniales, and Dinamebidales are phagotrophic; Dinophysiales and Gymnodiniales contain both phototrophic and phagotrophic species. Noctilucales contains 1 species with a vestigial chloroplast, and Dinamebidales has some species with a vestigial chloroplast. (Van Den Hoek et al, loc. cit.; Lee et al, loc. cit.)

Two evolutionary models have been put forward: the plate increase model (where thecal plate increase is considered apomorphic and plate decrease model (where the opposite is considered apomorphic), the latter being adopted by Van Den Hoek et al (loc. cit.) as comparative morphology and the fossil record support it. Thoracospherales and Pyrocystales are sometimes recognized also, separated from Peridiniales and Gonyaulacales, respectively, solely on differences in life cycle. Groups are distinguished by their thecal organization, and the thecal patterns (tabulations) are numbered and noted according to the Kofoid System.

Many fossil dinoflagellates are known – c. 2000 species in 425 genera. These are almost never fossils of the organism in its active state, but fossils of cysts, which enable dinoflagellates to survive periods when their environment dries up. One such class of cyst is called hystrichospheres, which are spiny balls with a long fossil record. Because dinoflagellates are sensitive to changes in ocean temperature, salinity, and nutrient levels, the dinoflagellate species that are present in a sediment sample are often used to determine the environmental conditions at the time and place that the sediment was deposited.

The fossil record of the taxon may extend into the Precambrian. Spherical organic-walled microfossils known as acritarchs, some of which may be dinoflagellate hystrichospheres, first appear in rocks about 1.8 billion years old. However, acritarchs lack a characteristic dinoflagellate feature, the archeopyle or excystment pore through which the dinoflagellate exits the cyst; they also lack the cingulum groove characteristic of many dinoflagellates. Various other algae can form cysts that look superficially similar. Exactly what the acritarchs were is not known with certainty; they probably included a number of clades of eukaryotic algae, and are thus a "form taxon," including all those spore-like fossils which have not been conclusively assigned to another group.

There is, however, geochemical evidence for the presence of dinoflagellates in the early part of the Paleozoic and late Precambrian, even though no cysts are known. This comes in the from of dinosteranes, which are derived from dinosteroids. Interestingly, the dominant modern coral group, the Scleractinia, a sponge taxon, took off at about the same time the dinoflagellates did. The zooxanthellas in these corals are dinoflagellates, and it is suspected that there may be a relationship between their diversifications.

The earliest known fossil that might possibly be a true dinoflagellate, *Arpylorus antiquus*, is from the Silurian, about 400 mya. Dinoflagellate fossils are rare in the Paleozoic, but abundant from the Late Triassic onward.

Tending to predominate in warm water communities dinoflagellates are particularly abundant in the Indian Ocean and Red Sea. Links to marine luminescence were demonstrated by Michaelis in 1830. At least 30 species in 5 autotrophic genera (*Gonyaulax, Protogonyaulax, Pyrodinium, Pyrocystis, Ceratium*) in the Peridiniarae complex, and 2 heterotrophic genera (*Noctiluca, Protoperidinium*) in the Gymnodiniarae complex, contain scintillons, particles/organelles that cause luminescence, a blue flash of 1/10 of a second duration, hence the name fire algae and Pyrrhophyceae for the entire class. Its purpose is unknown but it may be to ward off predators. The ocellus, which is present in all 7 genera of Warnowiaceae (Gymnodiniales) has a remarkable resemblance to the animal eye at the subcellular level. 90% of dinophyceans are marine, the other 10 % being fresh or brackish water, terrestrial (as snow algae), or intrazoic.

Some "blooms", dense growths of unicellular planktonic algae, cause red tides in which dangerous toxins are secreted. Genera responsible are especially the phototrophic *Gonyaulax, Protogonyaulax, Gymnodinium, Glenodinium, Dinophysis*, and *Prorocentrum*, and the heterotrophic *Noctiluca*. The red, but sometimes brown, colour is due to accumulation of carotenoids. Raphidophyceae, especially *Chattonella* and *Fibrocapsa*, also cause red tides but less frequently. (Van Den Hoek et al, 1995).

Zooxanthellas, dinoflagellates symbiotic in various animals, were described and named by Karl Brandt in the 1880s. Zooxanthellas are generally symbiotic, specifically mutualistic, with coral animals, so most coral reefs (see Chapter 18) are in shallow, tropical waters, because zooxanthellas prefer warm temperatures and rely on sunlight for photosynthesis, and the prodigious productivity of this ecosystem is because of this relationship. Dinoflagellates are symbiotic also in other cnidarians besides corals, and in forams, polycistines, and mollusks. Unicellular green algae that are endosymbiotes of freshwater animals, like the celenterate *Chlorohydra*, are called zoochlorellas. (Home-Douglas, 1991; Van den Hoek, 1995).

Euglenoids

The earliest classification of euglenoids was in Infusoria, as they were named by Ledermuller in 1763. Although described by O. F. Muller in 1786 in his "Animalcula infusoria," the first comprehensive treatment on infusorians, Ehrenberg's "Die Infusoriensthier" of 1836 was more influential. Flagellata were then established and were divided into 5 subgroups: Protomastigina, Polymastigina, Euglenoidina, Chloromonadina, and Chromomonadina. In 1892, Klebs configured the euglenoids into 3 families named by Senn 8 years later according to botanical nomenclature as Euglenaceae, Astasiaceae, and Peranemaceae. In 1898, Engler established the euglenoids as the order Euglenales. Senn, in 1900, divided Flagellata into 7 subgroups: Protomastigineae, Pantostomatineae, Distomatineae, Chrysomonadineae, Cryptomonadineae, Chloromonadineae, and Euglenineae, and in 1903, treated Flagellata as a phylum and the 7 groups as orders. Pascher, in 1931, elevated the euglenoids to phylum status as Euglenophyta. The phylum's 3 families grouped together the chlorophyllian, saprophytic, and holozoic genera and species, respectively, and the families were recognized as artificial. In 1933, G. M. Smith established the family Colaciaceae for the single genus *Colacium*, a chlorophyllian, nonmotile group forming into dendroid colonies in a gelantinous sheath and having a stalk system, and in 1938 created an order for it. In 1948, Skuja added 2 more families: Rhynchopodaceae and Rhizaspidaceae.

Leedale presented a scheme including 6 orders: Eutreptiales, Euglenales, Euglenomorphales, Rhabdomonadales (Menoldidae) (osmotrophic), Sphenomonadales (osmotrophic or phagotrophic), Heteronematales (phagotrophic), the 1st 3 being green or colourless, the 1st 2 being mostly with a stigma and flagellar swelling, and the last 3 being heterotrophic and without a stigma or flagellar swelling. It is now most commonly used but the orders are basically unclassified, and no families are recognized. Much better arrangements were done by Mignot (1966) and Johnson (1968). The former had 2 suborders and 7 families arranged like this:

Table 13-6. Mignot's 1966 taxonomy.

- Euglenina
  - Euglenidae (including Menoldidae)
  - Distigmidae
  - Eutreptidae
  - Euglenomorphidae
- Poranomina
  - Poranomidae
  - Petalomonadidae
  - Scytomonadidae

The latter had 3 suborders, the 1st suborder having 2 families, Euglenidae and Menoldidae, the 1st of these being divided into 3 subfamilies (Eutreptinae, Euglenomorphinae, and Eugleninae). Leedale (1978) did, however, present a putative phylogeny in which *Distigma* was the root and led to 2 lines, Sphenomonadales+(Eutreptales-Euglenomorphales+Euglenales) and Rhabdomonadales+Heteronematales.

Later, Kinetoplastida, Diplonemida (*Diplonema=Isonema*), Pseudociliata (*Stephanopogon*), and Hemimasigota were rightly added, because of fine structural affinities (corroborated by molecular methods, except possibly Hemimastigota, which are included because of pellicular similarites (see Foissner, Blatterer, and Foissner, 1988) and 2-fold rotational symmetry as in Percolomonada) forming an extended/expanded Euglenoida/Euglenophyceae or Euglenaria. Kinetoplastida, the (blood parasitic) hemoflagellates, is the largest order, with 2 families, 16 genera, and 600 species, created by Honigberg in 1963. The others all have only 1 genus, except for Hemimastigota, which has 4.

Often added also are the other discicristates. Polymastigota, a group of parasitic forms, contains 42 genera and some 600-800 species in subclasses Metamonada (orders Diplomonada [2 families], Retortomonada [1 family], and Oxymonada [5 families]) and Parabasalia (orders Trichomonada [4 families] and Hypermastigota [15 families]). Often the polymastigote subclasses are treated as classes and the suborders as orders. With the name Metamonada the clade is arranged as Preaxostyla (Oxymonada+*Trimastix*), Fornicata (Eopharyngia [Diplomanda, Retortomonada], *Carpediemonas*, and *Dysnectes*), and Parabasalia. Sometimes the name Discicristata is reserved for Euglenaria+Heterolobosa, and the name Discoba is reserved for Euglenaria+Heterolobosa+Jacobida.

Jakobids are unicellular, free-living, anisokontic, nonsexual, and bacterivorous, and have similar fine structure. There are 4 genera and 12 described species. *Jakoba* Patterson 1990 is typified by nonloricate trophic cells, each containing a branched mitochondrion and discoid cristas. *Reclinomonas* Flavin & Nerad, 1993 and *Histiona* M. Voigt 1902 are characterized by loricate trophic cells, each with an unbranched mitochondrion and tubular cristas. The latter 2 genera form Histionidae. *Andalucia* Lara et al 2006 was recently added. The feeding groove (excavate)-flagellar vane complex occurs in jakobids, *Malawimonas* O'Kelly & Nerad 1999, and some polymastigotes. Cercomonads have only the excavate. (O'Kelly, 1993; Jakobida-tolweb.org).

A cladistic analysis for euglenoids was performed by Shi Zhi-Xin (1996) using 33 OTUs (operational taxonomic units) and 35 characteristics. The resulting cladogram includes: the most primitive clade composed of phagotrophic, colourless euglenoids closely related to kinetoplastids; primitive green euglenoids; the most complex group, containing advanced green euglenoids and saprophytic, colourless euglenoids; and parasitic euglenoids with many flagella. He acknowledges 1 phylum, 1 class, and 5 orders, and identifies 5 subgenera of *Euglena* Ehrenberg 1830 as independent genera.

A molecular ML analysis for 22 Euglenaria genera was done by Busse and Preisfeld (2003) with the following results (in parentheses are the ML and MP bootstrap values):

Table 13-6. Euglenaria by Busse and Preisfeld.

sbph. unnamed (57, -)
- cl./ord. Diplonemida (100, 100) (1 gen.)
- cl./ord. Kinetoplastida (100, 100) (4 gen.)

sbph. Euglenoida (56, 53)
- cl./ord. Petalomonadida (1 gen.)(phagotrophic euglenoids)
- cl. Euglenoida (100, 100)
    - sbcl. Aphagea (osmotrophic euglenoids) (100, 100)

ord. Astasi-Distigmida (93, 88)(6 gen.)
ord. Distigmida (100, 100)(1 gen.)
sbcl. unnamed (76, 61)
ord. Peranemida (1 gen.)(phagotrophic euglenoids)
ord. Euglenida (100, 100)(8 gen.)(phototrophic euglenoids)

In this, *Astasia* is polyphyletic and *Distigma* is paraphyletic.

Aside from one possible Silurian fossil, a curious microfossil called *Moyeria*, the sparse fossil record of euglenoids is restricted to the late Tertiary.

Like dinoflagellates, Euglenoids have 3 chloroplast envelopes (chromistans have 4) and are mostly unicellular, and, like chromistans, the storage product is beta 1,3-linked glucan. The cells range in size from 15 microns to .5 mm. They occur world-wide and are found in mostly freshwater habitats, such as lakes, rivers, streams, ponds, and ditches. Marine forms, which are less common, occur in the open sea, tidal zones among seaweeds, and sandy beaches. "Blooms" occur in farmyard ponds, greenhouse tanks, agricultural drainage channels, dew ponds, and estuarine mudflats. Some forms favour very acidic environments such as peaty pools, sphagnum bogs, and sulfur lakes at pH 3-4. Terrestrial forms, which are rare, include those in snowfields, on trees, and in insectivorous plants.

Blackwell, W.H. 2009. Chromista Revisted. Phytologia 91: 191-225 (phytologia.org).

Bourelly, P. 1970. Les algues d'eau douce, Tome III. Boubée, Paris.

Busse, Ingo and Preisfeld, Angelika. 2003. Systematics of primary osmotrophic euglenids: a molecular approach to the phylogeny of *Distigma* and *Astasia* (Euglenozoa). IJSEM 53: 617-24.

Butcher, R. W. 1967. An introductory account of the smaller algae of British coastal waters, Part IV: Cryptophyceae. Fishery Invest., Lond., Series IV: 1-54 (as cited in Margulis et al, loc. cit.).

Encyclopedie Universalis. 1995.

Foissner, W., Blatterer, H., & Foissner, I. 1988. The Hemimastigophora (*Hemimastix amphikineta* nov. gen., nov. spec.), a new protistan phylum from Gondwanian soils. Europ. J. Protistol. 23: 361- 383.

Home-Douglas, Pierre (ed.). 1991. Oceans (How Things Works series). Time-Life, Alexandria, Virg.

Johnson, P.L. 1968. The taxonomy, phylogeny, and evolution of the genus *Euglena*. In: Buetow, D.E. (ed.), The Biology of Euglena, vol. I, pp 1–25. Academic Press, New York.

Kawachi, M., Inouye, I., Honda, D., O'Kelly, C.J., Bailey, J.C, Bidigare, R.R, and Andersen, R.A. 2002. The Pinguiophyceae classis nova, a new class of photosynthetic stramenopiles whose members produce large amounts of omega-3 fatty acids. Phycol. Res. 50: 31–47 (Abstract- Wiley.Com).

Kessel, Edward (ed.). 1955. A Century of Progress in the Natural Sciences. California Academy of Sciences, San Francisco (Archive.Org).

Lamoureux, Gisèle. 1985. Plantes sauvages du bord de la mer. Fleurbec, Montréal.

Lee, T. F. 1986. The Seaweed Handbook. Dover (republication; originally published in 1977 by Mariner's Press).

Leedale, G. F. 1967. Euglenoid Flagellates. Prentice-Hall, Englewood Cliffs, New Jersey.

Leedale, G. F. 1978. Phylogenetic criteria in euglenoid flagellates. Biosystems 10, 183–187.

Mignot, J. P. 1966. Structure et ultrastructure de quelques euglénomonadines. Protistologica 2: 51–140.

O'Kelly, C. J. 1993. The jakobid flagellates: structural features of *Jakoba*, *Reclinomonas,* and *Histiona* and implications for the early diversification of eukaryotes. J. Eukaryot. Microbiol. 40: 627-636 (Abstract – wiley.com).

Shi Zhi-Xin. 1996. Cladistic Analysis of Euglenoids. J. Syst. Evol. (formerly Acta Phytotaxonomica Sinica) 34: 265–275 (jse.ac.cn, abstract).

Chapter 14 Fungi

Fungi are the grand recyclers of the planet and the vanguard species in habitat restoration. - Paul Stamets, American mycologist – topfamousquotes.com.

Macroscopic funguses were, of course, known in Antiquity and, along with microscopic ones, including pseudofungi and slime molds, were first given regnal status by Necker (1783) and Fries (1832)(The Father of Mycology), the former as Regnum Mesymale, the latter as Regnum Mycetoideum.

15th century Italian Renaisssance humanist, diplomat, and professor Ermalao Barbaro, who did commentaries on Plinius and Dioscorides as well as important work on Aristotle, divided the Fungi, as he named name, into 8 groups: ovali, digitali, spongioli, poriginosi, pezicae, prunuli (along with spunuli and cardeoli), laciniae, and igniarii. It was in his Corollaries on the 5 Books of Dioscorides, published posthumously in 1516.

In 1682, in his "Methodus plantarum" and "Historia plantarum" (3 vols: vol. 1 in 1686, vol. 2 in 1688, vol, 3 in 1704), British botanist John Ray, who was one of the early parson-naturalists, did important work on higher plant taxonomy and is considered by many as the originator of natural classification, using various characteristics, especially the pileus (cap), recognized 5 divisions of what he also named Fungi: Pileati lamellati, Pileati lamellis carentes, Pileis destituti, Pulverulenti, and Subterranei, the last obviously refering to truffles. In the 2nd version in 1700 the funguses are preceded by Submarinae (the algae), and followed by Musci and Capillares (ferns). He also wrote on zoology and natural theology.

In 1719 in his "Catalogus plantarum," illustrated by himself, German botanist J.J. Dillenius distinguished Lamellosi, Porosi, and Echinati. Linneus recognized the 3 groups of Dillenius as the genera *Agaricus*, *Boletus*, and *Hydnum*, adding 8 more genera, *Phallus, Elvela, Clavaria, Clathnus, Peziza, Lycoperdon, Byssus*, and *Mucor*, but placing *Tremella* with the algae and the lichens with *Chara*, hepatics, and sponges. Linneus reunited in Cryptogamia the 4 orders already recognized by Ray: Filices, Musci, Algae, and Fungi. So the name Musci goes back at least to Ray and Fungi to at least Barbaro, even though Linneus is credited with them.

French botanist Sébastien Vaillant had disdain for funguses, as they are "flowers without flowers" and asexual, and saw them as a damned race and a diabolical invention upsetting nature's harmony. He nonetheless divided the genus Fungus (in those days the definition for a genus was not standardized) into 6 families in his "Botanicum parisense," published posthumously in 1727, conceieved in the same spirit as Dillenius, which were those with: caps having linings, caps having linings of small tufts or papillas, caps with linings of long tips like the needles of hedgehogs, caps lined with pipes, caps lined with branching nervations, and sheets. This was for fleshy funguses. The other funguses he called Fungoides.

Bernard de Jussieu, in 1759, placed the 11 Linnean genera of fungi in increasing order of complexity, rather than alphabetically as Linneus had done. De Candolle placed them in his Acotylédonés along with Algae, Hypoxylons, and Lichenes. De Jussieu's nephew, Antoine-Louis, adopted the same arrangement as de Candolle for the 1st 4 families of his Acotylédonés.

In 1809, German naturalist and botanist Johann H. F. Link, who studied under Blumenbach and taught Ehrenberg, distinguished the orders Epiphytae, Mucedines, Gasteromyci, and Fungi. The species he named, such as *Cordyceps*, *Fusarium*, *Phragmidium*, and *Myxomyces*, are still in use today. He wrote an abundant number of articles and books on a variety of scientific subjects, such as physics, chemistry, geology, botany, zoology, philosophy, paleontology, and early history. He was a graduate of the U. of Gottingen and affiliated to the universities of Rostock and Breslau, and a member of the Berlin Academy of Sciences, the Royal Swedish Academy of Sciences, and the Leopoldina Academy.

1825 saw the publication of French botanist Adolphe Théodore Brongniart's classification, which comprised 5 families: Urédinées, Mucidinées, Lycoperdacées, Fungi, and Hypoxylées, but which did not correspond to modern circumscriptions. Brongiart was president of the Académie des sciences, chairman of the Musée national d'histoire naturel, and recipient of the Wollaston Medal.

Genevan clergyman and botanist Jean Etienne Duby recognized the same groups as Brongniart but added Lichenes. He was president of the Société de physique et d'histoire naturelle de Genève in 1860–61 and a corresponding member of the Société de biologie de Paris and the Moscow Society of Naturalists.

A major advance came in 1801 with the much more systematic, extensive, and elaborate taxonomy of Christiaan Hendrik Persoon, South African-born mycologist working in Germany and France, in his "Synopsis methodica fungorum," who recognized 2 classes, 6 orders, 3 sections, and 71 genera, the 2 classes arranged as follows:

cl. Angiocarpi
 Sclerocarpi
 Sarcocarpi
 Dermocarpi (3 sections)
cl. Gymnocarpi
 Lytothecii
 Hymenothecii
 Nematothecii

The first version of Swedish mycologist (the Linneus of funguses and considered by some as the father of modern mycology) Elias Magnus Fries's system, appearing from 1821-1832, in his 3-volume "Systema mycologicum," a fundamental work in the history of mycology, distinguished the same orders as Link but renamed them Coniomycetes, Hyphomycetes, Gasteromycetes, and Hymenomyetes and elevated them to classes and each had 4 orders, each of these usually with 4 genera or 4 other subdivisions. He applied the philosophical ideas of Nees von Esenbeck (probably Theodore, the younger of the 2 botanist brothers, who authored Das System der Pilze in 1837, the older brother being named Christian) to the explanation of the 4 fundamental types represented in his 4 classes: a) determined by the preponderance of the reproductive effort itself, b) principally by air, c) by heat, and d) by light, respectively. Here is the taxonomy in more detail:

Coniomycetes (sporidia naked, without receptacles; all except part of order Entophytae corresponding to our present Fungi Imperfecti--this latter order contained also the rusts and smuts--they were not true organisms, according to Fries, but exanthemata of diseased plants).
Hyphomycetes (thallus floccose, the sporidia borne on or among the hypha; these, too, were mainly Fungi Imperfecti).
Gasteromycetes (the whole fungus closed, containing the sporidia in its interior; includes the present-day Gasteromycetes, Mucorales, Mycetozoa, and Pyrenomycetes).
Hymenomycetes (hymenium soon exposed, bearing the sporidia superficially).

His final version was in 1862 and included 6 families, each with 6 tribes, which were Hymenomycetes, Discomycetes 1836, Pyrenomycetes 1823, Gasteromycetes, Gymnomycetes, and Haplomycetes.

Fries was affiliated to Lund and Uppsala universities, director of the Botanical Gardens, member of the Swedish Academy, and foreign member of the Royal Society. His son Theodor Magnus was also a botanist. The genus *Friesia* (in Tiliaceae) was named after him by de Candolle in 1824 and some 100 fungal species are named for him.

Czech physician and mycologist August Corda, in his monumental 6-volume "Icones fungorum hucusque cognitorum," published from 1837 to 1842 and finally in 1854, adopted the same 4 classes as Fries's 1st version but used the name Myelomycetes for Gasteromycetes.

French physician and myclogist and U. of Paris graduate Joseph-Henri Léveillé, in his "Considérations mycologiques, suivies d'une nouvelle classification des champignons" from 1846, recognized 6 divisions: Basidiosporés (with basidiums), Thecasporés (with ascuses), Clinosporés, Cystosporés, Trichosporés, and Arthrosporés. The first 2 would survive in tact. It was he who coined the word "basidium", but his "theca" was replaced by "ascus" by Esenbeck.

Lewis David von Schweinitz was a German-American mycologist and Moravian cleric (his greatgrandfather, Count Niklaus Ludwig Zinzendorf, was the founder of the Moravian Church), is considered by some as the father of American mycology, and his herbarium had the largest private collection of plants in the US, with some 23,000 species. His system in his "Synopsis fungorum carolinae superioris" from 1832 follows Fries but rearranges the classes as follows:

Table 14-1.

A. Ascomycetes: bearing the sporidia in ascuses

Class I. Hymenomycetes (ascuses on an open receptacle)
Class II. Pyrenomycetes (ascuses within peritheciums)

B. Sporomycetes: bearing free sporidia, not in ascuses

Class III. Gasteromycetes (sporidia free within a peridium)

Coniomycetes of Fries, sporidia without peridium
Class IV. Hyphomycetes (sporidia borne directly on the thallus)
Class V. Gymnomycetes (sporidia borne on a sporodochium)

The British priest (probably Anglican) and botanist Rev. Miles Joseph Berkeley, considered the father of British mycology, described about 5,000 species, and his herbarium had some 10,000 species. He was a member of the Royal Society, the Linnean Society of London, and several other scientific organizations, and was winner of the Royal Medal. He also wrote on algae and mosses. Here is his classification, which was presented in his Introduction to Cryptogamic Botany in 1857.

Table 14-2.

Fungales

Sporidiiferi (sporidia in sacs)
Ascomycetes: asci formed from the fertile cells of an hymenium (same as our modern Ascomycetes)
Physomycetes (practically identical with our Mucorales)

Sporiferi (naked spores)

Hyphomycetes: spores variously seated on conspicuous threads which are rarely compacted; mostly small in proportion to the threads (mainly of Fungi Imperfecti, but including also Peronospora).

Coniomycetes: spores mostly terminal, seated on inconspicuous threads, free or enclosed in a perithecium (includes Uredinales and Ustilaginales, in addition to some of the dematioid imperfect fungi).

Gasteromycetes: hymenium enclosed in a peridium, seldom ruptured before maturity (includes, in addition to our present-day Gasteromycetes, also Mycetomycetes).

Hymenomycetes: hymenium free, mostly naked, or if enclosed at first, soon exposed (includes Tremellales (in the wider sense), Polyporales, and Agaricales).

Saprolegniales (in Oomycetes) was included by Berkeley among the Conferva group of algae (but with the doubt expressed that some genera may be molds, which they are, but related to other oomyctes), because their reproductive structures and manner, sexual and asexual (the latter by zoospores), were in his opinion of greater weight than their lack of starch and chlorophyll. This basically turned out to be the correct view as oomycetes are heterokontic algae.

The next important classification of the funguses was by German mycologist and microbiologist Heinrich Anton De Bary, in his 1866 textbook, Lehrbuch der Botanik. He divided the funguses into 4 orders: Phycomycetes, Hypodermii (smuts and rusts), Basidiomycetes, and Ascomycetes. He has no group set aside for what we call Fungi Imperfecti, which he rather looks on as nonsexual forms of Ascomycetes whose connections with the sexual stages have not been demonstrated. Eighteen years later, De Bary modified this classification by establishing 2 series as follows:

Table 14-3.

I. Ascomycetenreihe

1. Peronosporeen (nebst Ancydsteen und Monoblepharis; "together with Ancydsteen and Monoblepharis")
2. Saprolegnieen
3. Mucorineen oder Zygomyceten (Mucorales or Zygomycetes)
4. Entomophthoreen
5. Ascomyceten
6. Uredineen

II. Von der Ascomycetenreihe divergierende oder der Stellung nach zweifelhafte Gruppen (divergent from Ascomycetes or position as doubtful groups)

7. Chytridieen
8. Protomyces und Ustilagineen
9. Zweifeldafte Ascomyceten (doubtful Ascomycetes)(*Saccharomyces*, etc.)
10. Basidiomyceten

Groups 1-4, because of their presumed near connection with the algae, are brought together under the name Phycomycetes. In series II, groups 7 and 8 are to be treated in connection with Phycomycetes, 9 naturally with Ascomycetes, and 10 with 6. De Bary's opinion concerning Saprolegniales in his first article on the group in 1852 was apparently the same as Berkeley.

De Bary briefly practiced medicine in Frankfurt, then became Privatdozent in botany at the

University of Tübingen, succeeded the position of the well-known botanist Karl Wilhelm von Nägeli at Freiburg, where he established the most advanced botanical laboratory at the time and taught many students, including Oltmanns, Brefeld, and Francis Darwin (son of Charles), was botany professor at the University of Halle, editor of the pioneer botanical journal Botanische Zeitung, and botany professor at the University of Strasbourg, where he founded the Botanical Garden, and was the first rector (president) of the reorganized university.

German botanist Julius Oscar Brefeld, one of De Bary's students, learned methods from him and improved on his technique. He published a series of 15 Hefte (excercise books or journal issues) titled Botanische Untersuchungen (Botanical Research) from 1872 to 1912. Brefeld rejected the ideas of Sachs and De Bary, except the origin of Oomycetes from Siphoneae in the green algae, which for him does not represent true fungi. He regards the true funguses as begining with Mucorales, which he considers to have developed from green algae that produced zygospores (Conjugatae). He emphasizes that the algae retained their sexuality and evolved into the higher green plants. The primitive funguses (Mucorales) quickly began to magnify the importance and complexity of their asexual reproduction at the expense of their sexual reproduction, which soon disappears as evolution progresses toward the higher funguses. In the main line of fungal evolution, the basic group is class Zygomycetes (the class Oomycetes, in his view, comes to a dead end). This is his classification:

Table 14-4.

- Zygomycetes
    - Carposporangica
    - Exosporangica
- Mesomycetes
    - Hemiasci (asciform sporangia)
    - Homobasidii (basidiform sporangia)
    - Ustilaginae
    - Tilletiae
- Mycomycetes
- Ascomycetes (determinate sporangia)
    - Exoasci
    - Carpoasci
- Basidiomycetes (determinate conidiophores)
    - Protobasidiomycetes
    - Autobasidiomycetes

The classification of funguses followed in the first edition of Engler and Prantl (1897-1907) is, in its main features, the same as that of Brefeld. In the second edition (1926-1938) the main features are retained with some modifications made necessary by the cytological confirmation of the actual occurrence of sexuality in the higher fungi as well as in Phycomycetes.

French pharmacist and mycologist Narcisse Patouillard in 1887, in "Les Hyménomycètes d'Europe-Anatomie générale et classification des champignons supérieurs," and Brefeld in 1888, working independently, used microscopic structure as a basis for separating Basidiomycetes into groups with septate basidia and groups with nonseptate ones. It is Patouillard who coined the names Hétérobasidiés (also called Phragmobasidiomycetes) for the former and Homobasidiés (also called Holobasidiomycetes) for the latter. His doctoral thesis was "Des Hyménomycètes au point de vue de leur structure et de leur classification." In 1920 he became an honorary member of the British

Mycological Society. He is highly regarded for his taxonomic work in mycology; he described numerous genera and species, was the author of nearly 250 works, and was a leading authority on tropical mycology (over 100 of his publications involved studies of fungi from tropical locales).

The Lehrhuch der Botanik by German botanist Julius Sachs (1832-97) was of great influence in the development of botanical studies. This appeared in many editions and was translated into several languages. In his earlier editions he followed De Bary for the classification of the funguses. In his 4th edition (1874) he adopted a quite different arrangement. He places the plants below the group Bryophyta in the group Thallophyta. This he divides into 4 classes, each containing plants with chlorophyll and those without it. The main line of evolution he indicates goes upward in the chlorophyll-containing series (i.e., the algae), the chlorophyll-free organisms in each class being derived from those with chlorophyll in the same class. In other words, the funguses are polyphyletic and do not form a single phylum. The 4 classes are the following:

Table 14-5. Sachs' Classification of the Fungi.

I. Protophyta. No sexual reproduction
with chlorophyll (Cyanopliyceae, Palmellaceae (in part))
II. Zygosporeae. Sexual union of equal cells to produce a zygospore

with chlorophyll	without chlorophyll
Union of motile cells	
Volvocineae (Hydrodictyeae)	Myxomycetes
Conjugation of resting cells	
Conjugatae (including Diatomeae)	Zygomycetes

without chlorophyll (Schizomycetes (=Bacteria), Saccharomyces)

III. Oosporeae. Fertilization of egg to produce an oospore

with chlorophyll

Sphaeroplea
Vaucheria
Oedogonieae
Fucaceae

without chlorophyll

Saprolegnieae (Oomycetes)
Peronosporeae (Oomycetes)

IV. Carposporeae. Sexual reproduction results in the production of a spore body with chlorophyll

Coleocheteae
Florideae
Characeae

without chlorophyll

Ascomycetes (including the lichens)
Aecidiomycetes
Basidiomycetes

Sachs believed that Saprolegniales and then Peronosporales arose from algae closely related to *Vaucheria*. The branched, coenocytic, vegetative structure with cellulose cell walls, the production of zoospores in terminal segments of the hyphas, and the formation of large oogonia with antheridia usually arising nearby are characteristics common to *Vaucheria* and Saprolegniales.

The idea that Mucorales represent developments from Conjugatae in which the chlorophyll had been lost was adopted by Sachs from Brefeld, who emphasized the similarity in the formation of the zygospores in both groups. Although the suggestion of De Bary and Sachs that Saprolegniales are probably derived from *Vaucheria*-like algae had persisted in some quarters (Gaumann, 1949), some mycologists rejected the idea of the close kinship between these groups because of other factors: tubular, coenocytic hyphas in Mucorales, cellular hyphas with uninucleate cells in Conjugatae; cell wall mainly chitinous in the former, cellulosic in the latter; and abundant production of asexual wind-borne spores in the former, no special asexual cells in the latter.

Sachs was born in Breslau, Prussia (now in Poland), graduated from Charles U. in Prague, and was professor at the universities of Bonn, Freiberg, and Wurzburg. His famous students included Brefeld, Francis Darwin, Georg Klebs, Karl Prantl, C. E. Stahl, and Hugo de Fries, among others. His series of Keimungsgeschichten ("germination histories") laid the foundation of our knowledge of microchemical methods and the morphological and physiological details of germination. He revived the water-culture method and its application to the investigation of the problems of nutrition. He discovered that the starch grains in chloroplasts are the first visible product of their assimilatory activity. His later articles were almost exclusively published in the 3 volumes of the Arbeiten des botanischen Instituts in Würzburg (1871–88). Among these are his investigation of the periodicity of growth in length, in connection with which he devised the self-registering auxanometer by which he established the delaying influence of the highly refrangible rays of the spectrum on the rate of growth; his researches on heliotropism and geotropism, in which he introduced the clinostat; his work on the structure and arrangement of cells in growing points; the elaborate experimental evidence on which he based his imbibition theory of the transpiration current; and his exhaustive study of the assimilatory activity of the green leaf.

German botanist Friedrich Wilhelm Zopf was for many years Professor of Botany at the University of Halle. He made extensive studies on Chytridiales and other small aquatic funguses parasitic in algae and small animals. His textbook on fungi in 1890 was, next to that of De Bary, the outstanding work on the subject for many decades. He followed a classification similar in part to that of Brefeld, but placed Ascomycetes last. In this group he went from the simple forms, like *Saccharomyces*, *Endomyces*, Gymnoascaceae, to, at the peak, Pezizales. He recognized the formation of ascogonia in many ascomycetes and even the union of these in *Pyronema*, with club-shaped "pollinodia," but expressed doubt as to their real sexual function.

In the early part of the 20th century, French botanist and mycologist Pierre Clément Augustin Dangeard completely separates the funguses from the algae, thus forming 2 independent series. Both are assumed to have evolved from Flagellata. The algae became plants at the point of evolution where their flagellate, chlorophyll-containing ancestors lost the power of phagotrophy, and the funguses arose from the flagellates that lacked chlorophyll; likewise, at the point where they were no longer phagotrophic, the funguses are a kingdom like plants and animals. G. W. Martin apparently made a somewhat similar suggestion in 1932.

Dangeard worked at the Faculté des sciences de Caën, was professor of botany at the Université de Poitiers and Faculté des sciences de Paris (the leading higher ed science institution in France throughout its existence from 1811 to 1970), was a member of several scientific organizations including the Académie des sciences, the Société botanique de France (which he chaired from 1914-1918), Société mycologique de France, etc., and founded the magazine Le Botaniste in 1887, and received various honours such as a Légion d'honeur medal and a doctorate honoris causa from the University of Cambridge. He married Henriette Labrosse, the daughter of an admiral, with whom he had 4 children, including the phycologist Pierre Jean Louis Dangeard (1895-1970) and geologist Louis Dangeard (1899-1987).

Below is the system presented in Caratini (La vie de plantes, 1971) based mostly on Marius Chadefaud's 1960 system (the number of orders are in parentheses):

Table 14-6.

- Ascomycetes
  - Laboulbeniomycetes (1)
  - Discomycetes (5 spords.)
    - Lécanoriens (Lecanorales)
    - Helotiens (Hypodermales, Helotiales)
    - Leotiens (Ostropales, Leotiales, Sarcoscyphales)
    - Pezizéens (Pezizales)
    - Caliciens (Calicales)
  - Pyrenomycetes (3 spords.)
    - Dothydéens (Pleosporales, Dothoriales, Myrangiales)
    - Sphereacéens (Coronophorales, Diaporthales, Sordariales, Nectriales, Diatrypales, Xylariales, Hyponectriales, Glomerellales)
    - Clavicipiciens (1 ord.)
  - Plectomycetes (Discomycetes: Tuberales (truffles), Elaphomycetales (deer truffles); Pyrenomycetes: Erysiphales (powdery mildews), Eurotiales (blue and green molds))
  - Hemiascomycetes (Endomycetales, Dipodascales)
- Basidomycetes
  - Archeomycetes (Uredinales (rust), Ustilaginales (smut), Septobasidiales, Auriculariales)
  - Neomycetes
    - Heterobasidiés (Tremellales)
    - Homobasidiés hyménomycètes (Polyporales, Agaricales, Boletales)
    - Homobasidiés gastromycètes (Keratales (stinkhorns), Gastreales)

Sporobolomycetales (shadow or mirror yeasts)
Zygomycetes (Mucorales (bread mold), Entomophthorales)
Phycomycetes
Oomycetes (Leptomitales, Saprolegniales (white blister rust), Peronosporales (downy mildew))
Hyphomycetes (Hyphochytiales)
Chytridiomycetes (Blastocladiales, Chytridiales, Monoblepharidales)
Myxomycetes (Ceratiomyxales, Myxogastrales, Acrasiales)
Trichomycetes (Amebidales, Harpellales, Eccrinales)
Lichenes
Ascolichenes
Pyrenolichenes (Arthropyreniales, Verrucariales, Trypetheliales)
Discolichenes (Caliciales, Graphidales, Thelotremales, Cyanophilales, Lacideales, Lecanorales, Caloplacales)
Basidiolichenes (1)

The first to circumscribe the kingdom phylogenetically was Jeffrey in 1971, but the 4 subkingdoms and 9 divisions were not named (but Chytridiomycota was not included elsewhere and Oomycota was included in Chromobiota). Edwards in 1976 also delimited Kingdom Fungi phylogenetically and explicitly states the included taxons. Jeffrey in 1982 delimited the kingdom phylogenetically and named the included taxons. R.T. Moore gave the Latin diagnosis for the kingdom in 1980.

Below is the classification for phylum Eumycota presented in Parker (loc. cit.). The figures in parentheses are the families, genera, and species; the families are taken from Parker and the genera and species from the Penguin Dictionary of Botany.

Table 14-7.

Mastigomycotina
Chytridiomycetes (Chytridiales, Blastocladiales, Monoblepharidales) (14,110, 685)
Hyphochytridiomycetes (Hyphochytridiales)
Oomycetes (Saprolegniales, Leptomitales, Lagenidiales, Peronosporales)
Zygomycotina
Zygomycetes (Dimargaritales, Endogonales, Entomophthorales, Kickxellales, Mucorales, Zoopagales)(18, 85, 515)
Trichomycetes (Harpellales, Asellariales, Eccrinales, Amebidales)(6, 30, 105)
Ascomycotina
Hemiascomycetes (Endomycetales, Protomyceteales, Taphrinales)(no ascocarp) (6, 60, 330)
Loculomycetes (ascuses develop in locule (inner portion of ascocarp) and are double-walled, having an firm endoascous and an extensible exoascus (bitunicate condition))(58, 530, 2000)
Loculoplectascomycetidae (Myriangiales)
Loculoparenchemycetidae (Asterinales, Dothideales (sooty molds))
Loculoanoteromycetidae (Chetothyriales, Verrucariales)
Loculoedaphomycetidae (Hysteriales, Pleoporales)

Plectomycetes (Ascospherales, Gymnascales, Eurotiales,
Microascales, Elaphomycetales)(cleistothecial (no ostioles); ascuses scattered throughout the centrum)(10, 160, 2300)

Pyrenomycetes (Erysiphales, Meliolales (black mildews), Chetomiales, Melanosporales, Sordariales, Xylariales, Diaporthales, Hypocreales, Clavicipitales, Coryniales, Coronophorales, Laboulbeniales)(perithecial, with or without osteoles; ascuses form in hymenial layer)(20, 640, 6000)

Discomycetes (Ostropales, Phacidiales, Helotiales; Cyttariales, Medeolariales, Pezizales; Tuberales)(apothecial (ascocarp open); with operculate or inoperculate ascuses borne on disc and typically arranged in a palisade layer (hymenium) or tufts and with paraphyses (sterile filaments) apothecium commonly cup- or saucer-shaped giving common name of cup funguses)(27, 425, 3000)

Basidomycotina

Teliomycetes (Uredinales and Ustilaginales)(with resting spores (teliospores), haustoria (specialized structures sent into host cell), and sterigma (formed by meiosis and bearing haploid basidiospores))(10, 174, 5850)

Phragmobasidiomycetes (jelly funguses)(septate)

Heterobasidiomycetidae (Eutremellales, Septobasidiales)

Metabasidiomycetidae (Metatremellales)

Hymenomycetes (Exobasidiales, Agaricales)(basidospores forcibly discharged) (33, 675, 5000)

Gasteromycetes (Protogastrales, Hymenogastrales (false truffles), Podaxales, Gauteriales, Agaricogastrales, Melanogastrales, Keratales, Lycoperdales (puffballs), Tulostomatales (stalked puffballs), Sclerodermatales (earthballs), Nidulariales (bird's nest) (basidospores not forcibly discharged; with gleba (mass of spores and filaments) and often a capillitium (tangled mass of filaments among the spores))(21, 150, 700)

Deuteromycotina (Fungi Imperfecti, i.e., sexual stages absent or unknown)

Hyphomycetes (Melanconiales, Spheropsidales) (conidia (spores) borne on the substrate)(5, 930, 7500)

Celomycetes (Hyphomycetales, Stilbellales, Tuberculariales, Agonomycetales)(conidia borne formed in enclosed structures) (5, 870, 7000)

In Discomycetes, the first 3 orders are inoperculate (without apical lid), the second 3 are operculate (with apical lid), and Tuberlaes have lost the apical discharge mechanism.

Anders Tehler's (1988) cladistic classification in simplified form (Oomycetes and Hyphochtriomycetes not included; Basidiomycetes are not given in detail) is presented in Table 14-8. PAUP 2.4 and Fitch were used, there were 21 taxons, 51 characteristics (in morphology, chemistry, molecular biology, fine structure, ecology, and physiology), 73 steps, and there was a CI of .85. Oomycetes were used as the outgroup.

Table 14-8. Tehler's Fungal Classification from 1988 (names are those of Tehler; in parentheses are clarifications and the number of families, genera, and species).

Blastocladiomycota (Blastocladiales)(3, 7, 54)

- Unnamed 1
  - Chytridiomycota (Chytridiales)(9, 93, 550)
  - Amastigomycota
    - Zygomycotina (Zygomycetes)(25, 115, 620)
    - Unnamed 2
      - Dipodascomycotina (Dipodascaceae = Hemiascomycetes 1)(1 sp.)
      - Unnamed 3
        - Endomycotina (Endomycetaceae =Hemiascomycetes 2)(3 gen.)
        - Unnamed 4
          - Dipodascopsidomycotina (*Dipodascopsis* =Hemiascomycetes 3)(1 sp.)
          - Dicaryomycotina
            - Ascomycetes (*Myxotrichum*=(Eurotiales+Euascomycetes))(140, 1800, 13,300)
            - Unnamed 5
              - Protobasidiomycota (Saccharomycetaceae+Taphrinales=Hemiascomycetes 4)(2, 17, ?)
              - Basidiomycetes
                - Ustilaginaceae (17 gen.)
                - Unnamed 6
                  - Uredinales (5, 126, 5000)
                  - Unnamed 7
                    - Exobasidiales (1, 6, 20)
                    - Unnamed 8
                      - Tilletiaceae (13 gen.)
                      - Hymenomycetidae
                        - Homobasidiomycetes (63, 825, 5700)
                        - Unnamed 10
                          - Dacrymycetaceae (10, 70)
                          - Unnamed 11
                            - Auriculariaceae (20, 100)
                            - Unnamed 12
                              - Tremellaceae (20,100)
                              - Tulasnellaceae (3, 30)

In Parker, Tilletiaceae (bunt) is in Ustiliginales, Auriculariaceae in Eutremellales, and Dacrymycetaceae and Tulasnellaceae in Metabasidiomycetidae. Tehler designated Homobasidiomycetes as containing Aphyllophorales, Polyporales, Boletales, Agaricales, and Gasteromycetes. The first of these is an artificial group composed of ungilled Agaricales, gills (lamellas) being radiating, platelike structures on the underside of the cap that produce spores. About half of the 32 families in Agaricales in the Parker classification are without gills, and many polypores and several boletes lack them as well. Chanterelles (with the gomphoid group, the type genus being known as the scaley chantarelle sometimes separated as a family) are considered as not having true gills and are called sublamellate.

The classification by David Hibbett et al (66 authors)(2007) was a synthesis of various molecular taxonomies combined with input from various mycological taxonomists and is as follows (the numbers in parentheses are the order totals unless specified otherwise):

Table 14-9.

Microsporidia Balbiani 1882 (1)
Kickxellomycotina Benny 2007 (1 class)(4)
Zoopagomycotina Benny 2007 (1)
Entomophthomycotina Humber 2007 (1)
Blastocladiomycota James 2007 (syn. Allomycota Cavalier-Smith, 1981)(1)
Mucoromycotina Benny 2007 (1 class)(3)
Neocallimastigomycota-Chytridomycota
  Neocallimastigomycota Powell 2007 (1)
  Chytridiomycota Powell 2007
    Monoblepharidomycetes Hibbett et al 2007 (1)
    Chytridomycetes Sparrow 1958 (3)
Glomeromycota Walker and Schuessler 2001 (1 class) (4)
sbk. Dikarya Hibbett, James, and Vilgalys 2007 (syn. Carpomycetaceae Bessey 1907, Dikaryomycotina Tehler 1988, nom. nud., Neomycota Caval.-Sm. 1998)
  Ascomycota Berkeley 1857 (syn. Ascomycetes Berkeley 1857)
    Taphrinomycotina Eriksson and Winka, 1997
      Neolectomycetes Eriksson and Winka, 1997 (1)
      Pneumocystidomycetes Eriksson and Winka, 1997 (1)
      Schizosaccharomycetes Eriksson and Winka, 1997 (1)(fission yeasts)
      Taphrinomycetes Eriksson and Winka, 1997 (1)
    Saccharomycotina Eriksson and Winka, 1997 (1)(budding yeasts)
    Pezizomycotina Eriksson and Winka, 1997 (3 orders of uncertain position)
      Orbiliomycetes Eriksson and Winka, 1997 (1) (Discomycetes 1)
      Pezizomycetes Eriksson and Winka, 1997 (1) (Discomycetes 2)
      Dothideomycetes Eriksson and Winka, 1997 (8) (Loculomycetes 1 (Pyrenomycetes 2))
      Arthoniomycetes Eriksson and Winka, 1997 (1) (Loculomycetes 2 (Pyrenomycetes 3))
      Eurotiomycetes Eriksson and Winka, 1997 (7) (Plectomycetes 1 + Loculomycetes 3)
      Laboulbeniomycetes Engler, 1898 (2) (Pyrenomycetes 5)
      Lichinomycetes Reeb, Lutzoni, and Roux, 2004 (1)
      Lecanoromycetes Eriksson and Winka, 1997 (10) (Discomycetes 4)
      Unnamed 1
        Leotiomycetes Eriksson and Winka, 1997 (5) (Discomycetes 5 + Plectomycetes 2 (Pyrenomycetes 6)
        Sordariomycetes Eriksson and Winka, 1997 (16) (Pyrenomycetes 7)
  Basiodiomycota de Bary 1866 (2 classes (1 order each) of uncertain position)(syn. Basidiomycetes Bold 1957 ex Moore 1980)
    Wallemiomycetes Zalar et al, 2005 (1 genus)
    Entorrhizomycetes Begerow, Stoll, and Bauer, 2007 (1)
    Pucciniomycotina Bauer et al 2006
      Classiculomycetes Bauer et al 2006 (1)
      Cryptomycocolacomycetes Bauer et al 2006 (1)
      Mixiomycetes Bauer et al 2006 (1)
      Atractiellomycetes Bauer et al 2006 (1)
      Agaricostilbomycetes Bauer et al 2006 (2)
      Cystobasidiomycetes Bauer et al 2006 (3)

Pucciniomycetes Bauer et al 2006 (5)
Microbotryomycetes Bauer et al 2006 (4)
Ustilaginomycotina Bauer et al 2006 (1 order of uncertain position)
Ustilaginomycetes Bauer, Oberwinkler, and Vánky 1997, emend. Bergerov, Stoll, and Bauer 2007 (2)
Exobasidiomycetes Bergerov, Stoll, and Bauer 2007 (7)
Agaricomycotina Dowell 2001
Tremellomycetes Dowell 2001 (3)
Unnamed 2
Dacrymycetes Dowell 2001 (1)
Agaricomycetes Dowell 2001 (17)

They recognize 7 phylums, 10 subphylums, 35 classes, 12 subclasses, and 130 orders. Not included are 2 trichomycete orders, Eccrinales and Amebidiales, as they are indicated by molecular analyses to belong in Mesomycetozoa (aka Ichthyosporea), which is probably an artificial group as it has no synapomorphies (Paleobiology Data Base (eol.org; Encyclopedia of Life)), and is considered as sister group to Filozoa, which comprises Filasterea and Choanozoa+Metazoa, but probably has nothing in common with either, its similarities are with funguses, some molecular data indicate a fungal affinity (e.g., Sumathia, 2006), and most ichthyosporeans were placed either in Fungi or Haplosporidia (Lohr et al, 2010). Dermacysta and Corallochytria also belong in Fungi as early branching clades. Also, Nuclearida are considered as sister group to Fungi, based on molecular evidence, but have nothing in common with them, at least nothing in the phenotype that would indicate kinship.

Rozella, named by Cornu in 1872, occupies a deep-branching position in genotypic classifications of Kingdom Fungi, but bootstrap support is inconsistent and often weak in the most comprehensively sampled phylogenies by James et al in 2006 and Jones et al in 2011. The name "Rozellida" (renamed Cryptomycota by Jones et al in 2011) was informally coined by Lara et al in 2010 to accommodate the genus and several environmental sequences that form a distinct clade. (Jones et al, 2011).

Cryptomycotes exist in at least 3 morphologies in freshwater environments: uniflagellate zoospores, variably-shaped cells without flagella attached to other eukaryotic microscopic organisms (e.g., diatom hosts), and non-flagellate cysts. None of these stages were shown to possess a chitin or cellulose wall, although other life-cycle phases with a chitin and/or cellulose wall may remain undetected. (Jones et al, 2011).

The aphelids are a small group of intracellular parasitoids in planktonic algae with 3 genera, *Aphelidium* Zopf 1885, *Amoeboaphelidium* Scherffel 1925, and *Pseudaphelidium* Schweikert et Schnepf, 1996, and 10 valid species, which form along with related environmental sequences a very diversified group. They possess posteriorly-directed uniflagellate zoospores and form an endobiotic plasmodium. *Aphelidium* has lamellar cristas in zoospores and tubular cristas in cysts. The amebas of *Amoeboaphelidium* have lamellar cristas. A unique characteristic of aphelids is the intracellular, amoeboid trophic stage which engulfs the host cell contents. A similar stage is found in *Rozella* but is absent in the fast evolving Microsporidia and other funguses. Microsporidia, Rozellida, Aphelidea form the putative ARM clade, named Opisthosporidia by Karpov et al, which is sister group to eufungi. (Karpov et al, 2014).

Aphelidium and *Amoeboaphelidium* were placed in Cienkowski's Monadinea, comprised of "fungal animals" – organisms with a fungal-like life cycle but an ameboid trophic stage. In the 1950s and 1960s, they were included in the order Proteomyxida or subclass Proteomyxidia in the class Rhizopoda. They were subsequently and strangely completely forgotten in classifications in later years.

By the end of the 20th century much more was known about the life cycles, fine structure, and peculiarities of several species of aphelids. Gromov in 2000 established a new class, Aphelidea, for all 3 genera. Cavalier-Smith in 1998 suggested that *Aphelidium* belongs to the opisthokonts because of their posteriorly directed uniflagellate zoospores and flat mitochondrial cristae. Gromov in 2000 placed the class Aphelidea in the phylum Rhizopoda sensu lato on the basis of the amoeboid nature of the trophozoite stage. Later, Karpov transferred the class Aphelidea into the phylum Mesomycetozoa based on 18S rRNA data. In addition to these classifications based on the 18S rRNA marker, aphelids were placed in Ichthyosporea based on their parasitic nature by Shalchian-Tabrizi et al in 2008, and then, based on their morphology and lifestyle, as an order in class Rozellidea, in the new subphylum Paramycia of the phylum Choanozoa by Cavalier-Smith in 2013. (Karpov et al, 2014).

Jones et al caution that our current knowledge of the life stages of Cryptomycota is very incomplete and point out that cryptomycotes have some strong resemblances to chytridiomycotes in both structure (e.g., flagellar apparatus) and ecology, if not in cell wall chemistry. And James and Berbee (2012) point out that aphelids and cryptomycotes may well be reduced higher funguses so the ARM groups as early diverging may be an artefact of LBA.

Trichomycetes were apparently named by Duboscq, Léger, and Tuzet in 1948 (nhm.ku.edu-Chapter 12 Phylogeny) and were divided into Eccrinides (Eccrinales and Amebidiales) and Harpellides (Harpellales and Genistellales (these represent the 2 families of Harpellales in the standard classification). Asellariales were excluded. Neither Eccrinales nor Amebidiales have zygospores, trichospores, septa, nor typical fungal walls (they are composed of galactosamine-galactose), but Eccrinales do have sporangiospores that form from the apex downward, a feature unique to funguses (Moore et al, 21st Century Guide to Fungi, 2011, p. 415 (books.google)). Assembling the Tree of Life (2004), edited by Cracraft and Donoghue, p. 183 (books.google), treats them as incertae sedis.

Haplosporidians should also be included, as, like microsporidians and higher funguses, they are diplokaryotic (Perkins, 1991; Margulis and Schwartz, 1998). Desportes and Nashed (1983) suggest these 2 sporozoan groups are closely related because both proliferate with somatic stages alternating with sporulating ones, where sporulation is initiated by the differentiation of sporonts which divide into sporoblasts giving spores, and note that the spindle pole bodies and synaptonemal complexes in sporont nuclei are also similar in the 2 taxons.

Paramyxea, which contain 3 genera (*Paramyxa*, *Marteilia*, and *Paramarteilia*) have an organelle similar to haplosporosomes (considered homologous) in *Marteilia* and *Paramarteilia*, the 2 genera forming a family, which occurs also in Haplosporidia, which warrants the recognition of Ascetosporea, combining Haplosporidia with Paramyxea, established by Desportes and Nashed (1983). However, Paramyxea have singlet microtubules, like Apicomplexa. Haplosporidia should be placed in Fungi alongside Microsporidia, as they also have diplokaryosis.

Dipodascopsis and Dipodascaceae are in Saccharomycetes (taxonomicon.taxonomy.nl).

Pezizomycotina is the largest subphylum of Ascomycota with more than 32,000 described species. It approximately equates to Euascomycetes sensu Kirk et al from 2001 in Dictionary of the Fungi and Alexopoulos et al from 1996, and it includes all filamentous, sporocarp-producing species, with the exception of *Neolecta* of Taphrinomycotina. (Pezizomycotina-tolweb.org). Orbilariaceae, the sole family of Orbiliomycetes, is in Heliotales in Discomycetes in Parker. The 3 orders of uncertain position are Lahmiales, Medeolariales, and Triblidiales. Lahmiales (Pyrenomycetes 1), with the single genus *Lahmia* Körber 1861, is otherwise in Coronophorales (in Pyrenomycetes and Sordariomycetes), Medeolariales (Discomycetes 2) are otherwise in Discomycetes, and Triblidiales are otherwise in Hyphomycetes or Saccharomycetaceae (Triblidiales-Viquipèdia).

The class Dothideomycetes contains the majority of species with ascostromatic development and bitunicate ascuses that were placed in the Loculoascomycetes (named by Luttrell in 1955; also called Bitunicatae, named by Luttrell in 1951, and Loculoascomycetidae by Luttrell in 1981). The remaining Loculoascomycetes, the black yeasts, are placed in Eurotiomycetes, the latter containing in part the plectomycetes and being the cleistothecial ascomycetes. (Dothideomycetes-tolweb.org; Eurotiomycetes-Tolweb.org)

Lecanoromycetes are very diverse and, as recognized in this taxonomy, are the largest fungal class. It includes the about 90 % of all described ascomycote lichens (estimated to be over 13,500 species). A common characteristic for this class is their ascohymenial ascomal ontogeny, with a predominance of apothecial spores. In most lineages ascuses have a multilayered wall of which 2 layers are thick enough to be visible with light microscopy and display predominantly rostrate dehiscence. (Miadlikowska et al, 2006).

Sordariomycetes corresponds mostly with Pyrenomycetes. The term "pyrenomycetes" was used to unite funguses with perithecial ascomas and unitunicate ascuses by Luttrell in 1951. Its use was discontinued in molecular taxonomies based on the placement of perithecial species outside of the clade and the inclusion of species with prototunicate ascuses to avoid confusion. (Sordariomycetes -tolweb.org)

Arthoniomycetes, Lichinomycetes, and Leconoromycetes are the lichenized ascomycete groups.

A lichen is a stable, self-supporting association of a mycobiote and a photobiote, and more precisely, an ecologically obligate, stable mutualism between an exhabitant fungal partner and an inhabitant algal partner. 19% of all funguses and 42% of ascomycotes are lichenized; 13 of the 46 orders that include lichenous species are in Ascomycota, but only 4 orders are exclusively lichenous (Arthoniales, Lecanorales, Lichinales, and Verrucariales). There are 40 genera of algal partners: 25 eukaryotic algae, mostly green, and 15 cyanobacteria. (Hawksworth et al, 1996).

Wallemiomycetes, with the single genus *Wallemia* Johan-Olsen 1887, a xerophilic and halophilic mold causing farmer's lung in 1 species, based on parenthosome fine structure is linked to Filobasidiales in Tremellomycetes (Zalar et al, 2005).

Entorrhizomycetes, teliosporic root parasites, are otherwise placed in Ustilaginomycotina (Begerow et al, 2006; Bauer et al, 1997).

Pucciniomycotina is very diverse and contains about 7400 species in about 215 genera. More than 95% of the species and 75% of the genera are in Pucciniales (in Pucciniomycetes), which is the same as Uredinales. The next largest orders, Septobasidiales (in Pucciniomycetes) and Microbotryales (Microbotryomycetes), collectively constitute c. 5% of the species and 4% of the genera. (Pucciniomycotina-tolweb.org). The only classes based on new genera in Hibbett et al's taxonomy are the first 2 in Pucciniomycotina, the former from 2003 and the latter from 1990.

Agaricomycotina is Tehler's Hymenomycetidae and is Neomycetes but with Auriculariales added, and Agaricomycetes is Homobasidiomycetes but with Auriculariales added. Members of Agaricales s.l. are separated out as Polyporales, Boletales, Russulales, Telephorales, Corticiales, Hymenochetales, Gomphales, and Cantharellales, all in Agaricomycetes, the first 6, with Agaricales s.s., part of a less inclusive clade. The class has the most species of both edible and poisonous funguses. It contains about 16,000 described species (tolweb.org). Tulasnellaceae are in Cantharellales (taxonomicon.taxonomy.nl).

German mycologist Rolf Singer, in his classic and monumental The Agaricales in Modern Taxonomy, published in 4 editions, 1951, 1962, 1975, and 1986, used both Fries's macroscopic and Fayod's microscopic features, and his most recent classification included 230 genera in 18 families, and was arranged into 3 suborders: Agaricineae, Boletineae, and Russulineae (Agaricales-Wikipedia). He excluded Cantharellaceae and included the gilled polypores (book review of Singer's 1st edition, Mycologia 44 (jstor.org) and of the 3rd edition, Mycologia 68 (jstor.org)). The suborders have been corroborated by genotypic classifications but the 3rd not as directly related to the first 2. The 1st forms a clade, subclass Agaricomycetidae, with the 2nd , along with Atheliales. The 1st is now called Agaricales or the euagarics.

French mycologist Robert Kühner in 1980 divided Singer's Agaricales into 5 orders: Tricholomatales (including some gilled groups of Polyporales), Agaricales sensu stricto, Pluteales, Russulales, and Boletales (Methany et al, 2006). Three of the orders—Tricholomatales, Agaricales and Pluteales—conform mostly to Methany and colleagues' Agaricales (the euagarics) or Singer's Agaricineae. While Singer's taxonomy was influential, Kühner's was not widely recognized.

So genotypic results have mostly corroborated the traditional classification of funguses, as is the case with most other groups, and largely correspond to phenotypic cladistic results, as is also usually the case. And it is worthy of note that even before molecular studies, Lichenes, Plectomycetes, Gasteromycetes, and Deuteromycota were considered as artificial.

Notable macroscopic fungal forms include morels (*Morchella*), scarlet cup (*Sarcoscypha*), elf cup (*Tarzetta*) (Pezizales); truffles (*Tuber*)(Tuberales); swamp beacon (*Mitrula*), club mushrooms (*Leotia*)(Heliotales); chantarelles (Cantharellales); polypores (bracket funguses or shelf funguses) (Polyporales); agarics (*Agaricus*), amanitas (*Amanita*), honey mushroom (*Armillaria*), coral mushrooms (Clavariaceae), parasols (*Lepiota*), lepiotas (*Lepiota*), entolomas (*Entoloma=Rhodophyllus*), corts (*Cortinarius*), waxy caps (*Hygrophorus*), inky caps (*Coprinus*), milk caps (*Lactarius*), russulas or red caps (*Russula*), angel wings (*Pleurocybella*), jack o' lantern (*Omphalotus*), pluteus (*Pluteus*), pholiotas (*Pholiota*), fiber heads (*Inocybe*), boletes (*Boletus*) (Agaricales); puffballs (3 families), earthstars (*Geastrum*), splash cups (*Cyathus*) (Lycoperdales); bird's-nest fungus (*Nidularia*), cannon fungus (*Spherobolus*) (Nidulariales) (Lincoff, 1981); and in lichens, reindeer moss, pyxie cup, and British soldiers in *Cladonia*, Iceland moss (*Cetraria*), orchella (*Orchella*), cudbear (*Ochrolechia*) and orchil (*Roccella*) (used to make litmus dye), lungwort (*Lobaria*), and various *Parmelia* lichens (skull lichen, ashen stud lichen, wrinkled shield lichen, snakeskin lichen) (Lichens-Enc. Brit., 1970). The most beautiful ones are the gill mushrooms with the typical and familiar cap (pileus) and stalk (stipe).

Several species in Cantharellaceae are edible and the chanterelle (*Cantharellus cibarius*) and horn-of-plenty (*Craterellus cornicopoides*) are especially prized as esculents (edible mushrooms). The latter is also known as the black chanterelle, black trumpet, or trumpet of the dead, in allusion to its trumpet form, its blackish colour, and possibly the belief they were being played as trumpets by dead people underground. The fabulous cornucopia of Greek myth was the magical horn of the nymph Amalthea's goat (or herself in goat form) that filled itself with whatever fruit its owner requested and has become the symbol of plenty. Edible mushrooms also include certain species of agarics, boletes, coral mushrooms, club mushrooms, milk caps, russulas, morels, truffles, puffballs, etc. The field, white, or button mushroom (*Agaricus*) is probably the most popular.

The jack o' lantern is poisonous, but when gathered fresh and taken into a dark room the gills give off an eerie green glow (Lincoff, 1981). Other poisonous mushrooms include several species of agarics and milky caps, ergot (*Claviceps*), some coral mushrooms (*Ramaria*), devil's bolete, and the scaley chanterelle (*Gomphus*), inter alia (List of poisonous fungi-Wikipedia), all in Agaricales, except ergot, which is in Pyrenomycetes. Deadly species include several amanitas, conocybes (*Conocybe*), webcaps (*Cortinarius*), skull cap (*Galerina*), and dapperlings (*Lepiota*), inter alia (List of deadly fungi-Wikipedia), all in Agaricales. *Amanita* appears to have the most poisonous species, and over 90 % of mushroom-related deaths are caused by the aptly-named death cap (*Amanita*) (Parker,1982) so it is the world's deadliest mushroom.

Fungal luminescence (Foxfire: Bioluminescent Fungi-inamidst.com; Bioluminescent Fungi-

Mykoweb; Why bioluminescent fungi glow in the dark-news.mongaby.com; List of bioluminescent fungi-Wikipedia) is also called foxfire or less commonly fairy fire. The "fox" may derive from the Old French "fols," meaning "false", although the association with foxes is widespread and occurs in Japanese folklore (Foxfire-Wikipedia). In folklore, "fairy sparks" in decaying wood in forests indicated the place where fairies held their nightly revels, hence the other name. The beautiful green glow is caused by luciferase, an oxidative enzyme, which emits light as it reacts with luciferin, which is the same process in other bioluminescence. The intensity level and location in funguses varies by species, but typically occurs in the mycelium and/or basidiomes, but also in the mature spores. In many instances it is the hyphas present in decaying plant tissues that luminesce, resulting in the appearance of luminescent wood or leaves. The oldest recorded documentation of foxfire is by Aristotle in the 4th century B.C. and Pliny the Elder in the 1st century AD. Its cause was finally discovered in 1823; the glow emitted from wooden support beams in mines was examined, and it was found that the luminescence came from fungal growth. It has been reported in 71 species, all white-spored basidiomycetes traditionally placed in Tricholomataceae, but molecularly in 3 or 4 separate lineages but still in basidiomycetes, most (40) in the diverse and widespread genus *Mycena*. The other genera containing luminescent species include *Armillaria (*with 10), *Pleurotus, Gerronema, Panellus, Dictyopanus, Nothopanus, Neonothopanus, Filoboletus, Favolaschia, Roridomyces*, and *Xylaria*, making 13 genera.

The reason for fungal glow is unknown. It may offer some sort of selective advantage, but if so why is it not more widespread throughout the funguses? Such funguses emit light 24 hours a day (but can be seen only in the dark as the light is too faint to be seen otherwise), which must be an energy consumptive process. It has been hypothesized that it is to attract invertebrates that aid in spore dispersal, which may apply to those species with luminescent basidiomes, but not for those in which only the mycelium emits light. Additional hypotheses include the attraction of predators of mycetophagous invertebrates, and even the function of emitted light as a warning to nocturnal phagotrophs that might consume the fungus or its substrate, similar to warning colourations observed in other organisms. Experiments by Sivinski lend support to the idea of attraction of insects for spore dispersal.

It is also possible that the phenomenon in these funguses is only the by-product of some other metabolic process. Because the reaction is oxygen-dependent, it has been theorized that it may have evolved as a way to consume excess oxygen produced in the cells during other metabolic processes (i.e., as an antioxidant). Studies have suggested a link between lignin degradation and luminescence in funguses, with the oxygen-consuming reaction acting as a means of dealing with peroxides generated during the process. The phenomenon, then, may have arisen as a by-product of a beneficial metabolic process and subsequently been co-opted in a relatively small number of taxons for secondary use in attracting spore dispersal agents or deterring fungivores. And different species may glow for different reasons, especially dictated by which part of the mushroom glows.

Four specimens of 2 species of *Armillaria,* the humongous fungus, as dubbed by S.J. Gould (1992), are reputed to be the world's largest organisms, the 1st, *A. bulbosa (gallica),* being discovered by Myron Smith and colleagues (1992) in Michigan, which covers 27 acres (11 hectares), weighs in at 100 tons, and is 1500 years old, the 2nd, *A. ostoyae (solidipes),* found later that same year, by Shaw and Russell in Washington, the 3rd, *A. ostoyae*, in '98 in Oregon, which covers 2384 acres, and the 4th by Catharine Parks and colleagues in 2003, which covers 2200 acres and is 2400 years old. Most of these organisms are underground in the form of rhizomorphs.

The discoveries garnered much media attention and rekindled the debate as to what constitutes an individual organism. "It's one set of genetically identical cells that are in communication with one another that have a sort of common purpose or at least can coordinate themselves to do something," maintains mycologist Tom Volk (Strange but True: the Largest Organism in the World is a Fungus-scientificamerican.com, 2007), but this is also the definition of a clonal colony, which is what the funguses in question are, and it excludes unicellular organisms. Both the blue whale, which weighs 200 tons, and the humongous fungus fit this definition, and so does the 6,615-ton (6 mln.-kilo) colony of a male quaking aspen tree and its clones, interconnected by a single root system, that covers 107 acres (43 hectares) of a Utah mountainside. The redwood tree (*Sequoia*) is the largest organism by volume at 52,508 cu. ft. and is also the oldest at 3500 years. A clonal colony as a single, individual organism is questionable at best.

The largest families of funguses are Pucciniaceae (in Uredinales), with 17 genera and 4000 species in *Puccinia* alone (total for the family not stated), and Spheropsidaceae (Phomaceae), in Spheropsidales in Celomycetes, with 500 genera (the number of species is not stated, but for the order it is 6000)(Parker, 1982). As traditionally circumscribed, the largest orders according to number of families are Lecanorales, with about 40 families (the molecular version has 26), and Agaricales, also with about 40 families (in Hudson (1984), combining Agaricales and Aphyllophorales; in Parker it is 32; the molecular version has 33).

Information on the history of fungal classification was taken from Vuillemin (1912) and Kessel (1955) (for Patouillard's work it was Parker, loc. cit.), and biographical information was taken from Wikipedia.

Bauer, R., Oberwinkler, F., Vanky, K. 1997. Ultrastructural markers and systematics in smut fungi and allied taxa. Canadian Journal of Botany 75: 1273–314.

Begerow, Dominik; Stoll, Matthias; Bauer, Robert. A phylogenetic hypothesis of Ustilaginomycotina based on multiple gene analyses and morphological data. 2006. Mycologia 98: 906-16 (mycologia.org)

Desportes, I., Nashed, N.N. 1983. Ultrastructure of sporulation in Minchinia (Arvy), an haplosporean parasite *Dentalium entale* (Scaphopoda, Mollusca); taxonomic implications. Protistologica 19: 435-60.

Gould, S.J. 1992. A Humongous Fungus Among Us. Natural History 101: 10-14 (as cited in The Humongous Fungus-Ten Years Later – botit.botany.wisc.edu).

Hawksworth, D.L., Kirk, P.M., Sutton, B.C, Pegler, D.N. 1995. Ainsworth and Bisby's Dictionary of the Fungi (8th ed.). IMI (International Mycological Institute-CAB (Center for Agriculture and Biosciences), Oxon, UK, New York, NY.

Hibbett, David et al. 2007. A higher-level classification of Fungi. Mycological Research 111: 509-47 (usda.gov; umich.edu).

Hudson, H.J. 1984. Fungi. In: A Synoptic Classification of Living Organisms. Sinauer Associates, Sunderland, Mass.

James, T.Y, Berbee, M.L. 2012. No jacket required--new fungal lineage defies dress code: recently described zoosporic fungi lack a cell wall during trophic phase. Bioessays 34: 94-102.

Jones, Meredith; Richards, Thomas; Hawksworth, David; Bass, David. 2011. Validation and justification of the phylum name Cryptomycota phyl. nov. IMA Fungus 2: 173–175 (nih.gov).

Karpov, Sergey; Mamkaeva, Maria; Aleoshin, Vladimir; Nassonova, Elena; Lilje, Osu; Gleason, Frank. 2014. Morphology, phylogeny, and ecology of the aphelids (Aphelidea, Opisthokonta) and proposal for the new superphylum Opisthosporidia. Frontiers in Microbiology 5: 112 (nih.gov).

Kessel, Edward (ed.). 1955. A Century of Progress in the Natural Sciences. California Academy of

Sciences, San Francisco.

Lincoff, Gary. 1981. Audubon Society Field Guide to North American Mushrooms. Knopf.

Lohr, Jennifer, Yin, M., Wolinska, J. (2010). A *Daphnia* parasite (*Coullerya mesnili*) constitutes a new member of the Ichthyosporea, a group of protistans near the animal-fungi divergence. Journal of Eukaryotic Microbiology 57: 328-36 (unifr.ch).

Margulis, L, Schwartz, K. (loc. cit.).

Matheny, P. Brandon, et al. 2006. Major Clades of Agaricales: a multilocus phylogenetic overview. Mycologia 98: 982-95 (clarku.edu).

Miadlikowska, Jolanta, et al. 2006. New insights into classification and evolution of the Lecanoromycetes (Pezizomycotina, Ascomycota) from phylogenetic analyses of three ribosomal RNA- and two protein-coding genes. Mycologia 98: 1088–1103 (lichenology.info).

Parker, Sybil (loc. cit.)

Perkins, Frank. 1990. Haplosporidia (Chp. 2), in Handbook of Protoctista, Margulis, L., Melkonian, M., Corliss, J.O., Chapman, D.J. (eds.). Jones and Bartlett, Boston.

Smith ML, Bruhn JM, Anderson JB. 1992. The fungus *Armillaria bulbosa* is among the largest and oldest living organisms. Nature 356: 428–31, as cited in Strange but True: the Largest Organism in the World is a Fungus-scientificamerican.com, 2007)

Sumathia, Catherine, Raghukumara, S, Kasbekarb, D, and Raghukumara, C. 2006. Molecular Evidence of Fungal Signatures in the Marine Protist *Corallochytrium limacisporum* and its Implications in the Evolution of Animals and Fungi. Protist 157: 363-376.

Tehler, Anders. 1988. A cladistic outline of Eumycota. Cladistics 4: 227-277.

Vuillemin, Paul. 1912. Les Champignons: un essai de classification (gallica.bnf.fr).

Zalar, Polona; de Hoog, G. Sybren; Schroers, Hans-Josef; Frank, John Michael; Gunde-Cimerman, Nina. 2005. Taxonomy and phylogeny of the xerophilic genus *Wallemia* (Wallemiomycetes and Wallemiales, cl. et ord. nov.)(Abstract). Ant. van Leew. 87: 311-328 (springer.com).

Chapter 15 Animalia

Fossils

At midnight in the museum hall,
The fossils gathered for a ball,
There were no drums or saxophones,
But just the clatter of their bones,
Rolling, rattling carefree circus,
Of mammoth polkas and mazurkas,
Pterodactyls and brontosauruses
Sang ghostly prehistoric choruses,
Amid the mastodonic wassail
I caught the eye of one small fossil,
“Cheer up sad world,” he said and winked,
“It’s kind of fun to be extinct."

Ogden Nash, Carnival of Animals

Historical Overview

"Peri ta zoa historica" ("about the history of animals"; Latin transl., "Historia animalium;" English transl., History of Animals) by Aristotle (Aristoteles, "best purpose") divided animals into Anaima (without blood) and Enaima (with blood), corresponding to invertebrates and vertebrates, respectively. We now know that complex invertebrates do make use of hemoglobin, but of a different kind from vertebrates. He also distinguished between viviparous and oviparous. This system is maintained unchanged through the Middle Ages and into the 1600s, and continued in its basic form into the 1800s.

He believed in a hierarchical "Ladder of Life" (Scala naturae or Great Chain of Being), placing animals according to complexity of structure and function so that higher organisms showed greater vitality and ability to move, that creatures were arranged in a graded scale of perfection rising from plants on up to man. This was depicted by Émile Guyénot in 1941 as 14 grades or rungs: humans, mammals (viviparous quadripeds), birds, oviparous quadripeds (reptiles and amphibians), cetacea, fish, malacians, malacostracans, ostracoderms, entoma (insects), zoophytes, higher plants, lower plants, and inanimate matter (Strickberger, 1993). The basic concept was also Plato's and was adopted by Proclus and the Neoplatonists and continued in various forms into the 1800s.

He also believed that intellectual purposes, i.e., final causes, guided all natural processes. Such a teleological view gave him cause to justify his observed data as expressions of formal design. Noting "no animal has, at the same time, both tusks and horns," and "a single-hooved animal with two horns I have never seen," he suggested that nature, giving no animal both horns and tusks, was staving off vanity and giving creatures faculties only to such a degree as they are necessary. His system had 11 grades, arranged according "to the degree to which they are imbued with potentiality", expressed in their form at birth. Aristotle also held that the level of a creature's perfection was reflected in its form, but not preordained by that form.

He placed emphasis on the type or types of soul an organism possessed, asserting that plants possess a vegetative soul, responsible for reproduction and growth, animals a vegetative and a sensitive soul, responsible for mobility and sensation, and humans a vegetative, a sensitive, and a rational soul, capable of thought and reflection. Notable also is Aristotle's division of sensation and thought, which generally went against previous philosophers, with the exception of Alcmeon.

In modern times, the 3 most celebrated systems are those of Linneus, Cuvier, and Cuénot, but a few others are also notable. Linneus, in his classic "Systema naturae" of 1715, recognized 6 classes: Vermes, Insecta, Pisces, Amphibia, Aves, and Quadripedia.

Jean-Baptiste de Monet (Lamarck) (1801) recognized the Aristotelian categories naming them animaux avec vertèbres and animaux sans vertèbres but arranged them into 11 classes--7 invertebrate and 4 vertebrate (Poissons, Reptiles, Oiseaux, Mammaux). In 1807 he separated out the Infusoires from the Polypes and the Cirripèdes from the Crustacés, and after the work of Savigny on the ascidians he added the Tunicata class in 1816 (Encyclopédie Universalis, Zoologie, 1993).

Lamarck was interested in botany, especially after his visits to the Jardin du Roi, and he became a student under Bernard de Jussieu. Under Jussieu, he spent 10 years studying French flora. After his studies, in 1778, he published some of his observations and results in a 3-volume work, titled "Flore françoise." Lamarck's work was respected by many scholars, and it launched him into prominence in French science. Georges-Louis Leclerc, Comte de Buffon, one of the top French scientists of the day, mentored Lamarck, and helped him gain membership to the Academie des sciences in 1779 and a commission as a Royal Botanist in 1781, in which he traveled to foreign botanical gardens and museums. In 1788, he became keeper of the herbarium of the Jardin du roi. In 1790, at the height of the French Revolution, he changed the name to Jardin des plantes. Lamarck had worked as the keeper of the herbarium for 5 years before he was appointed curator and professor of invertebrate zoology at the Musée national d'histoire naturelle in 1793. The next year he was appointed to serve as secretary of the assembly of professors for the museum for a period of one year.

Lamarck began as an essentialist who believed species were unchanging, however, he grew convinced that transmutation, or change in the nature of a species, occurred over time. He set out to develop an explanation, and on May 11, 1800, he presented a lecture at the Musée in which he first outlined his newly developing ideas about evolution. In 1802, he published "Recherches sur l'organisation des corps vivants," in which he elaborated on his theory, and did 3 other works for it: "Philosophie zoologique," in 1809, and "Histoire naturelle des animaux sans vertèbres," in 7 volumes, from 1815–22.

Although he was not the first to advocate organic evolution, he was the first to develop a truly coherent theory for it. Like many natural historians, he believed that simple organisms arose through spontaneous generation, based on the common belief of his day, and on his own belief in a pre-Lavoisier, 4-essence chemistry. Lamarck believed in 2 forces underpinning evolution. The 1st was a tendency for organisms to become more complex, which he referred to as "Le pouvoir de la vie" or "la force qui tend sans cesse à composer l'organisation"--a driving force. The 2nd he referred to as "L'influence des circonstances"--an adaptive force that was the interaction of organisms with their environment, by the use and disuse of certain characteristics (inheritance by acquired characteristics).

The latter, now also called soft inheritance, was accepted in his time but later rejected based on genetics, but, in the field of epigenetics, there is growing evidence that it plays a part in the alteration of some organisms' phenotypes: it leaves the DNA unaltered but affects it by preventing the expression of genes. Some epigenetic changes such as the methylation of genes alter the likelihood of DNA transcription and can be produced by changes in behaviour and environment. Many epigenetic changes are themselves heritable to a degree. Thus, while DNA itself is not directly altered by the environment and behaviour except through selection, the relationship of the genotype to the phenotype can be altered, even across generations, by experience within the lifetime of an individual. (Epigenetics: DNA Isn't Everything-ScienceDaily.Com; Lamarck's Revenge-ExtremeTech.Com)

Epigenetics is the study of heritable changes in gene activity that are not caused by changes in the DNA sequence, that is, functionally relevant changes to the genome that do not involve a change in the nucleotide sequence. So that non-genetic factors cause the organism's genes to behave (or "express themselves") differently. Examples of mechanisms that produce such changes are DNA methylation and histone modification, each of which alters how genes are expressed without altering the underlying DNA sequence. An example of an epigenetic change in eukaryotes is cellular differentiation.

Georges Cuvier, in his "Règne animal distribué d'après son organisation, pour servir de base à l'histoire naturelle des animaux et d'introduction à l'anatomie comparée" of 1817, organized animals around 4 body plans to which the 4 *embranchements* (corresponding to phylums) correspond, which are Vertébrés, Mollusques, Articulés, and Zoophytes (animalia radiata)(Swainson, 1835), and in which both fossil and living forms were considered for the first time. In 1798 the "Tableau élémentaire de l'histoire naturelle des animaux" was published, which was an abridgement of his course of lectures at the École du Panthéon and may be regarded as the foundation and first and general statement of his natural classification of the animal kingdom.

Cuvier (1769-1832) was one of the most influential figures in science during the early 19th century. His work is considered the foundation of vertebrate paleontology. He strongly opposed St. Hilaire's theory that all organisms were based on a basic plan or archetype and that they blended gradually one into another, and argued instead that it was function, not hypothetical relationships, that should form the basis of classification. This issue was part of the famous Cuvier-St. Hilaire debate of 1830, which has often been interpreted in the retrospect of a post-Darwinian age as a debate over evolution, but it mostly revolved around the number of archetypes needed to categorize organisms.

Cuvier was an essentialist who believed that animals and plants were unchanging throughout their existence, so he was a strong opponent of Lamarck's theory of evolution, believing there was no evidence for evolution but rather evidence for successive creations after catastrophic extinction events, an idea he proposed in his "Essai sur la théorie de la terre" in 1813 (see Chapter 19).

In 1821, Cuvier made what has been called his "Rash Dictum": he remarked that it was unlikely for any large, unknown animal to be discovered. Many such discoveries have been made since then. He established many key concepts. He was able to convince his contemporaries that extinction was a fact, what had earlier been controversial speculation. His study of the Paris Basin with Alexandre Brongniart established the basic principles of biostratigraphy, and he was the first to demonstrate that different rock strata in the Basin held different mammalian fauna. He also documented that the lower the rock strata, the more different the fossils were from living species. Although he did not accept the idea of organic evolution, such findings ironically produced knowledge that would ultimately provide support for Darwin's evolutionary theories.

He held several academic positions: lecturer at the École Centrale du Panthéon, assistant and later professor of animal anatomy at the post-Revolutionary Musée national d'histoire naturelle, professor of natural history at the Collège de France, titular professor at the Jardin des plantes, commisssary of the Institut national, permanent secretary of the Institut, chancellor of the U. of Paris, supervisor of the faculty of Protestant theology (he was a Lutheran, son of a Hughenot Swiss Guard lieutenant), and Imperial U. councillor.

He also held several political positions: State Councillor, State Council president, Inspector-General, interim president of the council of public education, president of the Committee of the Interior, and Minister of the Interior.

He served under 3 different régimes—republican, imperial, and royal--serving 3 different kings. He was a member of the Institut de France (Académie des sciences after 1815), Académie des inscriptions et belles lettres, and the Doctrinaires (a group of liberal royalists during the Bourbon Restoration), was made grand officer of the Légion d'honneur in 1826, and Louis Philippe elevated him to the rank of "peer of France" in 1831 and was called Baron from then on.

British paleontologist Richard Owen (1804-92), famous for coining the word "dinosaur," in his "Remarks on Entozoa" from 1835, was the first to distinguish Metazoa as a kingdom (in 1859), recognized 5 groups by separating Cuvier's Zoophytes into Nematoneura and Acrita, the latter he considered equivalent to Lamarck's "animaux apathiques," which designated the Radiaux and Polypes (Ragan, 1997). He is also known for having helped create, with Benjamin Hawkins, the first life-size sculptures depicting dinosaurs as it was thought they might have appeared. Some models were initially created for the Great Exhibition of 1851, but 33 were eventually produced when the Crystal Palace was relocated to Sydenham in South London. He had nearly 2 dozen lifesize sculptures of various prehistoric animals built out of concrete sculpted over a steel and brick framework. He is known as well for his campaign to give the natural specimens in the British Museum a new home, which resulted in the establishment of the now world-famous Natural History Museum in London in 1881.

Thomas Huxley (1869) recognized 8 primary groups, as he called them, which were Vertebrata, Mollusca, Molluscoida, Celenterata, Annulosa (Insecta), Annuloida, Infusoria, and Protozoa, expressing doubts about Infusoria. The vertebrates he divided into Ichthyopsida (Pisces and Amphibia), Sauropsida (Reptilia and Aves), and Mammalia, noting that sometimes the 1st 2 of these 3 vertebrate groups are called Branchiata and the 3rd, Mammalia, Abranchiata.

Huxley (1825-1895) was a Fellow of the Royal Society and the Linnean Society and an English comparative anatomist who was known as "Darwin's Bulldog" for his advocacy of the theory of evolution. His famous debate in 1860 with Samuel Wilberforce was a key moment in the wider acceptance of evolution, and in his own career, even though he was slow to accept some of Darwin's ideas, such as gradualism, and was undecided about natural selection. Instrumental in developing scientific education in Britain, he fought against the more extreme versions of religious tradition. He was a member of the HMS Rattlesnake, which went on a voyage of discovery and surveying in 1847 and on which he made studies of marine invertebrates, especially celenterates. In 1864 he launched a dining club called the X Club for like-minded people working to advance the cause of science. In 1869 he coined the term "agnostic", describing his own views on theology. He had little formal schooling so was almost entirely self-taught, yet became perhaps the finest comparative anatomist of the late 19th century. He worked on invertebrates and later on vertebrates, especially the relationship between apes and humans, and believed birds evolved from small carnivorous dinosaurs, a theory generally accepted today.

Lucien Cuénot (1866-1951), another French scholar, designed the fundamental modern taxonomy of animals from 1940-1951, which became standard (Encyclopédie Universalis-Zoologie, 1993). His Y-tree model has bacteria at the base with eukaryotic algae and higher plants plus funguses forming a branch from them at the right, protozoans occupying also a low position on the trunk but to the left, sponges a little higher up on the right, celenterates occupying the middle part, 1 branch on the left (cnidarians) and 1 on the right (celenterates), and triplobastic metazoans the upper part, the left branch composed of deuterostomes, and the right composed of acelomates and protostomes.

As early as 1894, the Paris-born Cuénot, distinguished as the first French geneticist, and the most eminent of French biologists in the 1st half of the 20th century, was the only one in France to oppose Lamarckism and embrace Darwinism, but also the August Weismann theory of 1883, which postulated material support for heredity. With the rediscovery of Mendel's work in plants by Correns, De Vries, and Tschermak, Cuénot proved, in 1902, that Mendelism applied to animals as well; in that same year he promoted the idea of preadaptation; in 1903, he proposed a possible interaction between the "mnémon" (gene), "diastase" (enzyme), and "chromogen" (pigments); in 1905, he discovered the first case of a lethal mutation in animals; he discovered the first phenomenon of epistasy, in 1907, where several genes located at places different from the chromosome intervene in the same biochemical pathway; and in 1908, he discovered the first case of pleiotropism, where certain genes can act on several seemingly independent characteristics. Subsequently, he was the first person to describe multiple allelism at a genetic locus. It is thanks to him that his friend, Professor Philibert Guinier, wrote, in 1912, articles foreseeing what the laws of heredity and natural selection could bring to forest management settlements. He was also a member of the Académie de Stanislas.

The "Encyclopédie pour tous" in 1958 divided the animal kingdom into subkingdoms Protozoa and Metazoa, with the former containing Rhizopoda, Flagellata, Ciliata, Suctorians, and Sporozoa, and the latter containing Celenterata, Sponges, Echinodermata, Vermidia, Vermes, Mollusca, Arthropoda, Protochordata, and Vertebrata.

In the 1970 Britannica, in its Zoology article, Animalia is also divided into subkingdoms Protozoa and Metazoa, and is presented in simplified form as follows:

Table 15-1.

- Protozoa (1 phylum, 4 subphylums)
  - Flagellata
  - Sarcodina
  - Sporozoa
  - Ciliata
- Metazoa
  - Mesozoa
  - Parazoa
  - Eumetazoa
    - Radiata (Celenterata)(Actinozoa)
    - Bilateria (Triploblastica)
      - Acelomata
      - Pseudocelomata
      - Eucelomata
        - Lophophorata
        - Schizocela
        - Enterocela

The first to construe the kingdom phylogenetically (Choanozoa+Metazoa) may have been as late as 1993 by Wainwright et al in a Science article called "The monophyletic origins of Metazoa."

The general animal phylogeny is presented in Table 17-2, based mostly on Zrzavy et al (1998), Nielsen (1995), Neilsen et al (1996), Eernisse et al (1992), Halanych (2004), and Giribet et al (2007), as well as phylogenies for holometabolous insects, mollusks, vertebrates, extant mammals, and primates (Tables 15-3 to 15-8).

New information has come to light which extends Animalia to include 2 or 3 more lineages involving 3 protozoan genera besides choanoflagellates: *Spheroforma*, *Capsaspora*, and *Ministeria.* The 1st has an epidermal growth factor similarity in annotation domain structure to Metazoa (Shalchian-Tabrizi et al, 2008) and groups with *Capsaspora* and Metazoa in at least 1 molecular analysis (Ruiz-Trillo, 2006). The 2nd has 3 metazoan-specific TFs (transcription factors)(Runx, T-box, and NF-kappa B)(Sebé-Pedrós et al, 2011), possesses cadherins, which also occur in Choanozoa and Metazoa (Nichols et al, 2012), and has 5 annotation signaling or cell adhesion component similarities with Metazoa (Shalchian-Tabrizi et al, 2008). And the 3rd has 11 annotation signaling or cell adhesion component similarities with Metazoa (Shichlian-Tabrizi et al, 2008). Ubiquitin S30 fusion might be another synapomorphy for Animalia including these 3 other genera. It is unclear if *Ministeria* forms a lineage with *Capsaspora* as the molecular analyses are ambigious about it and there have been no classical analyses, but they do have untapered filose tentacles with a microfilamentous skeleton.

Biographical material was from Britannica.Com, Wikipédia, VictorianWeb.Org, UCMP.Berkely.Edu, and NewWorldEncyclopedia.org.

the Cambrian Explosion

The transition between the Proterozoic and Phanerozoic Eons, beginning 542 million years ago (mya), is distinguished by the diversification of multicellular animals and their acquisition of mineralized skeletons during the Cambrian Period, when the majority of animal phyla first appears in the fossil record, over half (29, 10 of these extant) the major animal groups emerged in this period and 5 times more than in the next most productive period, which was the Quaternary, with 6, is certainly one of the most important events in the history of life on Earth. But the causes have remained a mystery, and the question of what was the trigger for the emergence of multicellularity in animals has remained unanswered. Several theories have been proposed:

- warming trend allowing mineralization of tissues
- atmospheric accumulation of oxygen through photosynthesis reaching sufficient levels to permit mineralization as well as forming a protective blanket of ozone, that is, a change in oxygen concentration
- a change in the shape and extent of shorelines because of the hypothetical continental drift, thus profoundly transforming both climate and environment
- sea level changes that accompany glaciation
- new tidal effects caused by the moon
- evolution of eukaryotic sexual genetic exchange and/or regulatory genes that control multicellular development
- reaching of a threshold number of species
- cropping, which is when predators feed on the most abundant prey species thereby reducing their numbers and letting other species use resources formerly monopolized by the dominant prey; its evolutionary value extends also to predators through a feedback cycle, since the diversification of prey species leads in turn to the diversification of predator species.

None of these have any convincing evidence for them. The only plausible explanations have come only recently. Fernández-Busquets et al (2009) have put forward the calcium concentration theory, whereby the geologically induced increase of marine calcium might be the key for understanding the Cambrian Explosion. They have shown that the massive and sudden surge in the calcium concentration of the Cambrian seawater that is believed to be the result of volcanically active midocean ridges not only initiated the buildup of calcified shells, but was also mandatory for the aggregation and stabilization of multicellular sponge structures.

Sponges are the oldest extant Precambrian metazoan phylum and thus a valid model to study factors that could have unleashed the rise of multicellular animals. One such factor is the advent of auto-/allorecognition systems, which would be evolutionarily beneficial to organisms to prevent germ-cell parasitism or the introduction of deleterious mutations resulting from fusion with genetically different individuals. However, the molecules responsible for allorecognition probably evolved gradually before the Cambrian period, and some other (external) factor remains to be identified as the missing triggering event. Sponge cells associate through calcium-dependent, multivalent carbohydrate–carbohydrate interactions of the g200 glycan found on extracellular proteoglycans. Single molecule force spectroscopy analysis of g200-g200 binding indicates that calcium affects the lifetime (+Ca/-Ca: 680 s/3 s) and bond reaction length. Calculation of mean g200 dissociation times in low and high calcium within the theoretical framework of a cooperative binding model indicates the nonlinear and divergent characteristics leading to either disaggregated cells or stable multicellular assemblies, respectively. This fundamental phenomenon can explain a switch from weak to strong adhesion between primitive metazoan cells caused by the well-documented rise in ocean calcium levels at the end of Precambrian time. The authors propose that stronger cell adhesion allowed the integrity of genetically uniform animals composed only of "self" cells, facilitating genetic constitutions to remain within the metazoan individual and be passed down inheritance lines. The Cambrian explosion might have been triggered by the coincidence in time of primitive animals endowed with self-/non–self-recognition and of a surge in seawater calcium that increased the binding forces between their calcium-dependent cell adhesion molecules.

The work constitutes the first research where single-molecule-force spectroscopy studies have provided meaningful answers to the question of the origin of multicellular animals and might represent a milestone for both disciplines and an example of how multidisciplinary collaboration is an essential component of contemporary science, and is the result of a longstanding collaboration between the Institute for Bioengineering of Catalonia and Bielefeld U. together with the Friedrich Miescher Institute in Basel and the Marine Biological Lab at Woods Hole (Mass.) on biophysical single molecules and the effect of calcium on the interactions of cell adhesion molecules from marine sponges, considered a link between the single-cell dominated Precambrian and later multicellular organisms.

Considerable progress has been made in documenting and more precisely correlating biotic patterns in the Neoproterozoic–Cambrian fossil record with geochemical and physical environmental perturbations, but the mechanisms responsible for those perturbations remain uncertain. Peters and Gaines (2012) use new stratigraphic and geochemical data to show that early Paleozoic marine sediments deposited c. 540–480 mya record both an expansion in the area of shallow epicontinental seas and anomalous patterns of chemical sedimentation that are indicative of increased oceanic alkalinity and enhanced chemical weathering of continental crust. These geochemical conditions were caused by a protracted period of widespread continental denudation during the Neoproterozoic followed by extensive physical reworking of soil, regolith, and basement rock during the first continental-scale marine transgression of the Phanerozoic. The resultant globally occurring stratigraphic surface, which in most regions separates continental crystalline basement rock from much younger Cambrian shallow marine sedimentary deposits, is known as the Great Unconformity.

Although Darwin and others have interpreted this widespread hiatus in sedimentation on the continents as a failure of the geological record, this paleogeomorphic surface represents a unique physical environmental boundary condition that affected seawater chemistry during a time of profound expansion of shallow marine habitats. Thus, the formation of the Great Unconformity may have been an environmental trigger for the evolution of biomineralization and the Cambrian Explosion of ecological and taxonomic diversity following the Neoproterozoic emergence of animals.

Burgess Shale

The Burgess Shale (Briggs et al, 1994), one of the most important fossil sites in history, contains some strange and puzzling animals. The dozen or so localities of fossils are found in spectacular scenery, lying along the Cathedral Escarpment, which runs SE through Mt. Field and Mt. Stephen. Mt. Stephen is to the S of the Trans-Canada Highway which runs along Kicking Horse River in this area. To the W, Emerald Basin separates Emerald Lake from President Range. Walcott Quarry runs NW-SE from Waptia Mtn. to Mt. Field. Included also are Mt. Burgess and Burgess Pass. The town of Field is nearby. All are located in Yoho National Park in the Rockies of SE BC near the Alberta border. The fossils are in black shale and were discovered by Charles Walcott in 1909, although *Anomalocaris canadensis* and *Wiwaxia corrugata* were originally described by Whiteaves in 1892 and G.F. Matthew in 1899, respectively, from the *Ogygopsis* trilobite beds on Mt. Stephen. *Hallucinogenia* may be related to velvet worms (Onychophora) (One of Oldest Fossils Ever Found Is Finally Linked to Living Animals-The Huffington Post).

Walcott was an invertebrate paleontologist, and although he had not finished high school, was the foremost expert on the Cambrian. He was also Smithsonian Secretary, NAS President, AAAS President, USGS Director, and on the NRC, helped found the Carnegie Institute, the National Park Service, and the National Advisory Committee for Aeronautics (now part of NASA), was an advisor to Teddy Roosevelt and acquaintance of Louis Agassiz, and won the Thompson Medal from the NAS. He started his professional career discovering trilobite beds in upstate NY (the Walcott-Rust Quarry) and Vermont, and by selling collections. The Walcott Medal is awarded every 5 years for outstanding work in Cambrian and Precambrian life and history.

At Burgess Shale, Walcott found 65,000 specimens (which are at Harvard, the National Museum of Natural History, and the Royal Ontario Museum), yielding 125 genera and some 170 species. He described 78 of the genera and over 100 of the species. 92% or 114 of the genera are animals; the other 10 genera are all algae, 7 of which are definite--2 blue (*Marpolia* and *Morania*), 4 red (*Bosworthia, Dalyia, Wahpia*, and *Waputikia*), 1 green (*Margeritia*), and 1 possible red (*Spherocodium*), 1 possible green (*Yuknessia*), and 1 possible with no group designated (*Dictyophycus*). The groups with the most genera are arthropods, with 42, and sponges with 20. 18 of the genera have not been classified, although the 2 sclerite-bearing genera, which have been proposed to form Celosclerotophora as Chancelloridae and Wiwaxidae, along with other sclerite-bearers of the Lower Cambrian, Siphogonuchitidae, by Bengston and Missarzhevsky in 1981; 4 genera are considered family Anomalocaridae, but its position is unknown; and *Haplophrentis* is considered by some to be a phylum (Hyolitha)(e.g., Runnegar in 1980) but by others as a member of Mollusca (e.g., Marck and Yochelson in 1976). Anomalocarids are not unique to the area as they are found also in Sirius Passet, Greenland and Cheng Jiang, China, and other Burgess Shale animals or Burgess Shale-type faunas occur in some 30 other localities in Australia, Siberia, Spain, and Poland.

Walcott's taxonomic assignments were criticized especially by Stephen Jay Gould who made the site famous by bringing it to a popular audience in "Wonderful Life" from 1989.

There was also the Raymond Quarry begun by Percy Raymond of Harvard in 1930; Franco Rasetti's field work in 1947 and 1948; excavation of the Walcott and Raymond Quarries in 1966 and 1967 by Harry Whittington, an international expert on trilobites, and a Geological Survey of Canada team, which found 13,000 new specimens; and the research by Desmond Collins, Curator of Invertebrate Paleontology at the ROM, who found over 1000 new specimens starting in 1975.

Burgess Shale is a major Lagerstätte and was designated as a World Heritage Site by Unesco in 1981.

the Tully Monster

The Tully Monster first came to the attention of paleontologists in 1958 (Tully's Mystery Monster, Brian Switek-wired.com). While looking for fossils among the mining pits of NE Illinois, at the Mazon Creek fossil site, an important Lagerstätte, collector Francis Tully stumbled on an assemblage of marine organisms unlike any found elsewhere in the area. Especially perplexing were 6-

in., vermiform impressions found inside the numerous concretions that littered the pits. Soon other amateur fossil hunters began finding them.

When presented with some of these specimens by Tully, the professional paleontologists at Chicago's famous Field Museum were puzzled. The Tully Monster did not correspond to any other known animal. In his 1966 description, Eugene Richardson of the Field Museum dubbed them *Tullimonstrum gregarium*, honouring its discoverer, bizarre form, and the sheer number of individuals that had been discovered – but he refrained from giving it a precise place in the animal kingdom. "While this obscure but plentiful animal is being studied," Richardson stated, "I prefer not to assign it to a phylum."

Richardson published a more complete description of the beasts 3 years later with colleague Ralph Gordon Johnson. They still were not certain what it was. "There is no compelling reason to assign *Tullimonstrum* to any of the known phyla," they wrote, concluding, "It could be imagined as an aberrant member of one of several phyla but the critical evidence is not available." Nevertheless, examination of scores of specimens allowed the paleontologists to flesh out the anatomy of the monster.

The chief difficulty with studying the Tully Monster is that all the specimens are only impressions of the soft-bodied animals. No exoskeletons, no chitin plates, and no other hard parts were left behind. A few specimens that had begun to decay before they were buried allowed a blurry look at the organs, but Johnson and Richardson were mostly restricted to studying the external anatomy.

As reconstructed by Richardson and Johnson, the Tully Monster had segmented, semi-cylindrical bodies marked by 3 remarkable external traits. At the posterior end were 2 triangular tail fins arranged like the undulating side fins of squid. On the opposite end, there was a flexible proboscis tipped with a minutely-spiked grasping claw, and further back on the head were 2 stalks with cup-like depressions. A specimen in which pyrite preserved the form of these organs showed that these flexible stalks probably supported the eyes. Slight variations seen among various specimens suggests that the eye stalks could be angled forward or backward for different views. Richardson and Johnson were also able to say a little about their prehistoric habitat--the marine invertebrates lived in the warm, coastal waters of a 300 million-year-old ocean.

Over 4 decades later, we don't know much more about the Tully Monster. Merrill Foster, in his 1979 reassessment, considered it related to Gastropoda, which contains conchs, whelks, and limpets. Chen et al in a 2005 article in the Proceedings of the Royal Society, suggested it might instead be kin to the Cambrian invertebrate *Vetustovermis*, itself a problematic fossil of uncertain affinities. Although separated by about 200 million years, some have likened it to *Opabinia* of the Burgess Shale as they both had stalked eyes, a flexible proboscis tipped with a grasping appendage, and moved by way of flexible fins on the sides of their bodies. However, any kinship between the 2 was denied by Beall in 1991 (Delle Cave et al 1998).

Ediacara Hills

In 1946, Australian mining geologist Reginald Sprigg was exploring the Ediacara Hills in Flinders Range north of Adelaide, South Australia, when he stumbled onto fossilized imprints preserved generally on the undersides of slabs of quartzite and sandstone. Most were round, discoid forms that he dubbed "medusoids", from their seeming similarity to jellyfish. Appropriately, the name "Ediacara" comes from an aboriginal language meaning "veinlike spring of water" — the "spring," perhaps, from which complex animals have arisen. (ucmp.berkeley.edu).

Initially, Sprigg thought these fossils might be Cambrian, but it was not until the discovery of the frond-shaped fossil *Charnia* in England's Charnwood Forest in 1957 and the work of Martin

Glaessner that the Precambrian was seriously considered as containing life, and the detailed geological mapping by the British Geological Survey confirmed it sat in Precambrian rocks.

There are over 1,500 well-preserved specimens that have been collected from this locality, another of the most important fossil finds in history and one containing strange and puzzling animals, resulting in the naming of 40 genera and more than 60 species (Wikipedia-List of Ediacara Genera), of which 11 genera are discoid and of uncertain status as they might be microbial colonies, 2 are replaced by another genus, and 1, a pseudofossil, is rejected, so that only 28 are definite. Four of them belong to Proarticulata Fedonkin 1985, which belongs in Bilateria, and 2 to Trilobozoa Fedonkin 1985, which might belong in Cnidaria. Sprigg named 10 genera, in '47 and '49, but only 1 of these, *Dickinsonia*, is definite. Glaessner named 5 (1 rejected), Mary Wade (one of Glaessner's students) 5, Glaessner and Wade 8, and Gehling authored 2 and co-authored 6, the others naming only 1 each.

The type locality of Ediacara Hills, a major Lagerstätte, gave its name to what is refered to collectively as the "Ediacara Biota", which comprise 25 localities around the world, of the late Precambrian, however, Breandan MacGhabhan debunks the designation/concept "Ediacara Biota" showing it to be artificial and arbitrary as it cannot be defined geographically, stratigraphically, taphonomically, nor taxonomically. He points out that 8 particular fossils or groups of fossils considered "Ediacaran" have among them 5 taphonomic modes (preservation styles), occur in 3 geological periods, and have no phylogenetic meaning as a whole.

The "Ediacara Biota" fossils fall into 4 major categories according to Briggs et al (1994): circular impressions resembling celenterates; sessile, frond-like forms; benthic forms; and trace fossils. The 1st 3 are classified as Vendozoa, early animals, or Kingdom Vendobiota, unreleated to animals, by Adolf Seilacher (1989, 1992) and Buss and Seilacher (1994). The following is the classification for Vendobiota (peripatus.gen.nz-the Ediacara Assemblage):

Phylum Petalonamae Pflug 1972 (syn. Kingdom Vendobiota Seilacher 1992)
- Class Erniettamorpha Pflug 1972
  - Order(s) unknown
    - Family Erniettidae Pflug 1972 (*Ernietta, Swarpuntia*?)
    - Family Pteridiniidae Richter 1955 (maybe this family belongs with the Anthozoa or even Trilobozoa) (*Phyllozoan?, Pteridinium)*
- Class Rangeomorpha Pflug 1972
  - Order(s) unknown
    - Family Rangeidae (*Rangea*)
    - Family Charniidae (*Charnia, Charniodiscus?, Paracharnia*)

Gregory Rettalack and Mark MacMenamin regard the various genera as lichens, Zhuravlev as protistans, and Peterson et al as typical funguses. Martin Glaessner (1983, 1984) sees them as belonging to existing phylums, which is the majority view.

Most "Ediacarans" appear to have been very thin, which is why some propose that their tissues may have housed symbiotic algae. Mark McMenamin (1986, 1998) coined the term 'Garden of Ediacara' to encapsulate the concept. However, other scholars such as Bruce Runnegar point out that the hypothesis is both difficult to test and unlikely anyway, because some of the fossils appear to have been deposited below the storm wave base, about 330 ft. (c. 100 m.), where the intensity of sunlight is very much diminished, probably below useful limits for such organisms.

"Ediacarans" are commonly grouped into 3 types or assemblages: Avalon (refering to the type locality of the Avalon Peninsula in Newfoundland), Ediacara, and Nama (the type locality being

Namibia). Erwin in 2011, based on molecular clock dates, found 6 major clades: Rangeomorpha, Ernettiomorpha, Dickinsoniomorpha, Arboreomorpha, Triradiomorpha, and Kimberellomorpha, and 3 possible clades: Bilateriomorpha, Tetraradiomorpha, and Pentaradiomorpha.

These fossils are the focus of a version of the late bloomer hypothesis, the Garden of Ediacara hypothesis, proposed by Mark and Diana McMenamin (1990, 1998) who maintain that multicellular animals existed long before the Cambrian Explosion and that there was a Rodinia, Precambrian, low-nutrient, autotrophic phase dominataed by photo- or chemosynthetic animal-like organisms (the Ediacara biota), but the dismantling of the hypothetical Rodinia supercontinent caused a nutrient pulse which triggered a Gondwana (hypothetical supercontinent) high-nutrient phase dominated by large predatory animals. The idea is opposed to the latecomer (oxygen threshold and cell type) hypotheses.

The Hypersea Hypothesis, which combines Vernadsky geochemical, Gaia, and Darwinian theories, is proposed by the same authors, who regard the land biota as a biogeophysiological entity for which the above scenario was a prelude.

These ideas have not been well received and have been refuted by Ben Waggoner, Richard Fortey, and James Gehling.

Table 15-2. Animalia.

Here I used section (sct.) and series (srs.), which have been introduced due to the many ranks occuring in this kingdom. Branch and grade are also used but grade is not an acceptable term in phylogenetics and all are branches anyways; synapomorphies are in parentheses; figures in parentheses are generally approximations and are for classes, orders, families, genera, and species, in that order; extinct groups are designated with an asterisk. Also, to accomodate the new lineages and maintain the rank stability I use the ranks supersubkingdom and hypersubkingdom; Spongiae may be paraphyletic (Zrzavy et al, loc. cit.).

hpsbk. Spheroformazoa nom. nov. (1 gen.)
hpsbk. Cadherozoa nom. nov.
  spsbk. Filasterozoa nom. nov. (2 gen.)(possibly paraphyletic)
  spsbk. Choanozoa
 sbk. /phyl. Choanoflagellata Kent 1880 (syn. Craspedophyceae Chadefaud 1960, Choanomastigota, Choanozoa Cavalier-Smith 1989) (collar flagellates) (1, 2, 3, 42, 150)
sbk. Metazoa Haeckel 1877
 infk. /phyl. Archeocyatha* Bornemann 1884
 infk. Neozoa nom. nov.
   pvk. phyl. Spongiae (syn. Spongiaria de Blainville 1816, Parazoa Sollas 1884, Porifera Grant 1835) (3, 28, 100, 350, 9000)(possibly paraphyletic)
   pvk. Epitheliozoa nom. nov. (syn. Eumetazoa) (epithelium and belt desmosomes)
    mck./phyl. Placozoa Grell 1971 (only 1 species)
    mck. Euzoa (syn. Enterozoa Lankester 1878, Eumetazoa) (symmetry, organs, organ systems, gap junctions, basement membranes, and hemidesmosomes)
     nnk./phyl. Cnidaria (Celenterata 1) (4, 27, 236, 927, 9000)(including Myxozoa)
     nnk. Anazoa nom. nov. (multiciliate epithelium, acetylcholine as neurotransmitter, unidirectional synaptic impulse transmission, nerve fibers with insulating sheath)
      pck./phyl. Ctenophora Eschscholtz 1829 (Celenterata 2)
      pck. Artiozoa (syn. Bilateria Schimkewitsch 1891, Triploblastica Lankester 1900) (bilateral symmetry, 3 germ layers, well organized tisssue formation, CNS (central

nervous system), protonephridia, mobility, cephaly)
sbpck. Protostomia Grobben 1908 (monociliate epithelia, ventral longitudinal nerves paired or secondarily fused, protonephridia)
sptrph. Ecdysyzoa Aguinaldo et al 1997 (molting cuticle)
hprph/phyl. Cephalorhyncha (cls. Kinorhyncha Reinhard 1887 (syn. Echinodera Zelinka 1928), Priapula (Priapuloidea Sedgewick 1898), and Lorcifera)
hprph. Sensillozoa nom. nov. (sensillium) (or Nematozoa + Cephalorhyncha forming Introverta)
spph./phyl. Nematohelminthes Gegenbauer, 1859 (syn. Nematozoa) (Nematoda +Nematomorpha) (1000 gen., 12,000 sp.)
spph. Panarthropoda Neilsen 1985
phyl. Onychophora Grube 1851 (velvet worms)
phyl. Arthropoda von Siebold and Stannius 1845
sptrph. Spiralia Neilsen 1995 s.l. (spiral cleavage with 4-d mesoderm)
mgph. Platyzoa
phyl. Trochata (cl. Rotifera (syn. Rotatoria Ehrenberg 1838), cl. Acanthocephala Rudolphi 1808) (or with Ecdyzyzoa forming Aschelminthes Neilsen 1995)
phyl. Platymorpha nom. nov. (cls. Gastrotricha Metschnikoff 1864, Gnathostoma, Platyhelminthes (flatworms)) (2000 gen., 20,000 sp.)
mgph. Trochozoa (trochophore larva)
grdph. /phyl. Kamptozoa Cori 1929 (syn. Calyssozoa, Entoprocta Nitsche 1870) (+Cycliphora?)
grdph. Eutrochozoa
hprph./ phyl. Echiuroidea Sedgewick 1898 s.s. (cls. Echiura Savigny 1817 as Echiurida (spoon worms), Pogonophora Johansson 1937, and Chetifera)
hprph. Molluscoidia nom. nov.
spph/phyl. Nemertina Oersted 1844 (syn. Rhynchocela Schulze 1850)(ribbon worms)
spph. Molluscomorpha nom. nov.
phyl. Sipuncula Raffinesque 1814
phyl. Mollusca Linneus 1758 (syn. Malacozoa) (7, 34, 530, 3000, 50,000)
sbpck. Deuterostomia Grobben 1908 (syn. Enterocelia Hatschek 1811, Epineuralia Cuénot 1940, Notoneuralia Ulrich 1951, Oligomeria, Irregularia) (blastopore develops posteriorly, enterocely, and radial, indeterminate, and regular cleavage)
trph. Chetognatha Leuckart 1854 (arrow worms)
trph. Enterocela Lankester 1876 (syn. Trimera) (trimery (prosome, mesosome, metasome), dorsal CNS, upstream collecting larval ciliary bands)
ggph. Lophophorata Hyman 1959 (syn. Tentaculata Hatschek 1891; Polyzoa Thompson 1830)
phyl. Bryozoa Ehrenberg 1831 (syn. Ectoprocta Nitsche 1870, Zoophyta Cuvier 1812) (moss animals) (2, 10, 108, 300, 10,000)
phyl. Phoronozoa Zrzavy et al 1998 (cls. Brachiopoda Cuvier 1802, Phoronida Hatschek 1888 (syn. Pygocaulia Thompson 1927))
ggph. Epithelioneuria (axial complex, epithelioneural system)

mgph./phyl. Echinodermata Laske 1778 (6, 36, 145, 250, 7000)
mgph. Stomochordata (stomochorda)
grdph./phyl. Pterobranchia Lankester 1877
grdph. Cyrtotreta Neilsen 1995 (gill slits, dorsal neural tube, collagenous skeleton)
hprph./ phyl. Enteropneusta Gegenbaur 1870 (acorn worms)
hprph. Chordata Haeckel 1874 (syn. Chordonia auct.) (chorda, dorsal neural tube, true tail, dorsal longitudinal muscles, endostyle organ)
spph. /phyl. Urochordata (syn. Tunicata Lamarck 1816)
spph. Notochordata (chorda extending along the whole dorsal side, myomery, mesoderm segmentation, open capillary junctions, grey and white differntiaton in the neural tube)
phyl. Cephalochordata Owen 1846 (syn. Acrania Haeckel 1866) (lancelets)
phyl. Vertebrata Cuvier 1817 (syn. Craniata Lankester 1877) (?, 100, 800, 8,000, 44,000)

Table 15-3. Spcoh. Holometabola Martynov 1925 (Oligoneoptera auct., Endopterygota Sharp 1893)

coh Coleopterida Handlirsch 1908
ord Coleoptera Linneus 1758 s.s. (beetles)
coh Telomerida Boudreaux 1979
sbcoh Hymenopterida Handlirsch 1908 , emend. from Hymenopteroida
ord Hymenoptera Linneus 1758 (wasps, bees, ants)
sbcoh Meronida Boudreaux 1979
infcoh Neuropterida Handlirsch 1908, emend. from Neuropteroidea
spord Raphidida Handlirsch 1908
ord Raphidiodea Burmeister 1839 (snake flies)
spord Sialida Handlirsch 1908, emend. from Sialoidea
ord Neuroptera Linneus 1758 s.s. (syn. Plannipennes Latreille 1817) (lace-wings)
ord Megaloptera Latreille 1802 s.s. (alder flies and allies)
infcoh Panorpida Handlirsch 1908
hprord Mecopterida (=Siphonoptera + Antliophora, Hennig 1969) stat. nov.
spord Mecopterodea Boudreaux 1979)
ord Mecoptera Packard 1886 (scorpion flies)
ord Siphonaptera Latreille 1825 (fleas)
spord Hanstellodea Clairville 1798, s.s., emend. from Haustellata
ord Diptera Linneus 1758 (2-winged flies)
ord Strepsiptera Kirby 1813
hpord Trichopterida Boudreaux 1979 (syn. Amphiesmenophora Hennig 1969) stat. nov.
ord Trichoptera Kirby 1826 (caddis flies)
ord Lepidoptera Linneus 1758 (moths and butterflies)

Based on Hennig 1969, Boudreaux 1979, Kristensen 1991, and Whiting et al 1997. Antliophora, Mecopteromorpha, Dipteromorpha, and Amphiesmenophora are all strongly supported by both morphology and molecules (Whiting et al , 1997).

The amazing success and diversity of winged insects is attributable to water loss prevention (gas exchange system, spiracular closing mechanism, cuticular exoskeleton and impermeable waxy cuticle, laying of waterproof eggs, and Malpighian tubules), ability to fly, ability to bite and chew food, rapid generation, small size, malleable structures, and, in Holometabola, a specialized larval (pupal) stage, and, in beetles, chitinized armour to protect against predators and parasites.

In Holometabola are found the important pollinators (Coleoptera, Hymenoptera, Diptera, Lepidoptera), probably the most important being bees, which were long thought to have co-evolved with flowers, but new evidence--a comprehensive study of insect evolution (Sepkoski and Labandeira, 1993) and the discovery of bees' nests from the Triassic in the Petrified Forest of Arizona (Haciotis, 1995), which go back 220 mln. yrs. indicating an origin 140 mln. yrs. earlier than previously believed--strongly suggests that bees were already pollinating gymnosperms and that flowers evolved showy petals and scents to compete with them for their attention so that the tradtional view may be erroneous.

Boudreaux, H. B. 1979. Arthropod Phylogeny, with special reference to insects. Wiley.
Hasiotis, S.T. and Dubiel R.F. 1995. Termite (Insecta: Isoptera) nest ichnofossils from the upper Triassic Chinle formation, Petrified Forest National Park, Arizona. Ichnos 4: 126-130.
Kristensen. N.P. 1991. Phylogeny of extant hexapods. In: The Insects of Australia, I.D. Nauman et al, eds., pp. 125-40. University Press, Melbourne.
Sepkoski, J.J., Jr., and Labandeira, C.C. 1993. Insect diversity in the fossil record. Science 261: 310-15.
Whiting, M.F., Carpenter, J.C., Wheeler, Q.D., Wheeler, W.C. 1997. The Strepsiptera problem: Phylogeny of the holometabolous insect orders inferred from 18S and 28SrDNA sequences and morphology. Syst. Biol. 46: 1-68.

Table 15-4. Mollusca (Malacozoa) (synthesis of several morphological and molecular phylogenies, primarily Peel and Salvini-Plawen (Molluscan Phylogeny-paleos.com)).

sbph Scutopoda
  inph Chetodermomorpha (Aplacophora 1)
    cl Caudofoveta
sbph Adenopoda
  infph Neomenomorpha (Aplacophora 2)
    cl Solenogastres
  inph Testaria
    hpcl Polyplacophora
      cl Polyplacophora
    hpcl Conchifera
      spcl Pleistomollusca
        cl Tergomya (Monoplacophora 1) + Gastropoda
        cl Bivalvia (syn. Acephela, Pelecypoda)
      spcl Metamollusca nom. nov.
        cl Helcionelloida (Monoplacophora 2) + Cephalopoda
        cl Scaphopoda

Table 15-5. Vertebrata (based mostly on M.J. Benton (Vertebrate Paleontology, 1996, Chapman & Hall) (asterisks designate fossil forms).

Myxinoidia (Cyclostomata 1)(hagfish)
Neovertebrata
  Petromyzonti (Cyclostomata 2)(lampreys)
  Sclerovertebrata
    Anaspida*
    Osteodermata
      Pteraspidomorphi* (syn. Diplorhinia)(Heterostraci and Thelodonti)
      Metavertebrata
        Galeaspidia*
        Ambivertebrata
          Osteostraci*
          Gnathostomata Gegenbaur 1874
            Euconodonta*
            Eugnathostomata
              Placodermi* McCoy 1848
              Neognathostoma
                Chondrichthyes J. Muller 1846 (cartilaginous fish)
                Osteognatha
                  Acanthodii*
                  Kinocrania Hennig 1985
                    Actinoptergyi Cuvier 1828 (ray-finned fish)
                    Sarcopterygimorpha
                      Actinistia (syn. Celacanthini Agassiz 1843)(tassel-finned fish)
                      Choanata Owen 1846
                        Dipnoi
                        Metacrania
                          Porolepimorpha*
                          Neocrania
                            Osteolepida*
                            Tetrapodomorpha

Vertebrate phylogeny including fossils between Tetrapodomorpha and mammals and birds is a veritable quagmire of conflicting results and proliferation of ranks so I am leaving that part out and am not ranking the higher levels. There may be a dozen or 2 dozen or even about 40 extra rank levels and some groups are disputed. An example is *Panderichthyes*, which may be paraphyletic and, along with other forms, mostly sarcopterygians, is arranged into 9 levels as part of early Tetrapodomorpha. Also, for sarcopterygian fishes there are at least 4 competing theories (by Romer from '66, Rosen et al from '81, Janvier from '86, and Panchen and Simpson from '87) differing significantly--Dipnoi as basal, sister group to Porolepiformes, between Actinistia and Porolepiformes, and sister group to Tetrapoda). And often molecular evidence is used which won't necessarily resolve anything as explained in Chapter 3, and can even complicate matters as we see, for example, in molluscs. Sarcopterygii is divided into Crossopterygii (containing Rhipidistia ((Porolepiformes, Osteolepida) and Celacanthini), Dipnoi, Rhizodontia, and *Panderichthyes*. A simplified, practical system can be one that excludes fossil forms, but it can also be one that includes them, but with the grades, so is not cladistic. I am presenting the latter below, based on Benton (loc. cit.).

Table 15-6. Simplified System for Vertebrates.

Agnatha
Gnathostomata
  Chondrichthyes
  Osteognatha
    Actinopterygii (Osteichthyes 1)
    Neognatha
      Sarcopterygii (Osteichthyes 2)(paraphyletic)
      Tetrapoda

simplified system for tetrapods:

- Batrachomorpha
  - Protobatrachomorpha* (paraphyletic)
  - Amphibia
- Reptiliomorpha
  - Protoreptiliomorpha*(paraphyletic)
  - Amniota
    - Sauropsida
      - Anapsida (turtles)(paraphyletic)
      - Diapsida
        - Lepidomorpha*
          - Parapsida
          - Lepidosauria (lizards and snakes)
        - Sauropterygia*
          - Placodontia
          - Nothosauria
          - Plesiosauria
        - Ichthyosauria*
        - Archosauria
          - Crocodylotarsi
          - Dinomorpha (Ornithosuchia)
            - Pterosauria*
            - Dinosauria
              - Saurischia
                - Theropoda
                  - Prototheropoda* (paraphyletic)
                  - Maniraptora
                    - Protomaniraptora* (paraphyletic)
                      - Dromeosauria*
                      - Aves
                    - Neomaniraptora*
                - Sauropodmorpha*
                  - Prosauopoda (paraphyletic)
                  - Sauropoda
              - Ornithischia*

Cerapoda
Thyreophora
Synapsida
Pelycosauria*(Theromorpha)(paraphyletic)
Therapsida
Prototherapsida* (paraphyletic)
Cynodontia
Protocynodontia*(paraphyletic)
Mammalia

Table 15-7. Legion Mammalia (based on Benton and paleos.org)(only extant forms included).

ggcohort Metamammalia
mgcoh Monotremata
mgcoh Theria
hprcoh Metatheria (syn. Marsupalia)(1 order)
hprcoh Eutheria (syn. Placentalia)
spcoh Edentomorpha
ord. Edentata (syn. Xenarthra)(silky and giant anteaters, sloths, armadillos)
ord. Pholidota (pangolins [anteaters])
spcoh Epitheria
coh. Insectivora (1 order)(uncertain position)
spord Lipotyphla (shrews, moles, tenrecs)
spord Menotyphla (jumping shrews)
coh. Metaplacentalia
sbcoh. Glirarchonta
mgord. Gliromorpha new name
spord. Macroscelidia (elephant shrews)(1 order)
spord. Glires
ord. Lagomorpha (rabbits and pikas)
ord. Rodentia
mgord. Archonta
spord Volitantia
ord Dermoptera (flying lemurs)
ord Chiroptera (bats)
spord Primatomorpha
ord Scandentia (tree shrews)
ord Primates
sbcoh. Neoplacentalia
infcoh. Carnivora (1 order)(unceratin position)
infcoh. Ungulata
ggord. Paraxonia
spord Artiodactyla (1 order)
spord Cetacea (whales and dolphins)(1 order)
ggord. Metungulata
mgord. Tubulidentata (aardvark [anteater])(1 order)(uncertain position)
mgord. Neungulata

hprord. Perissodactyla (horses and rhinos)(1 order)
hprord. Penungulata
spord. Tethytheria (2 orders, including elephants)
spord. Hyracoida (hyrax or coney)(1 order)

Table 15-8. Primates (based on Benton).

subord Strepsirhini (lemurs and lorises)
subord Haplorhini
inford Tarsiformes (only living genus *Tarsius* [tarsiers])
inford Simiformes
hypfam Platyrhini (New World monkeys)
hypfam Catarhini
spfam Cercopithicoidea (Old World monkeys)
spfam Hominoidea (syn. Anthropoidea)
fam Hylobatidae (gibbons)
fam Hominidae (great apes)
sbfam Ponginae (orangatangs, 2 living species [Sumatran and Bornean])
sbfam Homininae
tribe Gorillini
gen Gorilla (2 sp., West and East)
tribe Hominini
gen Pan (2 sp., common or field and bonobo or pygmy chimps)
gen Homo (12 sp., only 1 extant, which has 2 sbsp, only 1 extant)

Delle Cavea, Laura; Insom, Emilio; Simonetta, Alberto Mario. 1998. Advances, diversions, possible relapses, and additional problems in understanding the early evolution of the Articulata. Italian Journal of Zoology 65: 19-38 (tandfonline.com).

Eernisse, D.J., Albert, J.S., Anderson, F. E. 1992. Annelida and Arthropoda are not sister taxa: a phylogenetic analysis of spiralian metazoan morphology. Syst. Biol. 41: 305-30.

Fernández-Busquets, Xavier; Körnig, André; Bucior, Iwona; Burger, Max; and Anselmetti, Dario. 2009. Self-Recognition and Ca2+-Dependent Carbohydrate-Carbohydrate Cell Adhesion Provide Clues to the Cambrian Explosion. Molecular Biology and Evolution 26: 2551(science daily).

Giribet, G., Dunn, C.M., Edgecombe, CG.D., Rouse, G.W. 2007. A modern look at the Animal Tree of Life. Zootaxa 1668: 61-79.

Halanych, K.M. 2004. The New View of Animal Phylogeny. Annu. Rev. Ecol. Evol. Syst. 35: 229-56.

Huxley, Thomas. 1869. An Introduction to the Classification of Animals (books.google).

de Monet, J-P (chevalier de Lamarck). 1801. Système des animaux sans vertèbres (lamarck.cnrs.fr).

Nichols, S.A., Roberts B. W., Richter, D., Fairclough, R.J. and King, Nicole. 2012. Origin of metazoan cadherin diversity and the antiquity of the classical cadherin/ß-catenin complex. Proc. Natl. Acad. Sci. U S A 109: 13046–13051.

Nielsen, C. Animal Evolution: Interrelationships of the Living Phyla. U. of Copenhagen Press/Oxford U. Press.

Nielsen, C., Scharff, N., Eibye-Jacobsen, D. 1996. Cladistic analyses in the animal kingdom. Biol. J.

Linn. Soc. 57: 385-410.
Peters, Shanan & Gaines, Robert. 2012. Formation of the 'Great Unconformity' as a trigger for the Cambrian Explosion. Nature 484: 363–366.
Ragan, M.A. 1997. A 3rd Kingdom of Life: the history of an idea. Arch. Protistenkd. 148: 225-43.
Ruiz-Trillo, I., Lane, C. E., Archibald, J. M., and Roger, A. J. 2006. Insights into the Evolutionary Origin and Genome Architecture of the Unicellular Opisthokonts *Capsaspora owczarzaki* and *Sphaeroforma arctica*. Journal of Eukaryotic Microbiology, 53: 379–384. wiley.com.
Sebé-Pedrós, Arnau; de Mendoza, Alex; Lang, Franz; Degnan, Bernard; & Ruiz-Trillo, Iñaki. 2011. Unexpected repertoire of metazoan transcription factors in the unicellular holozoan *Capsaspora owczarzaki* (oxfordjournals.org; nih.gov).
Shalchian-Tabrizi, Kamran; Minge, Marianne; Espelund, Mari; Orr, Russell; Ruden, Torgeir; Jakobsen, Kjetill; Cavalier-Smith, Thomas. Multigene Phylogeny of Choanozoa and the Origin of Animals. PLoS ONE 3: e2098.
Swainson, Wm. 1835. Treatise on the Geography and Classification of Animals (gallica.bnf.fr)
Zrzavy, J., Mihulka, S., Kepka, P., Bezdek, A., Tietz, D. 1998. Phylogeny of Metazoa Based on Morphological and 18S Ribosomal DNA Evidence. Cladistics 14: 249-85.

Chapter 16 Protozoans

Examining this water...I found floating therein divers earthy particles, and some green streaks, spirally wound serpent-wise...and I judge that some of these little creatures were above a thousand times smaller than the smallest ones I have ever yet seen, upon the rind of cheese, in wheaten flour, mould, and the like. (The first recorded observation of protozoa.) - Antonie van Leeuwenhoek, Letter to the Royal Society, London (Sept. 7, 1674), in John Carey, Eyewitness to Science (1997), p. 28 - todayinsci.com.

Animalcula was coined by Leeuwenhoek in 1676, Oken used the name Urtiere ("original animals" or "primitive animals"), and Protozoa (essentially the Greek form of the German) was named by Goldfüss in 1817 as a class divided into 4 orders: Infusoria, Phytozoa, Lithozoa, and Medusinae. Infusoria (German Aufgustiere, "infusion animals") were named by Martin Frobenius Ledermüller between 1760 and 1763, the full name being Animalcula Infusoria, and refers to life forms able to produce dessication-resistant stages and which can be reactivated by an infusion of water to hay or pepper contaminated with such resting stages. The Goldfüss classification had 4 families for this group: Monades, Vorticellae, Brachioni, and Polypi. "Monades" came from Leibniz's monads from 1714. This was replaced by Bütschli's familiar system (1880-82) containing classes Sarkodina Hertwig and Lesser 1874, Mastigophora Diesing 1865, Sporozoa Leuckart 1879, and Infusoria Ledermuller 1763, restricting the last group to ciliates and suctorians, which has prevailed well into the 1960s. Greene, in 1859, recognized Protozoa as a subkingdom. Doflein, in 1901, combined the 1st 3 of Bütschli's classes as Plasmodroma. In 1878, Lankester recognized Gymnomyxa and Corticata. Sarcomastigota, combining Sarcodina with Mastigota, was also recognized by some, with the latter becomng more commonly known as Flagellata.

Plasmodiophorae

Plasmodiophorans have an osmotrophic, plasmodial, unwalled stage and a walled, anisokontic, infective, zoosporic stage, tubular cristas, dessicant-resistant, resting spores (cysts), which form aggregations (sori), and uniquely possess cruciform mitosis. The group was once considered as related to either chytrids, myxomycetes, or heterokont molds. In 1878 Woronin described *Plasmodiophora brassicae*, parasitic in cabbage causing club root. In 1884, Zopf established family Plasmodiophoreen (Plasmodiophoraceae Engler 1892) for *Plasmodiophora* and *Tetramyxa.* Schroeter created a new order, Phytomyxini, in 1885, with 1 family, to include *Plasmodiophora, Phytomyxa,* and *Sorosphera,* in Myxomycetes, close to Myxogasteres, and in 1893 he created a new class, Phytomyxinae, placing it between Acrasiales and Myxogasteres in Myxomycetes. Marie and Tison, in 1909, adpoted the family name and stated that it should be considered as an entirely distinct group intermediate between Sporozoa and Myxomycetes. Sparrow in 1943 and Martin in 1950 placed it in Phycomycetes. A close kinship with Chytridiales was widely believed until the flagellar differences were discovered. Karling (1968) notes that many taxonomists continued to use the name Phytomyxinae or some alteration of it even though it has been evident since the beginning of the 20th century that *Phytomyxa* is no longer valid and relates to bacteria and mycorrhizal funguses. The phylum/kingdom is intracellularly parasitic chiefly in higher plants. *Plasmodiophora, Polymyxa, Spongospora*, and *Sorosphera* attack land plants; *Tetramyxa, Ligniera, Sorodiscus*, and *Membranosorus* chiefly aquatic seed plants; *Woronina* and *Octomyxa*, Oomycetes; and *Phagomyxa*, brown algae (Copeland, 1956; Margulis et al, 1990).

Spongomonada

Spongomonads have tubular cristas and 2 isokontic flagella, are phagotrophic, usually colonial, and the cells are embedded in a granular, mucilaginous matrix (Lee et al, 2000). They also have microtubules radiating in a semicircle perpendicular to the basal body with concentric rings of EDM (electron-dense material) intersecting the microtubules (Lipscomb, 1985). There is no information on their habitat in the usual sources, but it is mentioned that they are found in "Gewässer" ("waters") (Spongomonas-de.Wikipedia), in a water reservoir (Spongomonas-protist.i.hosei.ac.jp), the Mer Bleue (a bog in Ontario, Canada)(Spongomonas-You Tube), and bog water (Rhipidodendron-You Tube). As Amphimonadaceae they were once one of 9 or 10 families in Protomastigineae, a group established by Klebs in 1892 for forms that took in food at a specific place on the cell, and included also Euglenarian groups, choanoflagellates, Tetramitaceae (now in Vahlkampfidae), and Bicecaceae (Kessel, 1955). They are in Euglenaria (along with Pseudodendromonadidae) in Lipscomb (1989, 1991), and Karpov (1999) assigns them to Heterokonta. Pseudodendromonadidae, separated from Spongomonada, were assigned to Bicosecales by Karpov (2000) because of fine structure similarities (left BB orientation, concentric rings, and 8+3 R3).

Conosans

Pelomyxida (syn. Archamebae, Pelobiota, Karyoblastea) include 2 families: Pelomyxidae Schultze 1877 (*Pelomyxa* Greef 1874 and *Mastigina* Frenzel 1897) and Mastigamebidae (*Mastigella* Frenzel 1897 and *Mastigameba* Schulze 1875 (syn. for the latter: *Phreatameba* and *Dinameba*)). *Pelomyxa palustris* is known only from mud at the bottom of freshwater ponds, and was usually placed in Entamebidae. The number of species is unknown.

Cercomonads are unicellular, free-living, gliding, anisokontic ameboflagellates with filose pseudopods and tubular cristae, common in soil. The order was established as Cercomonadidea by Poche in 1913 as a replacement for Rhizomastigina, which was established by Bütschli in 1883, and contained only *Cercomonas*. Vickerman in 1984 added *Heteromita*. It was revised by Mylnikov in 1986. In Lee et al (2000) included are *Cercomonas, Heteromita,* and *Massisteria.* Mylnikov and Karpov (2004), on the basis of morphology, include 2 families: Cercomonadidae Kent, 1880 (=Cercobodonidae Hollande, 1942) and Heteromitidae Kent, 1880 emend. Mylnikov, 2000 (=Bodomorphidae Hollande, 1952). The former family has a variable body shape, habitual pseudopods, a microtubular cone, no kinetocysts, and has a plamodial stage. The latter family has a rigid body shape, temporary pseudopods, mostly no microtubular cone, kinetocysts, and no plasmodial stage. Cercomonadidae include *Cercomonas* Dujardin, 1841 and *Helkesimastix* Woodcock et Lapage, 1914. All species of *Cercobodo* are transferred to *Cercomonas*. Heteromitidae include *Heteromita* Dujardin, 1841 emend. Mylnikov et Karpov 2004, *Allantion* Sandon, 1924, *Sainouron* Sandon, 1924, *Protaspis* Skuja, 1939, *Cholamonas* Flavin et al, 2000, and *Katabia* Karpov et al, 2003. The names *Bodomorpha* and *Sciviamonas* are regarded as junior synonyms of *Heteromita*. Synonyms for Cercomonas are *Dimastigamoeba* Blochmann, 1894, *Cercomastix* Lemmermann, 1914, *Prismatomonas* Masssart, 1920, inter alia. *Proleptomonas* Woodcock, 1916 is excluded according to morphology but included according to gene analysis, and *Massisteria* Larsen and Patterson, 1988 is excluded on the basis of molecular evidence and different pseudopods. Karpov et al (2006) add *Eocercomonas* and *Paracercomonas* based on kinetid structure and genes. There are 70 species in all.

Apusomonads have 2 genera, *Apusomonada* and *Bodomorpha.* Thaumatomonads (6 genera, including *Protaspis;* type genus *Thaumatomonas* de Saedeleer, 1931) have been identified as Thaumastigaceae by Patterson and Zöllfel in 1991 (based on the type genus *Thaumatomastix*

Lauterborn, 1899), with Vørs in 1992 providing the zoological suffix. (Lee et al, 2000).

Myxobiotes, like most protozoans, are anisokontic, tubulocristate, and ameboid (with filose pseudopods), but have a walled and sporulating phase. Bacteria are the primary food source. Spore walls are chitinous or cellulosic. The plasmodium-sporophore complex sets them apart from other organisms. There are 8 orders, 19 families, 74 genera, and some 630 species. Usually recognized are subphylums Dictyostelia, including 1 class and order, 2 families, 4 genera, and about 50 species; and 2 noncellular slime mold classes: Protostelia, which count 1 order, 4 families, 14 genera, and 32 species, and Myxomycetes, which include 2 or 3 subclasses, 6 orders, 13 families, some 60 genera, and about 550 species. Phylogenetically, Protostelia are the most primitive and are paraphyletic, and Dictyostelia and Myxomycetes form a clade. Protostelia have a reproductive sporulating phase and an ameboid trophic phase. Dictyostelia comprise cellular slime molds which have 3 phases: ameboid, microscopic; multicellular and sporulating (and visible to the unaided eye); and an intermediary aggregation phase (the pseudoplasmodium formed by aggregating myxamebas). Myxomycetes alternate monoploid (haploid) and diploid generations. There are 3 stages: the reproductive, sporulating; a uninucleate, microscopic, ameboid (that may or may not be flagellated) trophic stage; and a plasmodial trophic stage, which is sometimes macroscopic and pigmented.

Eulobosa

Gymnolobosa (Gymnamebae), in Parker (1982), was arranged as 5 orders: Amebida, Schizopyrenida, Pelobiotida, Leptomyxida, and Stereomyxida. Amebida had 5 suborders (Tubulina, Thecina, Flabellina, Conopodina, and Acanthopodina) and 9 families. Tubulina had 4 families: Amebidae Ehrenberg 1838, Hartmanellidae Vohlkonsky 1931, Entamebidae Chatton 1925, and Vahlkampfidae Jollos 1917. In Margulis et al (1990) there are the same 5 suborders, the arrangement being modified from various sources, including Deflandre in Grassé, Levine et al, Page, and Margulis and Schartz, from 1953 to 1982. In Lee et al (1985) there are also the same 5 suborders and the same orders as in Parker (1982). Gymnolobosa is still recognized but emended and configured in Lee et al (2000) as 3 orders, Euamebida, Centramebida, and Leptomyxida, with 14 families (including 3 of uncertain affinities), 58 genera, and c. 200 species; Schizopyrenida and Pelobiotida are placed in a separate section.

Testalobosa had 2 orders: Arcellinida Kent 1880 (19 fams., about 40 genera) and Trichosida Moebius 1889 (1 genus, *Trichospherium*, and 3 marine and estuarine species). The former was divided into Eulobosina (6 fams.), which is paraphyletic, and Reticulolobosina (2 fams.) by Page in Parker (1982).

It is clear that actively moving amoebas form specific, differentiated structures valuable for species characterization (Smirnov et al, 2011). In 1926 Schaeffer noted that an actively moving ameba, despite minor variations, has a more or less dynamically stable shape (specific general outlines and traits like position of hyaloplasm, dorsal or lateral ridges, and flatness) that may be genus- or even species-specific. He introduced the term "locomotive form" to recognize this, arguing that defining this "shape" is the best way to characterize an ameba; this concept remains the basis of ameboid descriptions. He also noted the importance of nuclear structure and other traits and in the same year constructed a synthetic system of amoebas utilizing many of the above mentioned characteristics. Though criticized by some, his wisely multifaceted approach was successfully applied in many studies and became the most frequently cited. Attempts to create an alternative system continued, but none became widely accepted. Greeff, in 1866 and 1874, pointed out the importance of gross nuclear structure in ameboid descriptions; this characteristic was widely applied.

The nuclear division pattern was suggested as such a fundamental feature and a number of systems used it. This resurrected the approach of Glaeser in 1912, who stated, "the most reliable criterion for the classification of amebas is the division of the nucleus", which despite some support was strongly criticized, e.g., by Schaeffer in 1920, who wrote, "the classification based on nuclear characters would be a highly artificial system". Subsequent developments have shown that he was right.

Studies of diverse mechanisms of ameboid movement inspired Jahn and Bovee (1965) to use patterns of cytoplasmic flow in pseudopods to group amebas into higher taxons. They named the 2 patterns contractile-hydraulic (Hydraulea) and active sliding (Autotractea) in superorder Lobida. Testate and naked forms of lobose amebas were classified together, instead of the usual separate manner. Though widely ignored by other taxonomists these contrasting pseudopodial patterns are now corroborated by molecular data.

Analysis of morphotypes (Smirnov et al, 2011) and of the list of species belonging to each tells us that all lobose amoebas may be split into 3 basic groups: (A) those where the entire cell is always cylindrical or subcylindrical; (C) those where it is always flattened, being laterally expanded in cross-section, and (B) those able to alter their locomotive form from cylindrical to flattened under certain conditions. Species showing morphotypes of groups A or B all belong to the Tubulinea clade in molecular trees, while those in group C belong to Discosea. So there is a clear division of lobose amebas into two large groups: those with tubular pseudopods (or able to form them under certain circumstances), Tubulinea, and those generally with a flattened body, Discosea. The 22 morphotypes (Smirnov and Goodkov, 1999) are: polytactic, orthotactic, monotactic, rhizomonotactic, flabellate, branched, dactylopodial, flabellate (fan-shaped), paraflabellian, paramebian, lingulate, rugose, striate, lanceolate, mayorellian, lenticular, flamellian, acanthopodial, palmate, spineolate, vexilliferan, and reticulate.

Flattened amoebas, unified as Discosea, never form true pseudopods -- the flattened cell moves as a whole; liquid cytoplasm flows in streams separated by islands of gel-like cytoplasm. Such cytoplasmic flow is termed polyaxial by Smirnov and colleagues. In locomotion, they never form cylindrical nor sub-cylindrical pseudopods nor show clear monoaxial flow of cytoplasm.

Smirnov and colleagues caution, however, that, from analysis of the morphotypes, which are very diverse, and of the varied details of ameboid movement, that Discosea may not be monophyletic. Some molecular data weakly suggest this. At one point they say that all testate lobose amebas (Arcellinida, suborder Tectiferina in Himatismenida, and Trichosida) form a sister group to the Euamebidae clade, citing Nilolaev et al from 2005. But this would make Tubulinea polyphyletic also, and the testate clade may have evolved their shells from cuticular forms in Thecamebida. They add that the existence of amebas able to alter their locomotive morphology from flattened, expanded to tubular, subcylindrical in cross-section (i.e., the entire order Leptomyxida and the genus *Echinamoeba*) shows that these two types of cellular organization are not completely different and can be realized by the cytoskeleton of the same cell.

Table 16-1. Classification of Lobosa Presented in Smirnov et al (2011) (down to ordinal rank).

Class Tubulinea
Order Euamebida Lepşi 1960, em. Smirnov et al, 2011 (Amebidae and Hartmannellidae, 13 gen.)
Order Arcellinida Kent, 1880 (18 fams., 70 gen.)
Order Leptomyxida (Pussard and Pons, 1976) Page, 1987 (3 fams., 5 gen.)
Order Nolandida Cavalier-Smith 2011 (1 gen.)
Order Echinamoebida Cavalier-Smith, 2004 em., 2011 (2 fams., 2 gen.)
Class Discosea Cavalier-Smith in Cavalier-Smith et al. (2004), em. 2011

Subclass Flabellinia Smirnov et al., 2005, em. 2011
Order Dactylopodida Smirnov et al., 2005 (2 fams., 5 gen.)
Order Vannellida Smirnov et al., 2005 (5 fams., 11 gen.)
Order Himatismenida Page 1987 (3 fams., 5 gen.)
Order Stygamoebida Smirnov and Cavalier-Smith 2011 (1 fam. 2 gen.)
Order Pellitida Smirnov and Cavalier-Smith 2011 (1 gen.)
Order Trichosida Moebius, 1889 (1 gen.)
Subclass Longamoebia Smirnov and Cavalier-Smith 2011
Order Dermamoebida Cavalier-Smith, 2004, em. Smirnov et al 2011 (2 fams., 3 gen.)
Order Thecamoebida Smirnov and Cavalier-Smith 2011 (1 fam., 3 gen.)
Order Centramoebida Rogerson and Patterson, 2002, em. Cavalier-Smith, 2004 (2 fams., 3 gen.)

Filosa

Testafilosa comprises Euglyphida, with 3 families (Euglyphidae, Cyphoderidae, and Paulinellidae), 18 genera, and over 100 species, and Gromida, with 3 families (Gromidae, Amphitremidae, and Psammonobiotidae), 17 or 18 genera, and 88 species (Parker, 1982). Three families (Chlamydophryidae, Pseudodifflugidae, and Volutellidae) (with 22 genera) are added in Lee at al (2000) with Gromidae excluded. *Gromia*, is now regarded as related to the forams by some.

Gymnofilosa (Aconchulinida) has 2 families, Nuclearidae and Vampyrellidae, which are now considered orders by some. The former has discoid cristas and is placed with *Fonticula* as Cristidiscoidea, which have no mt support for their pseudopods, with *Fonticula* having aggregative behaviour (forming sori) like Dictyostelia and Acrasia, the latter being separated out based on fine structure (vesicular cristas, osmiophilic granules, and ribosome arrays) and combined with Schyzopyrenida (ameboflagellates)(2 families) to form Heterolobosa, with Dictyostelia considered as part of Myxobiota. Acrasians, as cellular slime molds, were once assigned to Dictyostelia since it is also a cellular slime mold taxon.

Retarians

Forams

In 1825, Alcide d'Orbigny established the foram order in "Tableau méthodique de la classe des céphalopodes." He was the first to study their mode of living and ecology. But the foram's unicellular nature was discovered by Félix Dujardin in 1835.

Alcide d'Orbigny, sometimes called the Father of Micropaleontology, was a French naturalist who made major contributions in many areas, including zoology (especially malacology), paleontology, geology, archaeology, and anthropology. In Paris he became a disciple of geologist Pierre Cordier and Georges Cuvier. He followed Cuvier's theory and opposed to Lamarckism.

D'Orbigny travelled on a mission for the Paris Museum between 1826 and 1833, visiting 8 South American countries, returning to France with an enormous collection of more than 10,000 natural history specimens. He described part of his findings, the other specimens being described by zoologists at the museum, in "La Relation du voyage dans l'Amérique méridionale pendant les années 1826 à 1833," published in 90 fascicles from 1824–47. His contemporary, Charles Darwin, lauded the book as "one of the great monuments of science in the 19th century". He had numerous interactions with Darwin, and named some species after him, for example, Darwin's rhea for *Rhea pennata*, a S. Am. bird.

In 1840, d'Orbigny started the methodical description of French fossils and published "La paléontologie française" (8 vols). In 1849 he published a closely related "Prodrome de paléontologie stratigraphique," intended as a preface to "Paléontologie stratigraphique," in which he described almost 18,000 species, and, with biostratigraphical comparisons, established geological stages, the definitions for which rest on stratotypes. He described the geological time scales and defined numerous geological strata, still used today as chronostratigraphic references, such as Toarcian, Callovian, Oxfordian, Kimmeridgian, Aptian, Albian, and Cenomanian. The chair of paleontology was created especially in his honour. The d'Orbigny collection is housed in the Salle d'Orbigny and is often visited by experts. In 1853, he became professor of paleontology at the Musée national d'histoire naturelle in Paris, publishing his Cours élémentaire that related paleontology to zoology, as a science independent of the uses made of it in stratigraphy. In all, he described some 1500 species, most new to science, and 6 species were named for him along with a genus-species and a genus.

Félix Dujardin was a largely self-taught French biologist. In 1834 he proposed that a new group of one-celled organisms be called "Rhizopoda"; the name was later changed to "Protozoa". In forams, he noticed an apparently formless life substance that he named "sarcode" (which gave the name Sarcodina), which was later renamed "protoplasm" by Hugo von Mohl.

In 1840 Dujardin was appointed professor of geology and mineralogy at the University of Toulouse and during the following year was a professor of zoology and botany at Rennes. Later in his career he became a member of the French Académie des sciences.

The classification for forams, traditionally an order in subclass Rhizopoda in class Sarcodina, at least in modern times, was 1st done by Glaessner in 1947 in "Principles of Micropaleontology," in which there were 7 super-families and 50 families. Loeblich and Tappan (1984) recognized it as an order and divided it into 5 sub-orders, 17 super-families, 96 families, and 25, 000 species. J.J. Lee (1990) states there are 1200 genera in 12 orders and 120 families. Tappan and Loeblich (1988) give the following putative phylogeny:

Table 16-2.

sbph. Allogromina (Allogromida)
sbph. Protopolythalamia nom. nov.
 cl. Fusulinea (extinct) (Fusulinida)
 cl. Miliodea
  spord. Miliodina+Silicoluculinida
  spord. Lagenida
  spord. Involutinida+Spirillinida
sbph. MetaPolythalmia nom. nov.
 cl. Textularina (Textularida)
 cl. Rotulina
  sbcl. Carterinia (Carterinida)
  sbcl. Rotulinia
   spord. Robertina (Robertinida)
   spord. Bilamella nom. nov. (Rotalida+Globigerinida)

According to J. W. Murray (Ecology and Paleoecology of Benthic Foraminifera, Longman Scientific & Technical, 1991, cited in paleopolis.rediris.es-cours traité de foraminifèrologie) the 6 possible functions of the test are:

to protect against predation;
to serve as a barrier against an unfavorable environnement;
as a receptacle for secreted matter;
to aid in reproduction;
to control movements;
to assist in growth.

The forams differ from typical rhizopods in 3 key characteristics: their protoplasm is included in a flexible envelope, which is calcareous in all marine species; their pseudopods are filose, branching, and anastomosed into a network, i.e., they are reticulate; they are without contractile vacuoles, except for a few freshwater species (imago-mundi-cosmovisions.com).

Sen Gupta in "Modern Foraminifera" (p. 7-36, Kluwer Academic, Dordrecht, 1999, as cited in paleopolis.rediris.es-cours traité de foraminifèrologie) gives a convenience classification based on the test:

Table 16-3.

Allogromida
mineralized
- agglutinating
  - Astrorhizida
  - Lituolida
  - Trochammida
  - Textualarida
- aragonitic
  - Involutinida
  - Robertinida
  - Fusolinida (microgranular and exclusively Paleozoic)
  - Miliolida (porcelanous)
- hyaline
  - Lagenida (monolamellar)
  - bilammellar
    - Rotalida (benthic)
    - Globigerinida (planctonic)

The World Register of Marine Species classification from 2013 is as follows:

Table 16-4.

cl. Globothalamea
ord. Carterinida
ord. Robertinida
ord. Rotaliida
sbcl. Textulariia
ord. Lituolida
ord. Loftusiida
ord. Textulariida

cl. Monothalamea
ord. Allogromiida
ord. Astrorhizida
genus *Nellya* Gooday, Anikeeva & Pawlowski, 2010
spfam. Xenophyophoroidea Tendal 1972
cl. Tubothalamea
ord. Miliolida
ord. Spirillinida

Actinopods

The traditional taxonomy is as follows:

Table 16-5.

sbph. Radiolaria Müller 1858
cl. Pheodaria Haeckel 1879 (7 orders; 6 named by Haeckel in 1887, 1 named by Cachon and Cachon in 1985)
cl. Polycystina Ehrenberg 1838 (Spumellarida, Nassellarida)
cl. Acantharia Haeckel 1881
sbcl. Holacantharia Schewiakoff 1926 (Holacanthida)
sbcl. Euacantharia Cavalier-Smith 1987 (Chauncanthida, Symphiacanthida, Arthracanthida)
sbph. Heliozoa Haeckel 1866
cl. Cryptoaxohelea (Desmothoracida and Taxopodida)
cl. Phaneraxohelea (Centrohelea Kuhn 1926) (Axoplastohelida, Centroplastohelida, Endonucleoaxoplastohelida, Exonucleoaxoplastohelida)

Ernst Haeckel referred to radiolarians as nature's works of art, and was an artist himself authoring and illustrating the first major treatise, a comprehensive report, in 1887, on these organisms based on samples collected from the renowned Challenger Expedition of 1873-76. He treated them as a class and divided it into 4 legions or tribes: Spumellaria, Acantharia, Nassellaria, and Pheodaria, with the 1st 2 grouped as subclass Porulosa and the latter 2 as Osculosa, based on the size and distribution of the pores in the central capsular walls. He split each legion into 2 sublegions, and recognized 20 orders, 85 families, and 4, 318 species in total. Although his system is not regarded as phylogenetic, and is not easily applied to species identification, it was one of the more thorough-going treatments of that century and has remained until recently the major source of information on actinopod diversity and taxonomy. "Kunstformen der Natur" (Art Forms of Nature) is a book of lithographic and autotype prints originally published in 10 sets of 10 between 1899 and 1904 and as a complete volume in 1904, and consists of 100 prints of various organisms done by Haeckel.

Petrushevskaya (1977) proposed the following scheme for Actinopoda: superclasses Heliozoa and Radiolaria, the latter divided into classes Acantharia Haeckel, 1881 and Euradiolaria, the latter in turn divided into subclasses Pheodaria and Polycystina Ehrenberg 1838, with Polycystina split into the superorders Nassellaria, Spherellaria, Collodaria, and Albaillellaria. Heliozoa were divided into Actinophryida, Centrohelida, Desmothoracida, and Taxopoda.

Contrary to what is often stated, based on genotypic data, most actinopods appear to be monophyletic. The MTOC is entirely surrounded by the nucleus in *Dimorpha* (in Heliozoa) as well as some other actinopods like the polycystines; the kinetocyst in Arthracantharia and Centrohelea has a spherical inner core and a sharp conical rodlet; and a hexagonal mt pattern is present in Chauncantharia, Arthrancantharia, Taxopoda (1 genus: *Stilolonche*), and Axoplastohelida. In fact, most of Heliozoa may be subsumed by Euacantharia, leaving only Desmothoracida and Exonucleoplastohelida (1 genus: *Tetradimorpha*), with the latter possibly going to Pedinellales, because of the shared triangular mt pattern with *Ciliophrys*. *Actinophrys, Actinospherium, Echinospherium* (might be same as *Actinospherium*) in Actinophryida, and *Ciliophrys* in Ciliophryida, primtive heliozoans, have been transfered to Pedinellales, which prior to this had only *Pedinella, Apedinella*, and *Pseudopedinella*. Plegmacantharia contains a single genus, separated from the heterogenous family Plegmacantharidae of Holoacantharia.

The genotypic data, however, have both Heliozoa and Radiolaria as monophyletic. The central capsule occurs in Polycystina, Pheodaria, and Acantharia, so is a synapomorphy for Radiolaria, and there are intracellular celestine crystals in each flagellum in Polycystina and Acantharia, and a dodecagonal mt pattern is present in Holacanthida and the polycystine Periaxoplastidata. Heliozoa, on the other hand, do not appear to have any unique shared derived features.

Apicomplexans

The traditional classification comprised class Sporozoa, established by Leuckart in 1879, which contained sucbclasses Telosporidia (orders Gregarinida, Coccidia, and Hemosporidia), Acnidosporidia (orders Sarcosporidia and Haplosporidia), and Cnidosporidia (orders Myxosporidia, Actinomyxidia, Microsporidia, and Helicosporidia). Microsporidians and myxozoans (myxosporidians) were separated out, going to Fungi and Animalia, respectively.

Two taxons, parasites of marine invertebrates, Haplosporidia and Paramyxea, have also been separated out because they have no structures as in Apicomplexa, so do not warrant placement there, and are considered of contentious position. Haplosporidia, established by Caullery and Mesnil (1899), contain 3 genera (*Haplosporidium*, *Minchinia*, and *Urosporidium*) and about 30 species, and possess haplosporosomes. They are regarded as closely related to Microsporidia (now known to be funguses, based on both phenotypic and genotypic data) by Desportes and Nashed (1983) because of telling similarities: diplokaryosis, as in higher funguses; proliferation with vegetative stages alternating with sporulating ones where sporulation is initiated by the differentiation of sporonts which divide into sporoblasts producing spores; similar spindle pole bodies involved in nuclear division; and presumably synaptonemal complexes in the sporont nuclei.

Paramyxea, which also contain 3 genera (*Paramyxa*, *Marteilia*, and *Paramarteilia*) have an organelle similar to haplosporosomes (considered homologous) in *Marteilia* and *Paramarteilia*, the 2 genera forming a family, which warrants the recognition of Ascetosporea, established by Desportes and Nashed (1983). However, Paramyxea have singlet microtubules, like Apicomplexa. Haplosporidia should be placed in Fungi alongside Microsporidia, as they also have diplokaryosis.

A phylogeny for the phylum is as follows:

- sbph. *Colpodella* (1 genus)
- sbph. Gamontomorpha nom. nov. (syn. Gamontozoa Cavalier-Smith, 1993)
  - spcl. Gregarinia Dufour 1828 (4 orders)
  - spcl. Coccidiomorpha Doflein 1901
    - cl. Coccidia Leuckart 1879 (4 orders)
    - cl. Hematomorpha nom. nov. (syn. Hematozoa Vivier, 1982)
      - ord. Hemosporidia Danilewsky, 1886

ord. Piroplasmida Wenyon, 1926

Colpodella is definitely an apicomplexan, as it has an apical complex and micropores, but is separated out for unknown reasons.

Apicomplexans are parasitic and include 14 orders, 120 families, about 300 genera, and some 5000 species.

Ciliophora

Traditionally called Infusoria the group was later subphylum Ciliophora and divided into classes Ciliata and Suctoria. Ciliata was split into subclasses Protociliata and Euciliata by Metcalf in 1918, the former containing the orders Holotricha, Spirotricha, Chonotricha, and Peritricha. Some 50 orders, 215 families, 1100 genera, and 7500 species are now usually recognized. Lynn recognizes 11 classes, 19 subclasses, and 52 orders. De Puytorac et al (1993) recognize 11 classes, 25 subclasses, and 70 orders. Corliss (1994) recognizes 8 classes.

The following classification combines the proposals for ciliate phylogeny by Small and Lynn (1992), de Puytorac (1994), and Baroin-Tourancheau et al (1992).

Table 16-7. Ciliata Perty 1852 (Ciliophora Doflein 1901)

sbph. Postciliodesmatophora Gerassimova and Seravin 1976
  cl. Karyorelictea Corliss 1974
  cl. Heterotrichea Stein 1859
sbph. Intramacronucleata Lynn 1996
 spcl. Spirotricha Bütschli 1889 (as class)
  cl. Hypotrichea Stein 1859 (as suborder)
  cl. Oligotrichea Bütschli 1887 (as suborder)
 spcl. Transversala de Puytorac et al 1993
  cl. Colpodea Small and Lynn 1981
  cl. Plagiopylea
 spcl. Filicorticata de Puytorac et al 1993
  cl. Litostomatea Small and Lynn 1981
  cl. Vestibulifera
 spcl. Epiplasmata de Puytorac et al 1993
  cl. Ciliostomatophorea de Puytorac et al 1993 (as superclass (Phyllopharyngea de Puytorac et al 1974)
  cl. Membranellophorea Jankowski 1975
  sbcl. Nassophoria Small and Lynn 1981 (as class)
  sbcl. Oligohymenophoria de Puytorac et al 1974 (as class)

The molecular analysis by Baroin-Tourancheau et al (1992) agrees very well with morphology at lower levels, with the grouping of Karyorelictea+Heterotrichea and with the Ciliodesmatophora-Intramacronucleata split. The results for Intramacronucleata are as follows: ((Hypotricha+Oligotricha) +Colpodea))+(Nassophora ((Oligohymnophora+Protostoma)+Litostoma)).

Baroin-Tourancheau, A., Delgado, P., Perasso, R., and Adoutte, A. 1992. A broad molecular phylogeny of ciliates: identification of major evolutionary trends and radiations within the phylum. Proc.

Natl. Acad. Sci. U S A 89: 9764–9768.

Caullery, M., and Mesnil, F. 1899. Sur le genre *Aplosporidium* (nov) et l'ordre nouveau des Aplosporidies. Comptes rendus des séances de la Société de biologie et de ses filiales, Série 1, 51: 789-91.

Corliss, J.O. 1994. An Interim Utilitarian ("User-friendly") Heirarchical Classification and Characterization of the Protistans. Acta Protozoologica 33: 1-51.

Desportes, I., and Nashed, N.N. 1983. Ultrastructure of sporulation in *Minchinia dentali* (Arvy), a haplposporean parasite of *Dentalium entale* (Scaphopoda, Mollusca); taxonomic implications. Protistologica 19: 435-60.

Hausman, Klaus; Hulsman, Norbert. 1996. Protozoology, 2nd ed. Georg Thieme.

Haeckel, E. 1887. Report on Radiolaria collected by H.M.S. Challenger during the years 1873-1876. In The Voyage of H.M.S. Challenger, Vol. 18 (Thompson, C.W., Murray, J., eds.), Her Majesty's Stationary Office, London.

Jahn, T.L., Bovee, E.C. 1965. Mechanisms of movement in taxonomy of Sarcodina. I. As a basis for a new major dichotomy into two classes, Autotractea and Hydraulea. Am. Midl. Nat. 73: 30–40.

Karpov, Serguei; Bass, David; Mylnikov, Alexander; Cavalier-Smith, Thomas. 2006. Molecular Phylogeny of Cercomonadidae and Kinetid Patterns of *Cercomonas* and *Eocercomonas* gen. nov. (Cercomonadida, Cercozoa). Protist 157, 125-158 (zoology.bio.spbu.ru).

Lee, J.J. 1990. Granuloreticulosa. In Handbook of Protoctista (Margulis, L., Melkonian, M., Corliss, J.O., Chapman, D.J., eds.). Jones and Bartlett, Boston.

Lee, J.J; Leedale, Gordon; Bradbury, Phyllis, eds. 1985/2000. Illustrated Guide to Protozoa, 2nd ed., Vol. 1, Society of Protozoologists.

Loeblich, A.R., Tappan, H. 1984. Suprageneric classification of the foraminifera (Protozoa). Micropaleontology 30: 1-70.

Lynn, D.H. and Small, E.B. 1997. A Revised Classification of the Phylum Ciliophora Doflein, 1901. Rev. Soc. Mex. Hist. Nat. 47: 65-78.

Margulis et al. 1990. Handook of Protoctista. Jones and Bartlett, Boston.

Mylnikov, A.P. and Karpov, S.A. 2004. Review of diversity and taxonomy of cercomonads. Protistology 3: 201-217 (researchgate).

Parker, S. (ed.). 1982. Synopsis and Classification of Living Organisms. McGraw-Hill, New York.

Pearse, A. S. 1949. Zoological Names: a List of Phyla, Classes, and Orders (prepared for Section F of the AAAS), 24 pp.

Petrushevskaya, M.G. 1977. On the origin of Radiolaria. Zoologicheskii Zhurnal 56: 10.

Petrushevskaya, M.G. 1977. Radiolaria. In: Zuse AP, editor. Atlas of the Microorganisms in the Bottom Sediments of the Oceans. Nauka, Moscow.

Puytorac, P. de.1994. Phylum Ciliophora Doflein, 1901. In: P. de Puytorac, ed., Traité de Zoologie, Tome II, Infusoires Ciliés, Fasc. 2, Systématique, p. 1-15. Masson, Paris.

Reshetnjak, V.V. 1981. Akantarii. Fauna SSSR, 123, pp. 1-210, Akad. Nauk SSSR, Zool. Inst., Nauka, St. Petersberg. .

Smirnov, A.V., Goodkov, A.V. 1999. Illustrated list of basic morphotypes of Gymnamoebae (Rhizopoda, Lobosea). Protistology 1: 20-29 (protistology.ifmo.ru).

Smirnov, Alexey; Chao, Ema; Nassonova, Elena; and Cavalier-Smith, Thomas. 2011. A Revised Classification of Naked Lobose Amoebae (Amoebozoa: Lobosa). Protist 162: 545–570 (cytspb.rssi.ru).

Tappan, H., Loeblich, A.R. 1988. Foraminiferal evolution, diversification, and extinction. Journal of Paleontology 62: 695-714.

Chapter 18 Statistics

Numbers have power. But not enough to supply all the energy for all your daily electrical needs. There's just not enough strength in numbers for all that. - Jarod Kintz, This Book is Not FOR SALE, 2011 – goodreads.com.

Totals for the major groups at the major levels are tabulated below.

Table 18-1. Taxon Totals for Kingdoms and Possible Kingdoms (many of the figures are approximations).

	phyla	classes	orders	families	genera	species
eukaryotes						
Animalia	25	84	300	3600	100, 000	1.1 mln.
Plantae	17	27	150	700	16,000	244,000
cormophytes	20	38	170	620	15,500	235,000
green algae	10	11	23	80	500	9,000
Fungi	7	17	80	250	5000	50, 000
Retaria	4	9	21	302	1681	11,850
Forams	1	2	5	270	1200	10,000
Heliocelida	1	5	9	14	400	1200
Pheodaria	1	1	6	17	80?	650
Taxopoda	1	1	1	1	1	?
Chromista	4	17	90	277	915	31,756
Heterokonta	1	14	83	260	814	14,000
Haptomonada	1	1	2	8	75	1550
Cryptomonada	1	1	4	8	21	200
Chlorarachnia	1	1	1	1	5-7	6-8
Alveolates	3	21	78	325	1530	17,000
Ciliophora	1	11	47	215	1100	8000
Apicomplexa	1	5	14	60	300	5000
Dinoflagellata	1	5	17	50	130	2000
Rhodobiota	1	2	17	67	600	4000
Eulobosa	1	4	9	44	125	2500
Conosa						
Myxobiota	1	2	8	14	61	600
Pelomyxida	1	1	1	3	6-7	?
Cercobiota	1	1	3	4	18	80+
Excavates	4	11	17	48	180	2365
Euglenaria	1	5	6	11	60	1700
Polymastigota	1	2	6	32	100	550
Heterolobosa	1	2	4	4	16-20	115
Jakobida	1	2	1	1	4	?
Filosa					60	

Euglyphida				3	18	?
Gromida				3	17-18	?
Nuclearida				1	20	?
Vampyrellida				1	6	?
Plasmodiophorae	1	1	1	2	11	?
Glaucobiota	1	1	1	1	3	12
Cyanidiobiota	2	2	2	2	2	6
Spongomonada	1	1	1	1	2	?
Cyanidioschyza	1	1	1	1	1	1
total	60	160	800	5200	132,000	1.5 mln.

bacteria

Eubacteria						
Gracilicutes	3	7	60	160	800	5000
Posibacteria	3	7	20	100	700	4000
Togabacteria	1	1	2	3	11	30
Metabacteria	4	5	9	17	120	1000
total	11	21	50	90	550	10,000

All totals are for known and valid extant groups. The major sources are Parker (loc. cit.), Barnes (1984), Lee at al (loc. cit.), Margulis et al (loc. cit.), Holt et al (loc. cit.).

The figures for number of species for the 4 bacterial groups are from Battistuzzi and Hedges (loc. cit.).

For plants Mabberley estimates 250,000 species, Prance et al, 300,000-320,000, and Govaerts and Bramwell 400,000 (Ungricht, 2004).

For fungal species the Enc. Brit. On-Line gave 50,000. The 1970 edition of the Enc. Brit. and the Penguin Dictionary of Botany from 1984 reported 50,000 "known" and "recognized", respectively, the latter stating the actual number may be between 100,000 and a quarter million. Barnes says 50,610, excluding chytrids. Hawksworth (1991) estimated about 70,000. Margulis and Schwartz (1998) said 60,000 have been described. Kirk et al in 2008 in Ainsworth and Busby's Dictionary of Fungi (Blackwell, 2011) and Corliss (1984) say nearly 100,000 described. Fries, in 1925, predicted the actual number, that is, factoring in unknown species, would be analogous to the species richness for insects and was thinking well in excess of 140,000, while Busby and Ainsworth in 1943 had estimated 100, 000, Martin in 1951 gave 260,000, and Hawksworth stated recent estimates ranged from half a mln. (by May in 2000) to 9.9 mln. (by Cannon in 1997), Hammond in 1992 and Rossman in 1994 giving 1 mln., Hawksworth giving 1.5 mln. (1991), a cautious total widely accepted by mycologists as a working hypothesis, and Pascoe implying 2.7 mln. in 1990 (Hawksworth, 2001). Blackwell (2011) says 5.1 mln. might actually exist, and Mora et al (2011) calculate 660,000 give or take 308,000, which would make as many as nearly 1 mln. The number of known genera is from the Penguin Dictionary of Botany.

Corliss (loc. cit.) makes a distinction between valid and described, the latter including both valid and invalid. And there is also the synonymy rate (Wortley and Scotland, 2004). Overestimates fail to factor in these elements. So, for instance, the number of described species for funguses excluding synonymy is probably about 80,000. Excluding invalid species it is probably about 50,000, which is the figure given by Mora et al.

The numbers of fossil species for protistans (Corliss, loc. cit.) are:

Forams	30,000
Chromista	16,000
Dinoflagellates	2,000
Actinopods	1,000?
Acritarchs	1000? (400 gen.)
Rhodophycota	750
Green Algae	450

Plantae has 13 phyla, 23 classes, and 40 orders that are extinct. In animals, there is 1 fossil phylum (Archeocyatha) and perhaps about 70 fossil orders. There are 250,000 fossil species in all (Raup, 1986), about 200,000 in plants and animals. Bacterial and fungal fossils are rare. Diatoms are estimated to have 50,000-100,000 recorded species but only 25,000 are valid, with 15,000 of these as fossils (Corliss, 1984). The number of extinct species in all is estimated at perhaps several 100s of mlns., but Discovery Channel (How many species have actually gone extinct?-discovery.com) reports an estimate range of 1-4 bln. The percentage of known fossils is said to be .001 by Stern et al (2008), which comes out to about 2.5 bln. If we take the highest figure, the normal or background extinction rate would be about 1 species per year. The average species life span is 5-10 mln. years (Lawton and May, 1995).

Estimates of the number of unknown extant species has ranged between 3 and 100 mln. (May, 2011), but is reasonably estimated at between 4 and 14 mln. (Stork, 1993; Odegaard, 2000), and Mora et al (2011) calculate 8.7 mln., give or take 1.3 mln.-- some 3/4 are insects.

The total number of individual bacteria is estimated at 3 decillion ($3x10^{33}$) (Whitman et al, 1998) and they are the majority. The estimate for the number of individual insects is 10 quintillion (Numbers of Insects-Encyclopedia Smithsonian (si.edu)). Viruses, between life and nonlife, are the most abundant forms; there are 10^{30} in the oceans alone (Witzgany, 2015).

Barnes, R.S.K., ed. 1984. A Synoptic Classification of Living Organisms. Blackwell, London.

Blackwell, Meredith. 2011. The fungi: 1, 2, 3 … 5.1 million species? American Journal of Botany 98: 426–438 (amjbot.org).

Corliss, J.O. 1984. The Kingdom Protista and its 45 Phyla. BioSystems 17: 87-126.

Hawksworth, D.L. 1991. The fungal dimension of biodiversity: magnitude, significance, and conservation. Mycological Research 95:641-655.

Hawksworth, D.L. 2001. The magnitude of fungal diversity: the 1.5 million species estimate revisited. Myco. Res. 105: 1422-1432.

Lawton, John, and May, Robert. 1995. Extinction Rates. Oxford U. Press.

May, Robert. 2010. Tropical arthropod species, more or less? Science 329: 41–42.

Mora C, Tittensor DP, Adl S, Simpson AGB, Worm B. 2011. How Many Species Are There on Earth and in the Ocean? PLoS Biol 9(8)(plosbiology.org).

Ødegaard, Frode. 2000. How many species of arthropods? Erwin's estimate revised. Biological Journal of the Linnean Society 71: 583–597 (si.edu; idealibrary.com).

Stern, K.R., Bidlak, J.E., Jansky, S.H. 2008. Introductory Plant Biology, 11th ed. McGraw-Hill.

Storks, Nigel. 1993. How many species are there? Biodiv. Conserv. 2: 215–232 (griffith.edu.au).

Ungricht, Stefan. 2004. How Many Plant Species Are There? And How Many Are Threatened with Extinction? Endemic Species in Global Biodiversity and Conservation Assessments. Taxon

53: 481-484.
Whitman, W., Coleman, D., and Wiebe, W. 1998. Prokaryotes: the unseen majority. Proc. Natl. Acad. Sci. USA 95: 6578–83.
Witzgany, Günther. 2015. Supplemental data to main article "The Agents of Natural Genome Editing" (researchgate.net).
Wortley, Alexandra and Scotland, Robert. 2004. Synonymy, Sampling, and Seed Plant Numbers. Taxon 53: 478-480.

Chapter 17 Ecology and Biogeography

The Sea

There is a pleasure in the pathless woods,
There is a rapture on the lonely shore,
There is society where none intrudes
By the deep sea, and music in its roar.

Lord Byron, from "Childe Harold," Canto IV.

The habitat distribution for the major taxons is tabulated below.

Table 17-1. Taxons Listed by Habitats (++ = predominant)

	oceanic	neritic	inttdl	brksh	frshwtr	terrestrial
<u>Eubacteria</u>						
Acidobacteria						+
Deinobacteria						+
Mollicutes						+
Chlamydiae						+
Rickettsiae						+
Heliobacteria						+
Proteobacteria	+	++	+	+	++	+
Actinobacteria		+			+	++
Endospora		+			+	++
Cyanobacteria		+	+	+	++	++
Spirochetes		+			+	+
Togabacteria	+	+			+	+
Chloroflexi				+	++	+
Saprospirae		+			+	+
Chlorobia	+	+	+	+	++	
Planctobacteria				+		
Aquificae		+			+	
Thermi		+				
<u>Metabacteria</u>						
Thermoplasmata					+	+
Methanobacteria	+	+	+	+	+	+
Sulfobacteria	+	+			+	+
Halobacteria		+			+	+
Archeoglobi		+				

Eukaryotes

Myxobiota						+
Acrasia						+
Plasmodiophorae						+
Cormophyta			+	+	+	++
Amastigomycota		+	+	+	+	++
Euglyphida		+	+	+	++	++
Chytridiomycota		+			++	++
Metazoa	+	+	+	+	+	++
Gromida		+	+	+	+	+
Eulobosa		+	+		+	+
Jakobida		+			+	+
Schizopyrenida		+			+	+
Vampyrellida		+			+	+
Polymastigota		+			+	+
Apicomplexa		+				++
Prerhodophyceae					+	+
Cercomonada				+	+	+
Euglenoida		+	+	+	++	+
Green Algae	+	+	+	+	++	+
Heterokonta		++	++	+	++	+
Rhodobiota		++	++	+	+	+
Cryptomonada	+	+	+	+	+	
Haptomonada	+	+	+	+	+	
Dinoflagellata	+	++	+	+	+	
Spongomonada				+		
Glaucobiota				+		
Chlorarachnia		+	+	+		
Pelomyxida		+		+		
Nuclearida		+		+		
Heliozoans		+		+		
Choanoflagellata		++		+		
Ciliata	+	+		+		
Forams	+	+				
Pheodarea	+	+				
Celestina	+	+				

(Data mostly from Margulis et al, van den Hoek et al, Lee et al, Holt et al, and Parker et al).

Plasmodiophorae and Apicomplexa are entirely parasitic and Polymastigota is mostly parasitic, the 1st in land plants and the latter 2 in animals; they are designated here according to the habitat of the host. And epizoic or epiphytic forms are designated according to the organism they live on, e.g., *Cryptochlora*, a chlorarachnian, was found on green algae in the supralittoral zone so is terrestrial.

Cosmopolitan are red algae, cryptomonads, haptomonads, euglenoids, dinoflagellates, plants, animals, funguses, cercomonads, myxobiotes, eulobosans, forams, and actinopods. In bacteria, most groups are widespread, and 9 groups are entirely or mostly parasitic: Saprospirae, Spirochetes,

Chlamydiae, Rickettsiae, Campylobacteria, Enterobacteria, Mollicutes, Endospora, and Actinobacteria.

There are 12 genera and 57 species of monocots (called seagrasses) in 4 families considered marine (Marine Meadows-botgard.ucla.edu), but some of these are actually estuarine. Estuaries are often considered marine (Odum, 1997) but are transitional as they are partially enclosed bodies of water (affected by tides and with a connection to the open sea), and contain brackish water (a mixture of both freshwater and seawater, which is .05–3% dissolved salt; freshwater is below this range and seawater is above it). They often include only river mouths, but also often include bays, sounds, lagoons, and fjords. And invariably in descriptions of groups there is a distinction made between brackish and marine. For example, 4 genera of raphidomonads are described as brackish or marine and one of them, *Chattonella*, is described as brackish and marine (Lee et al). The US Fish and Wildlife Service considers them separate from marine ecosystems (Mitsch and Gosselink, 2007; Keddy, 2010). A geomorphic classification of estuaries comprises drowned river valleys, bar-built estuaries, fjords, and deltaic formations (Dawes, 1998). The seagrass *Halophila* occurs in salt water and holds the remarkable record for depth for these monocots as it was found once at nearly 300 ft. (90 m.) in clear water (Marine Meadows-botgard.ucla.edu). Certain lakes and seas are also brackish, such the Caspian Sea (the world's larget lake), the Great Salt Lake, the Black Sea (the largest body of brackish water), and the Baltic; the latter 2 are almost entirely enclosed and are also non-tidal.

There are also 238 genera and 444 species of higher funguses that are considered marine (Species of Marine Fungi-usm.edu). Fungal species have been cultured from water samples as deep as >15,000 ft. (4610 ms.) and from sediment samples as deep as >11, 000 ft. (3425 ms.); .3% of funguses are marine and most are benthic (Dawson, 1966). According to Hawksworth et al (loc. cit.) there are 800-1000 sp. of marine funguses, which makes 1.6-2%. In all there are 200,000 species in the sea (Home-Douglas, 1991).

"Pelagic" is used for neritic and oceanic together and includes planktonic (surface-floating) and nektonic (free-swimming) life forms and excludes the benthic (bottom-dwelling) life forms, the 4 terms also used for lakes. Haeckel invented the nektonic and benthic concepts in 1866 and German physiologist Victor Hensen added the planktonic later on (Engel, 1962). The pelagic is divided into the epi- (down to 200 m.), meso- (200-1000 m.), bathy- (1000-4000 m.), abysso- (4000-6000 m.), and hadopelagic levels (Tait and Dipper, 1998). The benthic levels are intertidal, subtidal, bathyal (continental slope), abyssal (plains), and hadal (trenches), hadal being the lowest level (Barnes and Hughes, 1999). The lowest point is 35,800 ft. (nearly 11,000 m.) or nearly 6000 fathoms or about 7 miles, is in the Marianas Trench, also called Challenger Deep, and was discovered by the British oceanographic vessel Challenger II in 1951, and was visited by Piccard and Walsh in the 60 ft.-long, 50-ton Trieste bathyscaphe, launched in 1953, equipped with a 6½-ft. observation sphere attached to its underside, in 1960, in an historic 8 ½-hr. dive and 20-min. stay on the seafloor, and was written about in 1961 in the book "7 Miles Down" by Piccard and Dietz (Soundings, Sea-Bottom, and Geophysics-OceanExplorer.NOAA.Gov; Marx, 1990; Home-Douglas, 1991).

Benthic marine forms include red algae, green algae and angiosperms, funguses, and euglenoids. Planktonic marine forms include dinoflagellates, cryptomonads, haptomonads, and most protozoans. Only a few green algae are planktonic. Heterokonts are benthic (chrysophyceans, pheophyceans, xanthophyceans, diatoms) and planktonic (diatoms). Animals and bacteria are benthic, planktonic, and nektonic. (Dawson, 1966; Van Den Hoek et al, loc. cit., Lee at al, loc. cit.).

The trenches, which occur mostly in the Pacific Rim, are inhabited by animals, forams, eubacteria (mostly Proteobacteria), and metabacteria. The existence of life in the deep sea was established first by John Ross in a British Admiralty expedition to the Arctic in 1819 in which fauna was hoisted up from 6000 ft. (nearly 2000 m.). Charles Thomson, in the late 1860s on the surface vessel HMS Porcupine in the NE Atlantic, found life forms at c. 4000 m. (over 13,000 ft.) down. This was followed by the renowned round-the-world Challenger Expedition from 1872-76, which was also lead by Thomson and which was the first to physically sample the hadal zone, by collecting sediments from 7220 m. (nearly 24,000 ft.) down in the Japan Trench and dredged up marine life from 26,865 ft. (8188 m.) in the Kuriles Trench off Japan. From 1949 to 1976 the Russian research vessel Vitjaz sampled hadal depths during 20 expeditions to 16 Pacific and Indian Ocean trenches down to over 10,000 m. (nearly 33,000 ft.). The Danish Galathea Expedition, in 1951-52, sampled depths also of over 10,000 m., in 6 trenches, also in the Pacific and Indian Ocean. It was lead by Anton Bruun, who coined the terms "hadal" and "hadopelagic" in 1956 for depths and fauna of over 6000 m. (c. 19,000 ft.). In the Russian literature of the time, and still sometimes today, the term "ultra-abyssal" was used following Zenkevitch from 1954. (Marx,1990; Jameison, 2015).

In vent communities bacteria are symbiotic with tube worms, clams, and mussels, which have evolved mechanisms to avoid H_2S toxicity, so bacteria are as central to these communities as phytoplankton are in the main ocean food web. The first black smokers and chemosynthetic hydrogen sulfide bacteria were discovered in 1977 at 8000 ft. (2438 m.) in the Galapagos Rift, between the East Pacific Rise and Ecuador, by marine scientists John B. Corliss and John M. Edmond, in the world's most successful submersible, the legendary Alvin (named for its major supporter, Allyn Vine, a Woods Hole scientist), a Navy-owned, Woods Hole-operated, General Mills/Litton-built, 16-ton, 23 ft. long, 12 ft. high, 8 ½ ft. wide craft, launched in 1964. It has done 4,600 dives, taking more than 13,000 scientists, engineers, and observers to the seafloor, and is still in operation. It was also with Alvin that scientists discovered a green sulfur bacterium, dubbed GSB 1, related to *Chlorobium* and *Prosthecochloris*, in a vent community, which uses geothermal illumination to carry on photosynthesis instead of sunlight, the only organism known to do so. It has a crew capacity of 3, the pilot and 2 marine scientists (housed in a 7-ft. titanium sphere, enlarged in 2011-12), has typically 6-hr. dives, a maximum depth capability of 15,000 ft. (2.8 mi.)(increased to 21,000 in 2011-12), a cruising speed of 1 knot and maximum speed of 2 knots, and a sample capacity of 400 lbs. It has a conning tower to enter the sphere, a viewing port (three 4" plexiglass windows, increased to 5 in 2011-12), a manipulator arm to collect biological and geological samples, a hoisting bitt (to hoist it on and off its mother ship, Atlantis II), 2 ballast systems, 2 batteries, syntactic foam for additional bouyancy, and 4 thrusters and 2 propellers powered by electric motors. Similar deep-sea, 3-person submersibles are the Pisces III and IV, operated by HURL (Hawaiian Undersea Research Lab), the Cyana and Nautile, operated by Ifremer (Institut français de recherche et exploitation de la mer), and the Russian Mir 1 and 2. (Home-Douglas, 1991; MidOcean Ridges-Global Distribution of Hydrothermal Vents-WHOI.Edu; Alvin-Ocean Explorer.NOAA.Gov; Marx, 1990; Life on the Ocean Floor, 1977-2012 (the-scientist.com); Beatty et al, 2005).

Nearly 95% of marine prokaryotes are flagellated Gram-negative rods and nearly 70% are pigmented. Most are facultatively anaerobic, fewer are facultatively aerobic, and very few are obligately anaerobic or obligately aerobic. There are about 50,000 to 400,000 per ml. in near shore waters, some 2000 per ml. in waters 3-120 miles from land, and around 40 per ml. in open ocean waters. Most organisms live near shore because of higher nutrient levels there. Below a depth of 656 ft. (200 m.) bacterial populations become increasingly sparse until we reach the sea floor where vast numbers occur at all depths. 43% of prokaryotes are marine and are ubiquitous in this environment. (Dawson, 1966). At the epipelagic level in the North Pacific, Eubacteria cells outnumber Metabacteria cells about 100 to 1, at depths of 500 and 100 m. they slightly outnumber them, at a depth of 4000 m. Metabacteria slightly outnumber Eubacteria, while at a depth of 5000 m. they are even; both groups are more abundant at higher levels (Chp. 23 of Brock Biology of Microoganisms-marmara.edu.tr).

Red algae are 98% marine, dinoflagellates are 93% marine (mostly neritic), cryptomonads and haptomonads are about 50% freshwater and about 50% marine, green algae are about 13% marine (mostly ulvophyceans), which means probably about 80% are freshwater (mostly Volvocophyta, Gamophyta, and Charophyta), heterokonts are over 40% marine (mostly in pheophyceans (99% marine) and diatoms (about 50% marine)), c. 50% freshwater (mostly in chrysophyceans (c. 80% freshwater) and diatoms (c. 50% freshwater)), and around 5% terrestrial (mostly in xanthophyceans), and euglenoids are 3% marine and over 90% freshwater (Dawson,1966; Dawes, 1998).

The littoral zone is the coastal area and is divided up into supralittoral (also called the splash or spray zone), eulittoral (also called intertidal or mediolittoral), and sublittoral (also called neritic) (zonation and terminology by Lüning from 1990 (Dawes, 1998)). The sublittoral, which is about 260-330 ft. (80-100 ms.) deep, designates the sea above the continental shelf and beyond littoral, as opposed to oceanic, the sea beyond continental shelf waters. The supralittoral zone is dominated by lichens, especially *Verrucaria* (Pech and Regnault, 1992). The intertidal zone is only temporarily submerged (at high tide) but is considered marine and ranges from a few inches (Baltic Sea, Patos Lagoon in Brazil) down to 50 ft. (Bay of Fundy)(Dawes, 1998). The organisms living in the upper intertidal are not usually submerged so considering them marine is questionable; the remainder of the intertidal is dominated by brown algae (Lee, 1986). 88% of marine life forms are benthic and nearly all are littoral or neritic (Home-Douglas, 1991). T. A. and Anne Stephenson in 1949 in the Journal of Ecology identified the supralittoral, littoral, and infralittoral, and a supralittoral fringe which straddles the upper littoral and lower supralittoral, an infralittoral fringe in the lower littoral, and a midlittoral between the 2 fringes (Dawson, 1966).

Many red algae are sublittoral, and encrusting coralline red algae (Corallinales)(encrusting, that is, crustose, algae are called leprydophytes) are at the greatest depths of any algae, down to 880 ft. (268 m.)(in the Bahamas), well below the typical photic zone, which usually extends to a depth of about 650 ft. (c. 200 m.), which means very low light conditions, and this because of their red pigmentation, that absorbs light in the longest wavelengths (and lowest frequencies), about 500 to about 700 nm., compared to the c. 400 - c. 500 nm. of other algae, which are in shallower waters (Thomas, 2002).

Coral reefs are among the most productive ecosystems. The zones are the forereef, the crest or ridge, which is intertidal, and the inner reef, reef flat, or back reef. They are found in tropical and subtropical waters in a band between latitudes of about 30° N and 30° S and are estimated to cover about 230,000 sq. mi. (600,000 sq. km.). Some 60% are in the Indian Ocean and Red Sea, 25% in the Pacific, and nealy all the remaining 15% are in the Caribbean. There are 2 princpal developmental types: 1) primarily in the Pacific Ocean on slowly sinking volacnoes, 2) on Pleistocene platforms in the Caribbean. Coral reefs protect shorelines by absorbing wave energy, and many small islands would not exist without their reefs to protect them, and are economically important because of fisheries and tourism. They are made by colonial corals, which were thought to be plants until 1723 when French naturalist Jean André Peysonnel described them as animals, a view that was not accepted at first, but later it prevailed. Important information on coral reefs was collected by Captain James Cook aboard the Endeavour during the first modern maritime expedition for scientific exploration in 1768-76 (Engel, 1962). In 1842, Charles Darwin, in "On the Structure and Distribution of Coral Reefs," identified 3 major classes: fringing, barrier, and atoll, which are still recognized today, and proposed a theory of coral reef formation, involving volcanoes, which was confirmed in 1952 by scientists of the US Land-Air-Sea Special 1st Detachment drilling in the Eniwetok Atoll in the Pacific. A rival theory, by Canadian geologist Reginald Daly in 1919, probably explains the patch reef. Some of the most common members of the coral reef ecosystem are angelfish, butterflyfish, bannerfish, and the false Moorish idol, which are probably the most beautiful fishes, the first in Pomacanthidae and the other 3 in the closely related Chetodontidae, both in the large order Perciformes in the large superorder Percoidea. (Engel, 1962; Home-Douglas, 1991; Margulis et al, loc. cit.; Stafford-Deitsch, 1991).

Darwin noted in the same monograph that coral reefs harbour a great diversity of life forms, comparable to that of jungles, yet occur in clear, nutrient-poor (oligotrophic) tropical waters, and called them oases in an oceanic desert, which became known as Darwin's paradox. Rougerie and Wauthy have proposed geothermal endo-upwelling as an explanation. Sponges in cracks and crevices providing nutrients to the reef is another. Researchers have known for a long time that a variety of processes, such as nitrogen fixation, groundwater seepage, and the influx of nutrients from the oceans, enable reefs to thrive in nutrient-poor water. Nutrients are also thought to come from within the cracks and crevices that riddle the reef framework. (Rich Coral Reefs in Nutrient-Poor Water: Paradox Expalined?-NationalGeographic.Com; Rougerie and Wauthy, 1986).

The Great Barrier Reef is the world's largest reef system, stretching 1625 mi. (2600 km.) along the Queensland coast in the Coral Sea, covering an area of c. 135,000 sq. mi. (350,000 sq. km.) or over half the total area of coral reefs and is the only living structure visible from space. It is more or less coextensive with the GBR Marine Park and was made a World Heritage Site in 1981 by Unesco. It is one of the world's most popular tourist destinations, generating 5-6 bln. Australian dollars a year. It contains some 3000 individual reefs, about 800-900 islands, 400 species of coral, 4000-8000 mollusk species, c. 2200 species of vascular plants, 1500 species of fish, 1500 species of sponges, 500 algal species, 300-500 bryozoan species, and 330 species of ascidians. (Stafford-Deitsch, 1991; Great Barrier Reef-nationalgeographic.com; greatbarrierreef.org; Flora and Fauna of the GBR World Heritge Area-gbrmpa.gov.au).

European exploration of the reef began in 1770 with Captain Cook who ran his ship aground on it. The charting of channels and passages through the maze of reefs began with Cook and continued during the 19th century. The Great Barrier Reef Expedition of 1928–29 contributed important knowledge on coral physiology and coral reef ecology. A modern laboratory on Heron Island continues scientific investigations. (Great Barrier Reef-nationalgeographic.com).

The encrusting red algae *Lithothamnion* and *Porolithon* form the fortifying purplish red algal rim that is one of the Great Barrier Reef's most characteristic features, while the green alga *Halimeda* is found nearly everywhere (Great Barrier Reef-nationalgeographic.com). As mentioned in Chapters 11 and 13, red algae and dinoflagellates are important reef members.

Wetlands are lands that are permanently or seaonally saturated or inundated by surface water, groundwater, or precipitation because of poor or nonexistent drainage, have hydric, anoxic soil, are dominated by vegetation (hydrophytes), and have an average depth of 6.6 ft. or less. They can be classified as marshes (dominated by herbs) and swamps (dominated by trees), which mostly have surface flow as their main water source, and bogs and fens, which are peatland; variants are muskeg (tundra swamp), dambos (type of swamp in Africa), bayous, sloughs (stagnant wetlands), pocosins (type of marsh), prairie potholes, wet meadows, vernal pools (in grasslands), moors, mires, inter alia. They are transitional between terrestrial and acquatic. Sometimes included, but improperly, are lakes, ponds, rivers, and streams. Wetlands are mostly freshwater but can also be brackish (oligohaline) or saline; are mostly inland but can also be coastal; and can be minerotrophic (receiving water and nutrients from streams) or ombrotrophic (receiving water and nutrients from precipitation); oligotrophic, eutrophic, or dystrophic (nutrient levels); herbaceous, scrub, or arboreal (dominant life form); and biogeochemically open (abundant exchange of nutrients with surroundings) or closed (little movement across ecosystem boundary). Unlike terrestrial ecosystems, whose biogeochemical role is as a source, and acquatic ecosystems, whose role is as a sink, roles of wetlands are as source, sink, or transformer, which further attests to their intermediate status. (What Are Wetlands-Wetlands.Org; Wetlands Delineation Manual-el.erdc.usace.army.mil; Wetland Types-water.epa.gov; Keddy, 2010; Mitsch and Gosselink, 2007).

Detailed classifications for wetlands have been done by Cowardin et al in 1979 for the US FWS and by Brinson et al in 1993 (a hydrogeomorphic (HGM) classification). The former is based on landscape position (marine, riverine, lacustrine, and palustrine), cover type (e.g., open water, submerged aquatic bed, persistent emergent vegetation, shrub, and forested), hydrological regime (ranging from saturated to permanently flooded). Modifiers can then be added for different salinity or acidity classes, soil type (organic vs mineral), or disturbance activities (impoundment (damming), beaver activity, etc.). Thus, the Cowardin system includes a mixture of geographically-based factors, proximal forcing functions (hydrological regime, acidity), disturbance regimes, and vegetative outcomes, making 6 classes: emergent, forested, unconsolidated, scrub-shrub, acquatic bed, and moss-lichen.

The latter classification system is based on geomorphic setting, dominant water source, and dominant hydrodynamics, making 7 classes: riverine (in floodplains and riparian zones associated with stream channels), depressional (prairie potholes, vernal pools), slope (with groundwater as main source), mineral soil flats (in areas of low topographic relief (e.g., interfluves, relic lake bottoms, and large floodplain terraces) with precipitation as main source of water), organic soil flats (peatland), tidal fringe (salt marshes and mangrove swamps, with euhaline, mixohaline, and hyperhaline subclasses), and lacustrine fringe (along lake shores). (Methods for evaluating wetland condition #7: Wetlands Classification-epa.gov).

The major functions of wetlands include water purification, flood control, and coastal protection (shoreline stability). They are among the most biologically diverse and productive of ecosystems but also among the ones that undergo the most environmental degradation. Wetlands said to be the largest, such as the West Siberian Lowland, the Hudson Bay Lowland, the Amazon, Congo, Mackenzie, Mississippi, and Nile river basins, Prairie Potholes, the Pantanal in S. America, and the Magellanic Moorland also in S. America, are actually wetland complexes instead of individual wetlands, as attested to by their plural descriptions, for example, the West Siberian Lowland: "bogs, fens, swamps, and marshes." (Keddy, 2010). And the famous Everglades in Florida are a complex system of interdependent ecosystems that include sawgrass marshes, cypress swamps, and the estuarine mangrove swamps of the Ten Thousand Islands, hardwood hammocks (stands of trees), pine rockland, and Florida Bay. Coastal wetlands, which include salt marshes and mangrove swamps, and which are intertidal and of variable salinity, are sometimes considered marine.

The major terrestrial ecosystems or biomes are arctic and alpine tundra, forests, savannah (transitional between forest and grassland), grassland, and desert (Strahler and Strahler, 1976).

As mentioned earlier, Augustin de Candolle was the first to recognize biogeographical regions in 1820 and his system was based on plant distribution, and he was primarily interested in documenting the nature and floral composition of the plant "formations" (which we call biomes today). He recognized 20 areas of endemism (distribution confined to a given geographical area) and doubled this number in 1838. Engler's 4 realms, in 1879 and 1882, contained 32 regions and were based de Candolle's climatic and physiological criteria and were: Arcto-Tertiary (the temperate and cold areas of the northern hemisphere), Paleotropical (Old World tropics from Africa to northern Australia), Neotropical (most of Central and South America), and Ancient Ocean (coastal Chile, Tierra del Fuego, S. Africa, most of Australia, Tasmania, the islands of the S. Atlantic, the Indian Ocean, and Pacific). In 1908 the 4th realm was split into 3 realms by Diels: Antarctic, Cape, and Australia, and NZ he placed in the Paleotropical. Ronald Good, in 1947, 1953, 1964, and 1974, modified this scheme, recognizing 37 floristic provinces in 6 floristic kingdoms: Boreal (Holarctic), Neotropical, Paleotropical, Cape, Australian, and Antarctic. He returned NZ to the Antarctic. Also as mentioned previously, Takhtajan recognized 152 floristic provinces and 35 regions in 6 floristic kingdoms: Holarctic (Palearctic and Nearctic), Neotropical, Paleotropical (Ethiopian and various Oriental and Pacific regions), S. African (Cape), Australian, and Antarctic. In 2001, Cox deleted Takhtajan's Cape (S. African) and Antarctic kingdoms, assigning them to neighbouring kingdoms, and separating the Oriental and Pacific regions as Indo-Pacific. (Cox, 2001).

The earliest zoogeographers, such as Prichard in 1826 and Swainson in 1835, recognized as many as 6 continental regions. These were first formally recognized in the Philip Sclater system of 1858 for passerines and comprised 6 regions, which were later accepted by Wallace in 1876 for other animals especially mammals, and Wallace's map showed the following: Palearctic (North America), Nearctic (most of Eurasia and North Africa), Neotropical (South America), Ethiopian (Dark Africa), Oriental (tropical Eurasia including the western East Indies), and Australian (including the eastern East Indies). This scheme, mainly for mammals and birds, has become standard. (Cox, 2001).

Lee Dice in 1943 in his book "Biotic Provinces of North America," who introduced the concept of the biotic province, Ray Dasmann in 1973 in his book "Biotic Provinces of the World," and Miklos Udvardy (A Classification of Biogeographical Provinces of the World, 1975-cmsdata.iucn.org), subscribe to the idea of biotic or biogeographical provinces, defined by Dice as a continuous and considerable geographical area characterized by one or more ecological associations (an association being a group of organisms that live together in a certain geographical region, constituting a community with a few dominant species) that differ from those of the adjacent provinces. They have a synecological basis (community ecology, as opposed to autoecology), are characterized by features of flora and fauna, ecological climax, physiography, and soil, are subdivisions of the biome system, and are well suited for conservation purposes.

Udvardy recognized 8 biogeographical realms: Palearctic, Nearctic, Neotropical, Afrotropical (Paleotropical), Indomalayan (Oriental), Australian, Oceanian (Pacific), and Antarctic. In this there are no regions, only provinces, and the Wallace Line is included. It also includes lakes and lake systems as provinces. The Antarctic flora and fauna are poorly known so his provinces for it are provisional and are: Insulantarctica, Neozealandia, and, for Antarctica itself, Maudlandia and Marielandia. Antarctica itself is excluded in the Good and Takhtajan schemes. There are 146 provinces in all in the Udvardy taxonomy.

Phytogeographical regions are based mostly on angiosperms, and the geography of algae, funguses, and bacteria is mostly unknown. Zoogeography is based mostly on vertebrates, and the geographic analyses of terrestrial invertebrates, when attempted, show faunal entities often very different from those of vertebrates. (Udvardy, loc. cit.)

The 4 major differences between phyto- and zoogeographical areas is that zoogeography has no Holarctic, no Cape, and no Antarctic, and has the Wallace Line, separating east from west in the East Indies, so in Good (1947), for instance, there is a Malayan Archipelago region (in the Indo-Malayan subkingdom of the Paleotropical Kingdom, which is the only one to have subkingdoms, the other 2 subkingdoms in it being the African and the Polynesian), and Takhtajan, in his 1969 version, has 5 Paleotropical subkingdoms, adding Madagascan and New Caledonian (Udvardy).

Early work on marine zoogeography was "Natural History of European Seas" by Edward Forbes in 1859. More recent work was "Tiergeographie des Meeres" by Sven Ekman in 1935, followed by a revised English edition, "Zoogeography of the Sea" in 1965, in which he describes, for the continental shelf, regions in warm, temperate, and polar waters, their separation by zoogeographical barriers, and their endemism. In 1974, in the book "Marine Zoogeography," and 1995, various regions and provinces were established by J.C. Briggs. For continental shelves the 4 temperature zones of the world's oceans have usually been identified as tropical, warm-temperate, cold-temperate, and cold, and within each region provinces were recognized. But the distinction between tropical and warm-temperate is apparently inappropriate. The definition of provinces by 10% endemism has been generally accepted since 1976, but good arguments can be made for a criterion of 15 or 20%. (Briggs and Bowen, 2012)

From the beginning of the formation of biogeography there have been 2 major approaches: the landscape, based on quantitative analysis of the distribution of dominant life forms, creating the biological background for landscapes of different regions, and which considers only species and higher taxons that form large biomasses on vast areas; and the floristic-faunistic (or historical-genetic), based on qualitative analysis of the composition of floras and faunas of different regions and analysis of their origins, and considers all species and higher taxons (Golikov et al,1990). A landscape classification is presented by Golikov et al (1990) and is a nested heirarchy including the individual, monocene-monotope, population, democene-ecotope, biocenosis, ecosystem, association, facia, biome, and formation.

The concepts were introduced by the following:

association	Humboldt 1805
formation	Griesebach 1838
biocenosis	Möbius 1877
facia	Derjugin 1915
ecosystem	Tansley 1935
biome	Clements and Shelford 1939
monocene-monotope	Scherdteger 1968
democene-ecotope	Whittaker et al 1973

Biocenosis refers to a group of interacting organisms that live in a particular habitat and form an ecological community, especially a self-sufficient one occupying a specific biotope. "Biotope" is sometimes seen as synonymous with "habitat", but in some countries the subject of a habitat is a population, while the subject of a biotope is a community. "Biogeocenosis" is used in the German and Russian literature as a parallel term to ecosystem (Odum, 1997).

The plant ecologist Frederick Clements teamed up with the animal ecologist Victor Shelford to devise the biome system in their book "Bioecology." Biomes are the obvious subdivisions of the world's biota and are characterized by one prevailing dominant form of climax vegetation, meaning that which will develop if nature is allowed to take its course. The prevailing climate shapes the climax vegetation along with the soils and associated animal life. (Biogeographical Provinces, Dasmann, 1976-WholeEarth.Com)

The first definition for "ecotope" was provided by Tansley in 1939 and was essentially equivalent to "habitat", but later the term was used in relation to landscape. In 1948 Schmithüsen first used it in the landscape sense, employing a tile-and-mosaic metaphor and Carl Troll in 1950 first applied the term to landscape ecology, defining it as "the smallest spatial object or component of a geographical landscape", and can be comprehensively seen as the smallest, mappable, botanically homogeneous, ecological landscape unit (Ecotopes-Ecotope.Org).

The floristic-faunistic classification of the marine world is divided into 3 kingdoms and 15 regions as follows:

Table 17-2. Biogeographical Classification of the Marine World based on floristic-faunistic factors (Golikov et al, 1990).

Temperate/Cold Northern Hemisphere

Arctic
Pacific boreal
Atlantic boreal

Tropical

Mediterranean-Lusitanic subtropical
West-Pacific tropical
Indian Ocean tropical
Guinean tropical
Carribean tropical
Tasmanian subtropical
New Zealand subtropical
South African subtropical
Patagonian subtropical
Araucanic subtropical

Temperate/Cold Southern Hemisphere

Kerguelenic
Antarctic

These divisions hold even for bathyal and abyssal levels.

There is also the ecoregion marine classification by Spalding et al (2007) which contains 12 realms, 62 provinces, and 232 ecoregions. The 12 realms are:

Arctic
Temperate Northern Atlantic
Temperate Northern Pacific
Tropical Atlantic
Western Indo-Pacific
Central Indo-Pacific
Eastern Indo-Pacific
Tropical Eastern Pacific
Temperate South America
Temperate Southern Africa
Temperate Australasia
Southern Ocean

Ecoregions are defined by Spalding et al as "Areas of relatively homogeneous species composition, clearly distinct from adjacent systems. The species composition is likely to be determined by the predominance of a small number of ecosystems and/or a distinct suite of oceanographic or topographic features. The dominant biogeographical forcing agents defining the ecoregions vary from location to location but may include isolation, upwelling, nutrient inputs, freshwater influx, temperature regimes, ice regimes, exposure, sediments, currents, and bathymetric or coastal complexity." The term "ecoregion" was first proposed in 1962 by Canadian forest researcher Orie Loucks (Bailey, 2005).

American geographer Robert Bailey also did a marine taxonomy as follows:

Table 17-3. Ecoregions of the Oceans According to Bailey (1996).

Polar
- inner
- outer

Temperate
- poleward westerlies
- equatorward westerlies
- subtropical
- high-salinity subtropical
- jet stream
- poleward monsoon

Tropical
- tropical monsoon
- high salinity tropical monsoon
- poleward trades
- trade winds
- equatorial trades
- equatorial countercurrent

Bailey authored "Descriptions of ecoregions of the United States" in 1995, 2nd ed., originally published in 1978 as an unnumbered publication by the Intermountain Region, USDA Forest Service, to provide a general description of the ecosystem geography of the US as shown on the 1976 map "Ecoregions of the United States." It was reprinted in 1980 by the Forest Service as Miscellaneous Publication No. 1391. An explanation of the basis for the regions delineated on the map was presented by Bailey in 1983. and the technique was subsequently expanded to include the rest of North America by Bailey and Cushwa in 1981 and the world by Bailey in 1989. (US Forest Service-fs.fed.us).

In mapping ecoregions Bailey followed the concepts of Crowley, who mapped the ecoregions of Canada in 1967 based on macrofeatures of the climate and vegetation, and uses 20 principles for identifying ecoregion boundaries (Bailey, 2005).

Bailey refers to the levels as domains, divisions, and provinces. The domains are equivalent to the major climatic zones and are Polar, Dry, Humid Temperate, and Humid Tropical (Bailey, 1995, 1996, 2009), and they correspond to Trewartha's E and F, B, C and D, and A respectively. Whereas the Trewartha system has 15 subtypes, Bailey has 16 divisions, which correspond to them, adding 3 prairie divisions (in humid temperate), and removing icecap (Bailey, 1989). The ecoregions are defined as major ecosystems "resulting from large-scale predictable patterns of solar radiation and moisture, which in turn affect the kinds of local ecosystems and animals and plants found in them." They are similar in concept to the ecobiomes of Polunin, proposed in 1984. The domains and divisions are based largely on the major climatic zones of the Köppen system as modified by Glenn Trewartha in 1968 in his book "An Introduction to Climate." Three scales of ecosystems are identified: macro- (macroclimatic differentiation (ecoregions), c. 100,000 sq. kms.), meso- (landscape differentiation (landscape mosaics), c. 1,000 sq. kms.), and microscale (edaphic-topoclimatic (sites), c. 10 sq. kms.), and climate is the common attribute and primary controlling factor, and is emphasized at the broadest levels because of its overriding effect on the composition and productivity of ecosystems from region to region. The bases for ecosystem delineation are flora and/or fauna, soil, physiography, watersheds, and acquatic biota.

The first known climatic classification originated with ancient Greek scientists in the 1st or 2nd century B.C., and it recognized 3 zones: frigid, torrid, and temperate, with the 1st and 3rd having a N and S version, which are still basically recognized today but with modifications (McKnight, 1992). The most used in modern times is by Russian-born German climatologist and amateur botanist Wladimir Köppen in 1900 with several revisions until 1940 (Köppen climate classification-Britannica.Com). It uses a system of capital (for zones) and small (for types and subtypes) letters and is based on temperature and aridity. The Köppen-Geiger classification by Geiger and Pohl in 1953 is a modification (Strahler and Strahler, 1976). The zones are A (tropical, also called equatorial), B (dry, also called arid or subtropical), C (mild midlatitude, also called warm temperate), D (severe midlatitude, also called cold temperate or snowy), and E (polar, also called icy). It, however, omits important factors such as sunshine and wind, and there is apparently a rather poor correspondence with vegetational distribution in some areas of the world (Köppen climate classification-Britannica.Com). The Trewartha modification addressed some of these shortcomings, adding E (subarctic)(with E becoming F (polar)), and H (highland). Types for Koppen-Geiger (Strahler and Strahler, 1976) are as follows:

A: wet, savannah, monsoonal
B: temperate and subtropical desert and temperate and subtropical steppe
C: Mediterranean, humid subtropical, and marine west coast
D: humid continental and subarctic
E: tundra and ice cap

American zoologist and ethnographer C. H. Merriam, one of the original founders of the National Geographic Society in 1888, in an 1898 USDA publication titled "Life Zones and Crop Zones of the United States," developed the concept of life zones to classify biomes found in North America along an altitudinal sequence corresponding to the zonal latitudinal sequence from Equator to Pole. Leslie Holdridge (1967), American botanist and climatologist and father of Haitian-born cinema and TV composer, conductor, arranger, and musician Lee Holdridge (e.g, the magnificent Moonlighting theme)(L.R. Holdridge-digplanet; Lee Holdridge-IMDB), elaborated on life zones in 1947 and 1967, formulating a global, bioclimatic system based on 3 axes forming a triangle: latitudinal regions (polar, subpolar, boreal, cool temperate, warm temperate, subtropical, and tropical), correlated to potential evapotranspiration ratio; altitudinal belts (premontane, lower montane, montane, subalpine, alpine, and alvar), correlated to annual precipitation and biotemperature (heat effective in plant growth); and humidity provinces (superarid, perarid, arid, semiarid, subhumid, humid, perhumid, and superhumid). They form some 30 classes, which can be grouped as tundra, desert, forest, steppe, woodland, and scrub.

The abiotic areas are the physical geographic areas, called physiographic divisions, regions, and provinces, and the geological regions or provinces. There are about 60 physiographic divisions globally (Asia: physiographic regions of Asia-britannica.com; North America Physiographic Regions-harpercollege.edu; Europe Physiographic Regions-harpercollege.edu; Physiography, Geography, and Climate of Latin America-utdallas.edu; Physiographic Regions of Australia-clw.csiro.au; List of the Physiographic Regions of the World-Wikipedia). The largest ones are the North African Plateau (6 mln. sq. mi.;10 mln. sq. kms.), East Antarctica (5.5 mln. sq. mi.; 9 mln. sq. kms.), the Great European Plain (4.4 mln. sq. mi.; 7 mln. sq. kms.), which goes from western and northern France, through Benelux, Denmark, northern Germany, and into most of Eastern Europe, the Amazon Plain (4.4 mln. sq. mi.; 7 mln. sq. kms.), the Canadian Shield (3 mln. sq. mi.; 5 mln. sq. kms.), West Antarctica (3 mln. sq. mi.; 5 mln. sq. kms.), the South African Plateau, East African Highlands, Western Australian Plateau, Brazilian Highlands, Pacific Cordillera, which runs from Alaska through the Rockies and down to the Andes, and the Alpine-Himalaya mountain system, which runs from the Atlas Mtns. in NW Africa to the Pyrenees, Alps, Carpathians, and Himalayas in Central Asia.

There are also about 60 geological provinces globally, of which there are 5 types: shields, platforms, igneous provinces, orogens (mountain systems), and extended crust (thinned out crust). The largest are the Canadian Shield, Central and South African Shield, Brazilian Shield, the Interior Platform (in North America), East European Platform, Asian Platform, the East Siberian Orogen, Alpine-Himalayan Orogen, and the Pacific Orogen. (Geologic Province and Thermo-Tectonic Age Maps-USGS.Gov; Geological Regions-thecanadianencyclopedia.ca)

As with physiographical areas, geological provinces can be grouped into 4 terrestrial realms: Western Hemisphere, Afro-Eurasia, Australia, and Antarctica. The 2 types often overlap, e.g., the Canadian Shield and the Pacific Cordillera/Orogen, and the 2 approaches are based on geomorphology, landforms (which are the substrates or backdrop for life forms and habitats) in other words, so the bounds between them are often unclear.

In the marine world, 4 geological and physiographic realms can be recognized also, being the 4 oceans, but sometimes a 5th (Southern or Antarctic) Ocean is identified. By far and undoubtedly the largest geological and physiographic region is the MidOcean Ridge, the most prominent geological/geographic feature on Earth, a nearly continuous, globe-girdling volcanic mountain system (the length is given as 80,000 kms. (c. 50,000 mi.) by Birkeland and Larson (1989), 40,000 mi., which is 64,000 kms., in Home-Douglas, 60,000 kms., which is 36,000 mi., by the Woods Hole Oceanographic Institute (MidOcean Ridges-WHOI.Edu), Britannica.Com, and Science Daily; Ocean Explorer (What Is The Mid-Ocean Ridge?-NOAA.Gov) says 65,000 kms.). It is about 620 mi. wide (Birkeland and Larson, 1989), criss-crossed by many fracture zones, with 10% of it exposed as islands (Ocean Explorer), including Iceland, the Azores, the Canaries, Melanesia, Micronesia, and Polynesia, its highest peak being Pico in the Azores at 20,000 ft. high and 7544 ft. above sea level (Birkeland and Larson, 1989). The major components are the MidAtlantic, SW Indian, Central Indian, SE Indian, and Pacific-Antarctic Ridge, E Pacific Rise, Chile Rise, and Gakkel Ridge in the Arctic (MidOcean Ridges-Global Distribution of Hydrothermal Vents-WHOI.Edu). Abyssal plains are most common in the Atlantic and Indian Ocean and abyssal hills are most common in the Pacific, and the hills and plains together occupy 42 % of the sea floor (Birkeland and Larson, 1989)(in Home-Douglas it says abyssal plains occupy 40% of the Atlantic and about half of the Pacific and Indian Ocean basins), the largest plain possibly being the Sohm in the North Atlantic, which is 350,000 sq. mi. (906,000 sq. kms.) (Abyssal Plain-Britannica.Com), while trenches, seamounts, and tablemounts (usually called guyots, named after Arnold Guyot, 19th century Swiss associate of Louis Agassiz, and sometimes considered as a type of seamount) are most common in the Pacific (Birkeland and Larson, 1989; Home-Douglas, 1991).

Bailey, R.G. 1989. Explanatory Supplement to Ecoregions Map of the Continents. Environmental Conservation Vol. 16, Issue 4, Winter.

Bailey, R.G. 1995. Ecosystem Geography, 1st ed. Springer, New York (Amazon Look Inside).

Bailey, R.G. 1996. Ecoregions: Ecosystem Geography of the Oceans and Continents. Springer, New York (Amazon Look Inside).

Bailey, R.G. 2005. Identifying Ecoregion Boundaries. Environmental Management 34, Suppl. 1: S14–S26

Bailey, R.G. 2009. Ecosystem Geography: from Ecosystems to Sites, 2nd ed. Springer, New York (Amazon Look Inside).

Barnes, R.K.S., and Hughes, Roger. 1999. An Introduction to Marine Ecology. Blackwell Science (Amazon Look Inside).

Beatty, J. Thomas, et al. 2005. An obligately photosynthetic bacterial anaerobe from a deep-sea hydrothermal vent. Proc. Natl. Acad. Sci. U S A.102: 9306–9310 (nih.gov).

Birkeland, Peter, and Larson, Edward. 1989. Putnam's Geology, 5th ed. Oxford U. Press.
Briggs, J.C., Bowen, B.W. 2012. A realignment of marine biogeographical provinces with particular reference to fish distributions. J. Biogeogr. 39: 12–30 (academia.edu).
Cox, C. Barry. 2001. The biogeographical regions reconsidered. J. Biogeog. 28: 511-23 (desiguenza.net).
Dawes, Clinton. 1998. Marine Botany, 2nd ed. Wiley (Amazon Look Inside).
Dawson, E. Yale. 1966. Marine Botany: an Introduction. Holt, Rinehart, and Winston.
Golikov, A.N., Dolgolenko, M.A., Maximovich, N.V., Scarlato, O.A. 1990. Theoretical approaches to marine biogeography. Mar. Ecol. Prog. Ser. 63: 289-301 (int-res.com; researchgate.net).
Good, Ronald. 1947. The Geography of the Flowering Plants. Longmans, Green, and Co., London.
Hawksworth, D.L., Kirk, P.M., Sutton, B.C, Pegler, D.N. 1995. Ainsworth and Bisby's Dictionary of the Fungi (8th ed.). IMI (International Mycological Institute-CAB (Center for Agriculture and Biosciences), Oxon, UK, New York, NY.
Holdridge, L.R. 1967. Life Zone Ecology. Tropical Science Center, San Jose, Costa Rica (fs.fed.us).
Home-Douglas, Pierre (ed.). 1991. Oceans (How Things Work Series). Time-Life, Alexandria, Virg.
Jameison, Alan. 2015. The Hadal Zone. Cambridge U. Press (Amazon Look Inside).
Keddy, Paul. 2010. Wetland Ecology. Cambridge U. Press (Amazon Look Inside).
Lee, Thomas. 1986. The Seaweed Handbook. Dover.
Marx, Robert. 1990. The History of Underwater Exploration. (Dover republication of the original Into the Deep: the History of Man's Underwater Exploration from 1978 by the Van Nostrand Reingold Company).
McKnight, T.L. 1992. Essentials of Physical Geography. Prentice-Hall.
Mitsch, W.J. and Gosselink, J.G. 2007. Wetlands. Wiley, Hoboken, N.J (Amazon Look Inside).
Odum, Eugene. 1997. Ecology. Sinauer Associates.
Pech, Pierre, and Regneault, Hervé. 1992. Géographie physique (Collection premier cycle). PUF.
Tait, R.V., and Dipper, F.A. 1998. Elements of Marine Ecology, 4th ed. Butterworth-Heinemann (Amazon Look Inside).
Spalding, Mark and 14 other authors. 2007. Marine Ecoregions of the World: a Bioregionalization of Coastal and Shelf Areas. Bioscience 57: 573–583 (campam.gcfi.org).
Strahler, A.N. and Strahler, A.H. 1976. Elements of Physical Geography. Wiley.
Thomas, David. 2002. Seaweeds (Natural World Series). Smithsonian Institute Press.

Chapter 19 Written in Tablets of Stone

The flight of time is measured by the weaving of composite rhythms...the stratigraphic series constitutes a record, written in tablets of stone, of the lesser and greater waves of change which have pulsed through geologic time. - Joseph Barrell, 1917, quoted in "Principles of Stratigraphy," Michael Brookfield, 2004, p. 190.

Geology, geography, paleontology, and biology are sister sciences, together constituting natural history, as they are intimately interrelated, paleontology even being considered as either a part of biology or of geology, which is illustrated by geological time scales as in the table below.

Table 19-1. Modern Composite Geological Time.

eon	era	subera	period	subperiod	epoch		age starts
Phanerozoic (G.H. Chadwick, 1930)							
	Cenozoic (John Phillips, 1840) (Age of Mammals)						
			Quaternary (Giovanni Arduino 1760/ Jean Desnoyers 1829)				
					Holocene (F.L.P. Gervais, 1867)		12 tya
					Pleistocene (Charles Lyell, 1839)		2.6 mya
			Tertiary (J.G. Lehmann 1756/Giovanni Arduino1759/ Jean Desnoyers 1829)				
				Neogene (M. Hornes, 1853)			
					Pliocene (Lyell 1833)	1st hominids	5
					Miocene (Lyell 1833)		23
				Paleogene (M. Hornes, 1853)			25
					Oligocene (H.E. Beyrich 1854)		40
					Eocene (Lyell 1833)		55
					Paleocene (Schimper 1874)	1st primates	66
	Mesozoic (John Phillips, 1840) (Secondary J.G. Lehmann 1756/Giovanni Arduino 1759) (Age of Reptiles)						
			Cretaceous (Jean d'Omalius d'Halloy, 1822)				
					Late	appearance of hornworts	100
					Early	angiosperm radiation	146
			Jurassic (Alexandre Brongiart, 1829)				
					Late (White, Malm)		160
					Middle (Brown, Dogger)		175
					Early (Black, Liassic)		208
			Triassic (Friedrich von Alberti, 1834)				
					Upper	appearance of dinosaurs and mammals	230
					Middle		245
					Lower		250
	Paleozoic (Adam Sedgewick, 1838) (Primary Giovanni Arduino 1760)(Age of Fishes)						
		Upper Paleozoic					
			Permian (Roderick Murchison, 1841)				
					Late	appearance of conifers and mosses	265

Early		300
Carboniferous (W.D. Conybeare and John Phillips, 1822)		
Pennsylvanian (Upper)	1st reptiles	320
Mississipian (Lower)	1st winged insects	360
Devonian (Sedgwick and Murchison, 1839)		
Upper	1st liverworts, ferns, club mosses, and horsetails	385
Middle	1st amphibians	400
Lower		420
Lower Paleozoic		
Silurian (Murchison, 1835)		
Late	appearance of vascular plants (rhyniophytes)	420
Early		440
Ordovician (Charles Lapworth, 1879)		
Late	1st vertebrates, 1st corals	460
Middle		470
Early		500
Cambrian (Sedgwick, 1835)		
Upper		
Middle		
Lower		540
Cryptozoic (G.H. Chadwick, 1930)(PreCambrian, Joseph Jukes, 1862)		
Proterozoic (S.F. Emmons, 1888)		
Vendian (Late, Neo-) (Hadrynian)	1st acritarchs, amebas, metazoans, and green algae	1 bya
Riphean (Middle, Meso-) (Helikian)	1st red algae and heterokonts	1.6
Aphebian (Early, Paleo-)		2.5
Archeozoic (J. D. Dana, 1872)(Archean, J. D. Dana, 1872)		
Neoarchean		2.8
Mesoarchean	1st stromatolites	3.2
Paleoarchean	1st oxygenic bacteria	3.6
Eoarchean	1st bacteria, 1st rock (Acasta Gneiss)	4
Azoic (Murchison, de Verneuil, Keyseling, 1845)(Hadean, Preston Cloud, 1972)		
Early Imbrian (Zirconian)		4.1
Chaotic		
Nectarian		4.3
Basin Groups		4.5
Cryptic	1st mineral (zircon)	4.6

(Dates are approximate and vary according to source, but the differences are minor; there are 2 eons, 6 eras, 21 periods, 29 or 33 epochs, and 99 stages [sometimes called ages])

The names Ordovician and Silurian are based on the names of Welsh tribes; Cambrian comes from an old name for Wales; Permian is derived from Perm, located in the western foothills of the Urals in Russia; Jura comes from the Jura Mtns., a sub-alpine mountain range north of the Western Alps, mostly on the French–Swiss border, separating the Rhine and Rhône basins; Trias is based on the 3 layers found in Germany and NW Europe: red beds, chalk, and black shales; Devonian is from

Devonshire; Liassic is from French liais, "hard freestone" (fine-grained limestone)(leia [Old Saxon], "layered"; Cornish quarryman's term for 'layers'; leac [Gaelic], "flat stone"); Dogger is old provincial English for ironstone; and Malm is Middle English for sand. (Sissingh, 2012; Time Almanac 2001).

The Azoic was also called the Eozoic by Dawson in 1868 and suggested as a replacement for Azoic by Credner in 1872; Eobiotic or Eomorphic was suggested in place of Eozoic by C.H. Hitchcock in 1888; Hypozoic for Azoic was coined by Phillips in 1840 or 1855; Primitive and Presilurian were also synonyms for Azoic. Eo-, Meso-, and Neoproterozoic were coined by F.D. Adams in 1909. (Geological Time Classification of USGS – usgs.gov).

The Cambrian originally included the basal part of the Ordovician and some Archeozoic rocks and was arranged as 4 epochs designated A, B, C, and D (early to later), which probably correspond to Terreneuvian, Series 2, Series 3, and Furongian (Sissingh, 2012; International Chronostratigraphic Chart – stratigraphy.org).

The Jurassic was originally only a part of the modern defintion and its Middle period was divided into the Great or Bath Oolitic (sometimes called Oolite) or Bathonian, named by d'Omalius d'Halloy in 1843, and the Inferior Oolitic or Bajocian, named by d'Orbigny in 1849, corresponded mostly to the Oolitic, which was the previous name (Sissingh, 2012). The period was established and originally named Jura-Kalkstein (Jura-Limestone) by Alexander von Humboldt in 1792, first published in 1799, for a separate and distinct rock sequence in the Swabian Jura Mtns. of S Germany and the Folded Jura in N Switzerland and would represent the middle and upper parts of the Jurassic System as more comprehensively defined by Buch 10 years later (Sissingh, 2012). The French name, "jurassique", was coined by Brongniart in 1829 (Dubois et al, 2005). The English name came into the language in 1831 (Webster's 9th New Collegiate Dictionary, 1987). Oolitic probably comes from "oolithe" coined by Trévoux in 1752 and "oolithique" coined by German-Italian geologist Scipione Breislak in 1818 (Dubois et al, 2005). The English "oolite" was introduced in 1785 (Webster's 9th New Collegiate Dictionary, 1987). Dinosaurs, gymnosperms, and ferns are especially abundant in the Late Jurassic.

The Permian is sometimes divided into the Cisuralian, Guadalupian, and Lopingian (early to later).

The Carboniferous, which originally included the Devonian, is also refered to as Coal Measures, and "measures" was a miner's term for sequences of sedimentary rock which could be measured in bore-holes or mine shafts. It is occasionally used as a synonym for "beds" or "series" of strata, an example being Barren Red Measures (the Upper Coal Measures of Scotland) at the Carboniferous-Permian junction in Europe. The Coal Measures Group consists of the Upper, Middle, and Lower Coal Measures Formations. The group records the deposition of fluvio-deltaic sediments, which consists mainly of clastic rocks (claystones, shales, siltstones, sandstones, conglomerates) interstratified with the coal beds, clastic rocks being ones built up from fragments of pre-existing rocks. (Whitten and Brooks, 1985; Coal measures-Wikipedia).

The Silurian originally included most of the Ordovician. The Upper Silurian is usually divided into the Ludlovian and Pridolian (early to later) epochs and the Early Silurian into the Llandoveryan and Wenlockian (early to later) epochs (Series and Stages of the Silurian System, C.H. Holland – stratigraphy.org).

The Lower Triassic is also called Bunter or Buntsandstein (Gm., coloured sandstone), Middle Triassic the Muschelkalk (Gm., shell limestone), and Upper Triassic the Keuper (old miner's term); these are Germanic lithostratigraphic stage names corresponding more or less to the epochs of the Triassic, and hence the name, as mentioned earlier (Time Almanac 2001).

It is unclear when fossils were first seen as extinct species, but Lo Han, Chinese writer of the 4th c., may have been the first to do so (Sissingh, 2012). But the discovery of fossils goes back to the Stone

Age, as perhaps the oldest evidence of human contact with fossils comes from the Gobi Desert, where in a site dating back to the Paleolithic, parts of a dinosaur egg shell were found drilled with holes, evidently used in necklaces (Bret-Surman et al, 2012).

Native Americans, long before European contact, had many earth science concepts. As Douglas Wolfe, leader of the Zuni Basin Paleontology Project, observed, "It's all there in one elegant myth: evolution, extinction, climate change, deep time, geology, and fossils" (Mayor, 2005; Bret-Surman et al, 2012). Publishers' Weekly declares in a review at Amazon of Adrienne Mayor's book: "Indian notions of "deep time," changing landforms and climates, and the descent of contemporary species from fossilized ancestors anticipate the insights of present-day geology and evolutionary theory, she contends, while Inca legends of extinction by "fire from heaven" prefigure modern theories of extinction by asteroid impact." And Booklist states in a review of the same book also at Amazon, "Though tribal myths actually anticipate key Darwinian concepts of species change, Native American traditions have too often been dismissed as mere superstition by orthodox scientists."

Nicolaus Steno (Niels Steensen), 17th century Lutheran geologist, who later became a physician to the Grand Duke of Tuscany and Catholic bishop, saw fossils as remains of dead organisms and demonstrated the orderly succession of stratified deposits, but, because of religious beliefs, did not admit deep time. He formulated 4 fundamental laws of temporal relationships in strata: superposition (the layer below is older), original horizontality (layers are horizontal and rarely tilted, and when they are it is subsequent), lateral continuity (unless abutting to other, older rocks layers are deposited globally without interruption), and cross-cutting (a body of rock cross-cut by another is older). He also established that the angle between corresponding faces in a crystal is fixed (a consequence of molecular arrangement), which is sometimes called Steno's Law. He expounded his ideas in *De solido intra solidum naturaliter contento dissertationis prodromus* (Preliminary discourse to a dissertation on a solid body naturally contained within a solid), called simply *Prodromus,* in 1667.

Steno's work is traditionally considered to mark the beginning of modern geology and encouraged further studies of this key subject. The growing interest in fossils was prompted by, among other things, an accelerating accumulation of fossil evidence throughout the 18th century, the result of field work, explorations, and the rapidly increasing rate of construction projects, particularly rock quarries and canals. (The Early Development of Modern Geology – viu.ca).

However, Drake (2007) points out that Robert Hooke, great 17th c. polymath, inventor, and architect, and a founder of the Royal Society, was not only aware of the principles of superposition and original horizontality, the organic nature of fossils, and the process of fossilization, but he discounted Noah's Flood, seeing the earth as much older than was admitted at the time, and considered fossils as perhaps helping us to establish a chronology, presaged Lamarck and Darwin with ideas of the extinction of species and evolution of new ones as a result of environmental changes, and his geological concepts were almost identical to those of Hutton some 2 centuries later, who did not mention him, and he not only recognized Steno's Law but also saw that it was an expression of the internal arrangment of particles. Hooke proposed the rock cycle (also called the geochemical cycle) before Hutton did (the rock cycle being magma, solidification [as igneous rocks], denudation [weathering, erosion, and transport], sedimentation, diagenesis and lithification [as sedimentary rocks], metamorphosis, and anatexis [regeneration of magma]). Hooke's series of lectures to the Royal Society starting in 1667 and lasting over 30 years were published posthumously in 1705 by Richard Waller and titled Lectures and Discourses of Earthquakes and Subterraneous Eruptions.

The first divisions of geological time were formally introduced by Italian abbot, geologist, naturalist, and professor of philosophy and rhetoric Anton Moro in 1740 in "De' crostacei e degli altri marini corpi che si truovano su' monti" ("On crustaceans and other marine bodies that are found on mountains"), as Monti Primari and Monti Secondari. The idea that the earth has undergone many tumultuous changes in the course of its existence was readdressed in an original manner by him. He claimed that these major formative events, as well as their resulting major sequences of exposed rock, were all volcanic in origin (which is why he is described as an ultraplutonist). According to him, "the stratification of the fossil-bearing deposits is not due to an aqueous agent, as shown by the chromatic and density differentiation of the stratal successions. The Deluge had not occurred at all and the sea had not repeatedly risen by itself and thereby frequently flooded the land in previous times," as assumed by many others. "Instead, the marine fossil-bearing rocks, as exposed in the land masses, are testimony for the origin of the land by tectono-volcanic uplift of the sea floor. All continents and islands had emerged out of the sea by volcanic action. This activity occurred locally at a variety of times and left marine strata on the mountain flanks and over their adjacent plains (involving resedimentation related to erosion of raised areas)." (Sissingh, 2012).

German geologist J.G. Lehmann designed one of the first modern geological time scales in 1756, which included Primary (possibly corresponding to PreCambrian), Secondary (Upper Carboniferous and Permian [the late Mesozoic]), and Tertiary (Cenozoic, which he called gebirgen ["mountain chains"]), with the Quaternary added by 18th century geologist Giovanni Arduino, the Father of Italian Geology, 3 years later, and refined in 1774 (he termed his units ordini). (Sissingh, 2012; Giovanni Arduino-Encyclopedia.Com).

Abraham Werner (The Early Development of Modern Geology-viu.ca; Sissingh, 2012) was inspector of mines and professor of mining and mineralogy at the Freiberg Mining Academy and has been called the "Father of German Geology". In his *Kurze Klassifikation und Beschreibung der Verschiedenen Gebirgsarten* ("short classification and description of various mountain chain categories")(36 pp.) in 1787 and lectures, he advocated that rocks should be arranged in chronological order, that is, based on the age of their formation, instead of mineral types (the traditional method), thereby placing reconstruction of historical processes at the center of geological inquiry, and he demonstrated chronological succession in rocks, for which he is remembered. The groupings are as follows:

Primitive	(Precambrian and Cambrian)
Transitional	(Ordv. to Lower Carb.)
Floetz	(Upper Carb. to Quaternary)
Vlanc	(Tert. to Quat.)
Alluvial	(Tert. to Quat.)

But he is especially known as originating Neptunism, which held that the earth was originally covered in an ocean and the majority of rocks were precipitated out of this vast, receding ocean, and in a definite order to form the current landscape, and that granites are the oldest rocks. A specific point of contention was the origin of basalt as sedimentary. Neptunism did accept uniformitarianism but depended on certain causes (for the sudden movement of the oceans) which no longer operated.

His most famous student at Freiberg was Alexander von Humboldt, Franco-Prussian explorer, polymath, and polyglot, Neptunist turned Plutonist, and Copley Medalist, and he and German contemporary, Karl Ritter, who wrote *Die Erdkunde* ("Earth Science"), are considered the founders of modern geography. One of Ritter's most famous students and supporters was Arnold Guyot, who was later professor of physical geography and geology at Princeton (then the College of New Jersey) from 1854 to 1880, and who we encountered in the preceding chapter. (About.Com; Columbia Viking Desk Encyclopedia, 1960).

James Hutton, 18th century Scottish geologist, chemist, naturalist, and farmer, known as the Father of Modern Geology, was an influential backer of uniformitarianism, the doctrine which postulates uniformity of geological processes over time, in other words, that the geological processes that operated in the past still operate today, so that the present is the key to the past, which has become the most important working principle of historical geology and biology. But the notion was not original to Hutton as the renowned genius, scientist, inventor, and artist Leonardo da Vinci had anticipated it 300 years earlier during the Italian Renaissance in the 16th c., which, as earlier stated, was formulated by Robert Hooke in the 17th c., and had been formulated by 18th century French physician Georges Füchsel. Hutton is also famous for saying, "no vestige of a beginning and no prospect of an end", but he may have been refering to the rock cycle instead of the Earth. (Britannica.Com; Foundation of Modern Geology–Plutonism-publish.illinois.edu; The Early Development of Modern Geology-viu.ca; Birkeland and Larson, 1989).

Plutonism, the currently accepted view as well, was first proposed by Moro, but Hutton was responsible for expounding it to the Royal Society of Edinburgh and the general scientific community and laid the foundation for its success. It holds that processes which created and arranged rocks into the current landscape are driven by heat concealed in Earth's interior, and that granite and basalt are igneous, hence the alternate name of Vulcanism. It was, of course, opposed to Neptunism, which is now discarded. Hutton elucidated his insights in "Theory of the Earth" in 1785. (Foundation of Modern Geology–Plutonism - publish.illinois.edu).

Georges Cuvier's (The Early Development of Modern Geology-viu.ca) study of the composition and fossil record of the different layers of the Paris Basin led him to propose a series of sudden inundations and retreats of the sea. For Cuvier the only satisfactory explanation for such evidence was a series of huge catastrophes, considering present processes produce effects too small and slow and volcanic action as too local. In short, it proposed repeated creations following catastrophic events as the dominant forces in geological processes. It was supported by notable geologists such as Elie de Beaumont, Alcide d'Orbigny, and Louis Agassiz.

Cuvier's work, "Essai sur la théorie de la terre" of 1813 was translated and published in England in the same year, where the catastrophic theory became very popular, not least of all because part of it gave strong support to the notion that the last such catastrophic flood could be dated at roughly the same time as the one described in Genesis. To make this point more persuasively, Cuvier had included a survey of mythological accounts of a great flood from many cultures and had traced them back to an approximate common date.

The Scriptural geologist William Buckland drew on Cuvier's idea, which he supported with the discovery in 1821 of the Kirkdale Caves in Yorkshire, which contained the bones of many extinct animals which had apparently died in a sudden flood, to corroborate the Biblical account.

This version of catastrophism was particularly helpful in "resolving" the then growing problem of the age of the earth. A literal interpretation of Scripture strongly suggested that Creation had taken place in about 4000 BC. Cuvier's theory dated the last great flood at around that time. Hence, invoking the notion that the Mosaic account described the last of several creations, the earlier ones having involved different lands and alternative life forms, geologists could reconcile their time requirements with Scripture.

William Smith, a civil engineer and surveyor who worked on the construction of canals in England, kept an extensive record of the fossils he observed in his tireless inquiries, and made some fundamental discoveries: the strata in SE England occured in the same order, different layers contained distinctive assemblages of fossils, and the distribution of fossils in that area could be represented on a map. Such a map he produced in 1815, which was a classic. It was the first geological map of Britain, the first colour geological map, and the first national-scale geological map. Today, geological mapping is one of the most important techniques of the geologist. In "Stratigraphic System of Organized Fossils", in 1817, he proposed that sedimentary rock units could be classified according to the fossil remains they contained (the law of strata identified by fossils), and that, with the aid of the fossil record, geologists could assign relative ages to rocks. The identification and classification of fossils at that time provided the only sure means of comparing the ages of different rocks (other than the law of superposition). As it so happens, Cuvier and Brongniart made the same discovery, fossils for the determination of geochronology, independently at about the same time in France. Like Leeuwenhoek and Walcott he was self-taught. (The Early Development of Modern Geology - viu.ca; Whitten and Brooks, 1985; Moore, 1956).

Charles Lyell, 19th century British baronet, lawyer, geologist, and Copley Medalist, was prominent in promulgating uniformitarianism and Plutonism, including them in his textbook, "Principles of Geology," from 1830 (1st Vol.), 1832 (2nd Vol.), and 1833 (3rd Vol.), which went through 12 editions, all in the 1800s, allowing the success of the theories in his century. His "Elements of Geology" went through 6 editions, also all in the 1800s. What is contradictory about Lyell is that he supported the older ideas that earth and water trade substances and shape each other, maintaining some kind of long-range balance (the steady-state Earth), and that time and life are cyclical. It was conceivable to him, because the Earth undergoes periodic climatic changes and animals are adapted to certain climates, that animals and humans could all become extinct, only to be replaced in a subsequent creation, followed, in some distant age, by a new creation, which actually sounds a lot like catastrophism, and, though he was a mentor and friend of Charles Darwin, he was reluctant to endorse biological evolution. He was cautious about contradicting anyone, had a desire for social acceptance, and was apparently a master of the art of coming down squarely on both sides of an issue. (Sissingh, 2012; Britannica.Com; stephenjaygould.org; Millar et al, 1996; Wikipedia).

Uniformitarianism was opposed to catastrophism, of course, but the former has come to admit the rates of geological processes have varied--sometimes there is a faster rate. For example, the movement of 2 blocks in the crust along a fault in opposite directions might occur over a span of 10 mln. years, but the total displacement is the result of a succession of small, rapid displacements; a canyon is the result of a succession of relatively small, catastrophic floods of local extent; and beach erosion is primarily accomplished during catastrophically short episodes when ocean current and wave action are at peak intensity, such as during cyclonic storms. And some catastrophes are large-scale rather than local (mass extinctions, ice ages) but are considered rare. (Birkeland and Larson, 1989).

Lyell's other scientific contributions included an explanation of earthquakes, the theory of gradual "backed up-building" of volcanoes, and classification in stratigraphy. He identified 2 systems of formations:

- System I
  - Pyroid Formations (Lower Precambrian)
  - Primordial Formations
    - Agalysian
    - Hemilysian Formations (Upper Precambrian to Carboniferous)
  - Secondary Formations
    - Ammonian Formations (Permian to Cret.)
    - Tertiary Formations (Tert. and Quaternary)
- System II
  - Plutonian Formations (Precambrian) (Pyroid and Agalysian)

Neptunian Formations (Phanerozoic or Permian to Quaternary)

Four of the 6 Cenozoic epochs were established by Lyell and this was mostly on the basis of the percentage of extinct marine molluscs in the sediments. The Eocene had over 90% of the species extinct, the Miocene 60-80%, and the Pleistocene only 10%. (Birkeland and Larson, 1989).

James Dwight Dana (Sissingh, 2012; Wikipedia) was a 19th century American geologist and mineralogist, and Copley, Wollaston, and Clarke Medalist, who is known for "Manual of Mineralogy" from 1848, which became a standard college textbook, and has gone through 23 editions, the latest in 2007, "System of Mineralogy" from 1837, which went through 8 editions, the latest in 1997, "Manual of Geology" (4 eds.), an influential textbook from 1862, "A Text-book of Geology," a more elementary work, from 1864, and "Corals and Coral Islands" from 1872, which was the climax to his notable studies of coral reefs, begun on the Wilkes expedition. In explorations of the South Pacific, the U.S. Northwest, Europe, and elsewhere, he made important studies of mountain building, volcanic activity, sea life, and the origin and structure of continents and ocean basins. He divided the geological past into 4 times (modern stratigraphic equivalents are in parentheses):

Archean Time
- Azoic Age (Archeozoic)
- Eozoic Age (Proterozoic)

Paleozoic Time
- Silurian Age or Age of Invertebrates (Cambrian-Silurian)
- Devonian Age or Age of Fishes (Devonian)
- Carboniferous Age or Age of Coal-Plants, Age of Amphibians (Carboniferous)

Mesozoic Time
- Secondary or Age of Reptiles (Triassic-Cretaceous)

Cenozoic Time
- Tertiary or Age of Mammals (Tertiary)
- Quaternary or Age of Man (Quaternary)

He also divided the Archean into Laurentian and Huronian and the Paleozoic into the Paleozoic Section (Cambrian and Ordovician) and the Neopaleozoic Section (Sil., Dev., Carb., and Perm.).

Haeckel in 1866 used the names Archolithic (Primordial Time), Paleolithic (Primary Time), Mesolithic (Secondary Time), Cenolithic (Tertiary Time), and Anthropolithic (Quaternary Time); in 1876 he called them epochs instead of times, used an i instead of an o in Archolithic, and called the Carboniferous the Coal Period and the Cretaceous the Chalk Period, which, of course, is what the usual names refer to. (Sissingh, 2012).

Old Red Sandstone is a facies (sum total of features that characterize a sediment as having been deposited in a given environment [Whitten and Brooks, 1985]; rock unit with specified characteristics that forms under certain conditions of sedimentation, reflecting a particular process or environment [Wikipedia]) of continental rocks correspondinig mostly with the Devonian, occuring in the British Isles, Norway, Greenland, and E North America. It is made mostly of sandstone and mostly red (due to iron oxides). The term was first used for deposits in the Welsh Basin and described as part of the Lower Carboniferous. Naturalist Robert Jameson coined the term in 1820, Murchison was the first to realize the ORS was a separate entity giving it the status of system, which for the next century was broadly correlated with the Devonian that he went on to describe with Sedgwick, and Conybare synonymized it with the Devonian in 1822. With modern stratigraphical methods it has merely become a classic term describing the facies of the Wenlockian in the Silurian to the Lower Mississipian in the Carboniferous, but is still commonly split into its Lower (Wenlockian to Lower Devonian), Middle (Middle Devonian), and Upper (Upper Devonian and Lower Mississipian) subdivisions. It is believed to have originated in the Caledonian Orogeny (orogeny or orogenesis being mountain building). ORS deposits have been extensively studied in Britain, where the facies is accorded supergroup status and where local and regional stage names have been applied. Plant and animal fossils occur in these deposits; the fossil fauna is characterized by primitive, often armoured fishes, and late in this succession came the first tetrapods. The New Red Sandstone corresponds to the Permian and was named the Dyassic by Jules Marcou in 1853. (OldRedSandstone.Com; Britannica.Com; Wikipedia; Sissingh, 2012).

Geological or geochronological and stratigraphic units correspond as follows: eon/eonothem, era/erathem, period/system, epoch/series, and age/stage. Lithostratigraphic units are, in increasing size, bed (synonomous with stratum), member (part of a formation), formation (primary or basic and mappable unit), and group (2 or more formations), with sub- and super- prefixes being sometimes applied.

A biostratigraphic unit, called a zone or biozone, is an interval of geological strata defined on the basis of its characteristic fossils. The major types are: range (acrozones)(which represent the occurrence range of a specific fossil taxon, based on the localities where it was recognized), interval (in which the top and base are defined by horizons that mark the first or last occurrence of 2 different fossils), assemblage (where a stratigraphic interval is characterized by an assemblage of 3 or more coexisting fossils that distinguish it from surrounding strata), abundance (in which the abundance of a particular taxon [or group of taxons] is significantly greater than neighbouring parts), lineage (which contain fossils that represent parts of the evolutionary lineage of a particular fossil group; a special case of range zones), and peak zones (epiboles)(where a fossil reaches its acme). Unlike lithostratigraphic units, they have no hierarchical significance and are not based on mutually exclusive criteria. A subbiozone or subzone is a division of a biozone, and superbiozone or superzone is a grouping of 2 or more biozones with related biostratigraphical attributes. A biochron is the length of time represented by a biostratigraphic zone. (Whitten and Brooks, 1985; Brookfield, 2004; Mathur, 2008).

The Chaldeans, noted astronomers of the ancient Near East, maintained the Earth had existed since 2.15 mln. yrs. (215 myriads)(Holmes, 1913). The ancient Chinese had the figure at 88.64 mln. yrs (Drake, 2007). The earliest scientific estimate for the age of the Earth in modern times was by G.-L. Leclerc (comte de Buffon) in 1774 (in the 1st supplementary volume of his "Histoire Naturelle") and 1778 (in the 5th volume), was based on cooling rate, and was 75,000 yrs. One of the best estimates in the modern era but prior to the advent of radiometry was made by Haughton in 1865, which was 2.3 bln. yrs. (Azoic starting at 2.3 bya and Zoic starting at 1 bya), and it was based, like Leclerc, on the cooling of the Earth. In 1871, his estimate was based on sedimentology and was 2.5 bln. yrs. He made another estimate in 1878, on the same basis, with a far different result: 100 mln. yrs. (this time he used 3 divisions: Azoic, Paleozoic, and Neozoic). Again in 1878 and also based on sedimentology his result was 200 mln. yrs. From 1860 and 1909 19 estimates based on sedimentology were done by 16 different geologists, and ranged from 3 mln. yrs. (Winchell in 1883) to 2.6 bln. yrs. (McGee in 1893) averaging 210 mln. yrs. A famous estimate of the time was by William Thomson (Lord Kelvin) in 1883, who suggested 100 mln. yrs. based on the decrease in Earth's rotational speed, the Sun's energy output (heat received from the sun), and cooling of the Earth. These were unreliable methods as pointed out by Arthur Holmes in 1913 in his classic book "Age of the Earth," who was a supporter of the radioactive method. His estimate was c. 2-3 bln. yrs. The current, widely accepted age of 4.6 bln. yrs. was determined by Cal Tech geochemist C. C. Patterson in the 1950s, using several meteorites including ones from Canyon Diablo, however, a radioactive dating of galena resulted in an estimate of 5-5.4 bln.

yrs. (Whitten and Brooks, 1985). Holmes, in accord with Thomson, found no support for uniformitarianism, at least as it was formulated in the 1800s (he did see the doctrine as an advance). (Sissingh, 2012).

Geochronology concerns itself with the crucial issue of measurement of geological time and 2 types of methods are used: relative (materials are compared in age only relative to each other; does not do actual dating but is periodization considered without actual age, in other words, measures duration), which is essentially stratigraphic and based on superposition of beds and fossil character, with specialized techniques (tephrachronology [based on volcanic ash], varve chronology, ice core chronology, and fluorine analysis [determines whether vertebrate fossils from the same horizon of the Quaternary are contemporaneous]), studies sedimentary rocks, has been around for about 300 years, and is done typically by visual observation; and absolute, which determines the actual age in years (before the present), studies igneous and metamorphic rocks, and requires expensive and complex analytical equipment. (Dating Methods Study Guide-enote.com; Wikipedia; Geological Time Classification of the USGS, M. Grace Wilmarth, 1925).

A varve is in the wide sense a layer of sediment deposited in a single year but in practice refers to sediments deposited in melt-water lakes, which consist of a coarser layer, representing summer deposition, and a finer layer, representing winter deposition. Counting and correlating these varves has (notably by De Geer and his students in 1940) has led to the development of a detailed Pleistocene chronology for the N Hemisphere. (Whitten and Brooks, 1985).

Fluorine dating determines whether vertebrate remains from the same horizon of Quaternary deposits are contemporaneous by comparing relative amounts of fluorine compostion in skeletal remains. A higher fluorine composition indicates a bone has been buried for a longer perod of time. (Dating Methods Study Guide-enotes.com).

Annual layers of snowfall in ice cores can be counted as easily as tree rings, allowing precise dating of events. Distinct annual layers stand out because, in snow that falls in summer, crystals are larger and acidity higher than in winter snow. Ice cores have revealed that global climate, long thought to change only very gradually, can shift with great speed, sometimes in only a matter of years. One such quantum leap occurred about 12,000 years ago, as the last Ice Age, the Pleistocene, was giving way to our current, warm, interglacial epoch, the Holocene. The Younger Dryassic (usually called Younger Dryas) Cold Event (sometimes called Episode), as it is known, lasted about 1,300 years before it returned, just as suddenly, to the temperatures typical of the time immediately preceding it. (Nova-pbs.org).

The Younger Dryassic is the latest and longest of 3 stadials (intervals between pauses [interstadials] in the advance of ice in a glaciation) that resulted from typically abrupt climatic changes that occurred over the last 16,000 years. It was preceded by a warmer stage, the Allerød Oscillation, which in turn was preceded by the Older Dryassic around 14,000 yrs. ago that lasted c. 200 years. The Older Dryassic, in turn, was preceded by another warmer stage, the Bølling Oscillation, that separates it from a third and even older stadial, the Oldest Dryassic, which lasted about 15,000 yrs. ago and lasted about 330 years, according to the GISP2 ice core from Greenland. The interval is named for an indicator genus, the magnificent, white, alpine-tundra wildflower *Dryas octopetala* of the rose family. The nomencalture comes from the Byltt-Sernander classification of northern European climatic phases. (Younger Dryas-PediaView.Com).

There have been at least 5 ice ages (major glaciations) (Birkeland and Larson, 1989; Ice Ages – Nature.Ca):

Pleistocene	2 mya – 12 tya	
Karoo	350 – 250 mya	Carboniferous-Permian

Andean-Saharan	450 – 420 mya	Ordovician-Silurian
Cryogenian	800 – 600 mya	Late Proterozoic
Huronian	2.4 – 2.1 bya	Early Proterozoic

Two methods can be used for both absolute and relative dating. Magnetic dating compares iron particle orientation with known movements of the magnetic North Pole over geological time. Its range is 100s of 1000s to mlns. of years. Obsidian hydration dating was introduced by Irving Friedman and Robert Smith in 1960 and is used only in archeology. (Chronological Methods 10 and 11-ucsb.edu).

In the absolute type of geochronological measurement there are 4 modern methods:

radiometric or isotopic (used extensively since the '50s)
incremental or floating chronology (dendrochronology [the counting of tree rings, used since the '20s and invented by A.E. Douglass, range: 11, 000 yrs.], lichenometry [first used by Knut Faegri in '33, range: 10,000 yrs.])
amino acid racemization (relates changes in amino acids to the time elapsed since they were formed; first done by Hare and Mitterer in '67, range: a few 1000 years)
cation-ratio dating (ratio of the more mobile cations to titanium in rock varnish decreases with time; used mostly for archeological artifacts and limited to arid lands)

Previous absolute methods were based on calculations of development of life through the ages, amounts of salinity in the oceans, deposition rates, earth temperatures, denudation rates, and heat received from the sun. (Dating Methods Study Guide-enotes.com; Wikipedia; Geological Time Classification of the USGS, M. Grace Wilmarth, 1925).

Radiometric dating (Radiometric Dating-waikato.ac.nz); Geochronology-Wikipedia; Cosmogenic Nuclide Dating - AntarcticGlaciers.Org; Birkeland and Larson, 1989; paleobiology.si.edu; about.com) is the one used for deep time as the other modern methods do not have sufficient ranges. After the discovery of radioactivity by French physicist Henri-Antoine Becquerel in 1895, the discovery of radioactive decay by British physicist and chemist Ernest Rutherford in 1902, and the discovery of isotopes by British radiochemist Frederick Soddy, Rutherford proposed the method in 1905, with an analytical chemist with the very alliterative name of Bertram B. Boltwood announcing the first such date--140 mln yrs.--for a rock from Connecticut in 1907. With the invention of the mass spectrometer in the '40s, the '50s heralded the modern age for the technique.

Uranium–lead dating is considered superior because there are 2 radioactive uranium atoms (those of mass 235 and 238), so 2 ages can be calculated for every analysis. The age results or equivalent daughter–parent ratio can then be plotted one against the other on a concordia diagram. If the ages concur, the result is concordant, and a closed-system unequivocal age is established. Any leakage of daughter isotopes from the system will cause the two ages calculated to differ, and data will plot below the curve, and the result is said to be discordant. (Dating (geochronology)-Britannica.Com).

The presence of 2 radioactive parents provides a second major advantage because, as daughter products, lead atoms are formed at different rates, and their relative abundance undergoes large changes as a function of time, so the ratio of lead 207 to lead 206 changes by about .1 % every 2 mln. yrs., which is easily calibrated and reproduced at such a level of precision that errors as low as c. 2 mln. yrs. at a confidence level of 95 % are routinely obtained for lead 207–lead 206 ages. By contrast, errors as high as c. 30 to 50 mln. yrs. are usually quoted for the rubidium–strontium and samarium–neodymium isochron methods (an isochron being a line on an isotope ratio diagram).

It is based on 4 assumptions: a closed system [an isolated system in which no parent or daughter isotope can be added or lost], the same initial ratio [a known amount of the daughter element present initially], the same age of a geological system [shown by an isochron], and a constant rate of decay.

Two other techniques were added later on, so there are 3 classes:

I. radioactive decay and isotope ratios: 3 subclasses: a) uranium-lead, thorium-lead, potassium-argon, rubidium-strontium, and samarium-neodymium, b) uranium series disequilibrium (range of up to c. 1/2 mln. yrs.), c) cosmogenic nuclides, which are C 14-N 14 (range up to about 80, 000 years), beryllium-10 and -11, aluminum-26, chlorine-36, helium-3, neon-21, argon-38, and krypton-81; U-Pb, K-Ar, and Rb-Sr have a range up to 4.6 bln. yrs.; rhenium-osmium, lutetium-hafnium, and iodine-xenon, in subclass a), have only limited potential.

II. fission tracking: uses tracks produced by U nuclear fission; developed in '65 by Price, Walker, and Fleischer, range up to 2 bln. yrs.).

III. luminesence: OSL (optical stimulated luminescence) was developed in '84 by David Huntley and colleagues at Simon Fraser in BC, range to 500,000 yrs.; IRSL (infrared stimulated luminescence); TL (thermal luminescence) is good only for dating pottery.

Practically all radiometric dates have been obtained from igneous and metamorphic rocks, whereas fossils, the basis for the relative time scale, are almost exclusively restricted to sedimentary rocks, which usually cannot be radiometrically dated. However, it is possible to bracket the age of the associated fossil-bearing layers and place absolute ages on the relative time scale by dating igneous rocks interlayered with sedimentary ones, dating rock bodies that cut across sedimentary sequences, or dating igneous and metamorphic rocks that uncomformably underlie sedimentary rock sections. This interweaving has allowed the construction of an absolute age framework for the relative time scale resulting in the modern composite time scale. (Birkeland and Larson, 1989).

Birkeland, Peter, and Larson, Edwin. 1989. Putnam's Geology, 5th ed. Oxford U. Press.

Brett-Surman, Michael K.; Holtz, Thomas; Farlow, James (eds.). 2012. The Complete Dinosaur (Life of the Past Series), 2nd Ed. 1st Chp. Indiana University Press (Barnes and Noble Read Sample).

Brookfield, Michael. 2004. Principles of Stratigraphy. Blackwell (books-google).

Drake, E.T. 2007. The geological observations of Robert Hooke. In Four centuries of geological travel (GSL SP 287), Wyse Jackson, Patrick N., ed., Chp. 3, The Geological Society of London (books-google).

Dubois, Jean; Mitterand, Henri; Douzat, Albert. (eds.) 2005. Grand dictionnaire étymologique et historique du français, 2me ed. Larousse, Paris.

Holmes, Arthur. 1913. Age of the Earth (Archive.Org).

Mathur, S.M. 2008. Elements of Geology. Prentice-Hall of India (books-google).

Millar, David, Ian, John, and Margaret. 1996. The Cambridge Dictionary of Scientists. Cambridge U. Press.

Mayor, Adrienne. 2005. Fossil Legends of the First Americans. Princeton U. Press (Amazon Look Inside).

Moore, Ruth.1956. The Earth We Live On. Knopf, New York.

Sissingh, Wim. 2012. Rocky Roads from Firenze: History of Geological Time and Change 1650-1900 (Utrecht Studies in Earth Sciences, Vol. 20) (dspace.library.uu.nl).

Whitten, D.G.A., Brooks, J.R.V. 1985. Dictionary of Geology. Penguin.

Diagnoses

Sporobacteria - eubacteria with spores present or absent; inclusive of Cyanobacteria, Proteobacteria, and Mollifirmicutes.

Cyanidioschyza – alga unicellulare cum uno nucleo, uno chloroplasto, uno mitochondrio, chlorophyllum *a* et C-phycocyaninam; chloroplastus cum due membranae; sine flagella, spora, vacuoli, nec acidi trienoicae; autotrophica necessarie; in duas partes discessu propagatur; loca thermalia acidaque incolit.

Cyanidiobiota – algae cum genome medius et magnitudo ADN chloroplastus medius et chlorophyllum *a* solum.

Glaucobiota - algae cum chlorophyllum *a* solum et flagella; cum protein vel cellulose in murus.

Rhodobiota - algae sine flagella, cum chlorophyllum *a* solum (previously nom. nud. Jeffrey 1971).

Primiphyta – plantae cum radices flagellares cruciatae et chlorophyllum a et b; chloroplastus cum due membranae.

Noviphyta - plantae cum radices flagellares unilateratae et chlorophyllum *a* et *b*; chloroplastus cum due membranae.

Apicophyta - plantae cum apicalibus incrementum, C6-C3 phenylpropanoid-derivatur flavonoid compositum, et similitudines in machinationes mitoticae.

Stomophyta - plantae cum stomata, cellulae custodiae, sporangium collumatum, et D-methionine.

Coscinophyta - plantae cum leptoidium et tamisium-elementum, gametophyta axiti, gametangia extremi, sporophyta pertinax et discernatae interne.

Stelophyta - plantae cum stele, ratio vascularis, cellulae se gero acqua, crassitudines annulares-helicales, tracheidi, et lignum.

Neochromista - algae cum epsilon-carotena, mastigonemata formatur in involucro nucleare et reticulo endoplasmatico, et cum mastigonemata tripartitae.

Dikaryomycotina - fungi unicellulares vel filamentosi, flagellis carentes, saepe stadium dikaryoticum includentes (previously nom. nud. Tehler 1988).

Keratales - fungi cum volva subterranea ubi novella, consistento ex peridium membranose tegento lamina glutaminosa et continento gleba; habiunt odor putridus; type gen. *Keras* nom. nov., gleba ferris in apex stirpes in membrana precaepuus formata conus. (possible syn. *Muscamyces)*

Celestina - actinopods with celestine (also called celestite or strontium sulfate).

Heliocelestina - actinopods with hexagonal axonemal microtubular pattern.

Eulobosa - protozoans with lobose pseudopods, tubular cristas, and no flagella; some shelled, some not; exclusive of Pelomyxidae, Entamebida, and Schyzopyrenida.

The diagnosis for Cyanidioschyzophycota, Cyanidioschyzophyceae, Cyanidioschyzales, Cyanidioschyzaceae is the same as for Cyanidioschyza. Deuterobacteria, Paratoga, Deinothermi, Thermi, Proteinomurus, Mollifirmicutes, Ubiquinonae, Neosoma, Anisokonta, Myxobiota, and Dinociliata are described earlier.

Appendices

Table A1. Data Set – Bacteria (all of the states are 0 absent and 1 present except where otherwise indicated; in TNT the taxons are numbered starting with 0 as are the data).

morphology

flagella

0. proterokonty
1. (proterokontic) position 0 polar 1 lateral 2 peritrichous 3 internal
2. no. of poles 0 monopolar 1 bipolar
3. no. of polar flagella 0 single 1 multiple
4. (proterokontic) BB 0 with inner rings 1 with outer rings
5. (protk.) pass through outer membrane 0 absent 1 present
6. (prtk.) (rotation) 0 with rt-handed helix 1 with left-handed helix

general

7. cell shape 0 coccoid 0 bacillary 0 coryneform 0 vibrioid 0 spiral 1 oval 1 ellip.
8. 0 coccoid 0 bacillary 0 coryneform 0 vibrioid 0 spiral 0 oval 1 ellip.
9. 0 coccoid 0 oval 0 ellip. 1 bacillary 1 coryneform 1 vibrioid 1 spiral
10. 0 coccoid 0 oval 0 ellip. 0 bacillary 0 coryneform 0 vibrioid 0 spiral 1 coryneform
11. 0 coccoid 0 oval 0 ellip. 0 bacillary 0 coryneform 1 vibrioid 1 spiral
12. 0 coccoid 0 oval 0 ellip. 0 bacillary 0 coryneform 0 vibrioid 1 spiral
13. capsule
14. fimbrias
15. toga
16. prosthecas
17. sheathed filaments
18. carboxysomes

chemistry

cell wall

19. peptidoglycan 0 pep 1 psdpep 2 -
20. heteropolysaccharides
21. protein 0 - 1 + 1 glycoprotein
22. 0 - 0 + 1 glycoprotein
23. murein 0 N-acetylated 1 N-glycolated
24. peptide cross linkage 0 A anchors at subunit position 3 and 4 1 B at pos. 2 and 4
25. peptide bridge for A 0 none 1 all
26. 0 1 monocarb 2 dicarb
27. 0 1 polymerized subunits
28. peptide bridge for B 0 L-amino acid 1 D-amino acid
29. amino acids at pos. 3 0 lysine 1 ornithine 1 DAP acid

30. 0 lysine 0 ornithine 1 DAP acid
31. peptidoglycan layer 0 thin 1 thick
32. teichoic acid
33. mycolic acid
34. cell wall outer membrane
35. KDO (ketodeoxyoctanate)
36. LPS (lipopolysaccharide)
37. cell wall with glycoprotein hexagonal array

storage products

38. type 0 PHB (polybetahydroxybutyrate) 1 a-1-4 amino acids
39. LSP (lysine sunthesis pathway) 0 with DAP (diaminopimelic) acid 1 with AAA (aminoadipic acid)
40. glutamine I

proteins (general)

41. cytochromes 0 c 0 a1 0 aa3 0 d 1 b 1 o
42. 0 c 0 a1 0 aa3 0 d 0 b 1 o
43. 0 c 0 b 0 o 0 d 1 a1 1 aa3
44. 0 c 0 b 0 o 0 d 0 a1 1 aa3
45. 0 c 0 b 0 o 0 a1 0 aa3 1 d
46. flagellar protein 0 flagellin 1 pilin-like
47. HSP (heat shock protein) 90
48. actin-tubulin folding chaperonins, no. of subunits 0 7(Group I) 1 8 (thermosome)(Group II) 2 9 (Group II)(TRIC (TCP1 ring complex)(CCT)(chaperonin-containing TCP-1))
49. actin-tubulin folding chaperones 0 GROES 1 GROEL/GIMC (2 subunits) 2 GROEL/GIMC (6 subunits)
50. proteasomes
51. protein secretion mech. 0 post-translational 1 co-translational
52. protein secretion chaperone 0 SecA 1 SecB

enzymes

53. citrate synthase with N-terminal helix
54. citrate synthase sensitvity/inhibition to NADH (reduced nicotinamide adenine dinucleotide)
55. citrate synthase NADH inhibition AMP reactivation
56. citrate synthase inhibition by alpha-oxoglutarate
57. citrate synthase size 0 small 1 large
58. STK (succinate thiokinase) size 0 small 1 large
59. catalase
60. tyrosine kinases
61. oligosaccharyl transferases
62. split glutamate synthase
63. FDP (fructose diphosphate) aldolases
64. FDP-activated lactate dehydrogenase
65. serine proteases 3-D structure

66. SOD (superoxide dismutase) 0 FeMn 1 CuZn coenzymes
67. factor 420

lipids

68. methanopterin lipids
69. membrane lipids 0 straight chain fatty acids with ester bond 1 branched chain aliphatic acids with ether bond
70. diether/tetraether ratios 0 high/low 1 low/high
71. triterpenes 0 - 1 hopanoids 2 sterols
72. carotenoids 0 a-car. 0 A 0 M 0 gm-car. 1 H
73. 0 A 0 M 0 gm-car. 0 H 1 a-car.
74. 0 a-car. 0 M 0 gm-car. 0 H 1 A
75. 0 a-car. 0 gm-car. 0 H 0 A 1 M
76. 0 a-car. 0 M 0 H 0 A 1 gm-car.
77. quinones
78. quinone types 0 benzoquinones 0 anthraquinones 0 anthracyclinones 1 naphthaquinones
79. 0 benzoquinones 0 anthracyclinones 0 naphthaquinones 1 anthraquinones
80. 0 benzoquinones 0 anthraquinones 0 naphthaquinones 1 anthracyclinones
81. sphingolipids
82. sulfonolipids
83. archeol
84. caldarcheol
85. unsaturated fatty acids 0 monoenoic 0 polyenoic 1 cyclopropane 1 10-methyl
86. 0 monoenoic 0 polyenoic 0 cyclopropane 1 10-methyl
87. 0 monoenoic 0 cyclopropane 0 10-methyl 1 polyenoic
88. fatty acid pathway 0 anaerobic 1 Desaturase I 2 Desaturase II
89. fatty acid synthase 0 type II 1 type I
90. long chain diabolic acids
91. waxes
92. PI (phosphatidylinositol)
93. PIM (phosphatidylinositol mannoside)
94. lipid formation pathway 0 malonate 1 mevalonate
95. DOXY pathway

<u>physiology</u>

general

96. cellulose accumulation
97. chitin
98. glycosylation 0 O 1 N
99. S deposition 0 ext 1 int
100. DPA (dipicolinic acid)
101. EMP (Embden-Meyerhoff-Parnas) pathway 0 without PFK (phosphofructokinase) 1 with PFK
102. EMP pathway reversal
103. metabolism 0 fermentation 1 respiration

104. type of respiration 0 anaerobic 1 aerobic
105. carbon sources 0 CH O (heterotrophic) 1 CO (autotrophic)
106. energy sources 0 chemical compounds(chemotrophic) 1 light(phototrophic)
107. electron or H donors 0 organic compounds(organotrophic) 1 inorganic compounds and C(lithotrophic)
108. hyperthermophily 0 + 1 -
109. TCA (tricarboxylic acid) (citric acid cycle)(Krebs cycle) cycle 0 incomplete 1 complete
110. CO fixation (assimilation) pathway 0 hydroxyproprionate 1 reverse TCA 2 Calvin-Benson
111.
112. sulfur compound oxidation
113. sulfate reduction
114. iron oxidation
115. CO oxidation
116. hydrogen oxidation
117. nitrate reduction
118. methanogenesis
119. clastic system
120. adenylate ADP-ATP exchange system

photosynthesis

121. chlorophyllian photosynthesis
122. reaction center pigments 0 bacteriochlorophylls 1 chlorophylls
123. bacteriochlorophylls 0 c 0 a 0 b 0 d 0 g 1 e
124. 0 c 0 e 1 d 1 a 1 b 1 g
125. 0 c 0 e 0 d 1 a 1b 1 g
126. 0 c 0 e 0 d 0 a 0 g 1 d
127. 0 c 0 e 0 d 0 a 0 d 1 g
128. chlorophylls 0 a 1 b 2 c
129. 0 a 0 b 1 c
130. phycobilins
131. carotenoid structure 0 aryl 1 aliphatic 2 alicyclic
132. antenna pigments 0 bchl a or b 1 bchl c, d, or e 2 phycobilins and chl a
133. photosynthetic system 0 chlorosomes 1 cytoplasmic membrane 2 phycobilisomes

general

134. metabolic type 0 anoxygenic 1 oxygenic
135. electron donours 0 H2 1 S or H2S 2 HO2
136. ALA (aminolevulate acid) 0 glycine succinyl Co A 1 L-glutamate

reproduction

137. cell division 0 with septum 1 without septum
138. sporulation
139. sporulation position 0 external 1 internal
140. myxospores
141. budding movement
142. motility

143. gliding
144. longitudinal gliding motility
145. magnetotaxis

molecular biology

ribosomes

146. small subunit 0 without bill, lobe, gap, or platform split 1 with bill only 2 with bill + lobe, gap, and platform split
147. large subunit lobe
148. LSU (large subunit) filled gap
149. LSU bulge
150. 70S subunit association 0 tight 1 loose 5S secondary structure
151. no. of helices 0 4 1 5
152. helix IV base loop nucleotides 0 3 1 4
153. potentially coaxial helices 5S tertiary structure
154. helices 0 4 1 5
155. region E loop
156. G region 0 long 1 short
157. 5S rRNA 5' termini with 0 monophosphate 1 triphosphate ribosomal A protein
158. C- terminal region
159. N-terminal region
160. valine content 0 high 1 low
161. size (no. of residues) 0 large 1 small r-proteins
162. no. 0 54-65 1 60- 65 2 70-80
163. acidity 0 low 1 high
164. r-subunit protein LX
165. IF (initiation facor) hypusine
166. mol. mass of hypusine 0 low 1 high
167. EF (elongation factor)-1 aminoacyl tRNA-to-ribosome catalysis 0 EF-Tu 1 EF-alpha
168. EF-1 G affinity
169. EF-1 inserts 0 4-amino acid 1 11-amino acid
170. EF-2 0 without diphthamide 1 with diphthamide
171. peptidyl tRNA translocation 0 EF-G 1 EF-2
172. EF-2 compatibility 0 + 1 -
173. subunit compatibility

RNA

174. RNAP type 0 simple 1 multicomponent
175. no. of RNAP subunits 0 4 1 8 or more
176. RNAP A
177. RNAP A structure 0 split 1 integral
178. RNAP B
179. RNAP B structure 0 split 1 integral
180. RNAP H

181. RNAP G
182. protein encoded nuclear genes with 0 cis-splicing of miniexons 1 trans-splicing
183. SRP (signal recognition particle) size 0 4.5S 1 7S
184. SRP helices 1-4
185. protein-RNA mass ratio 0 low 1 high
186. tRNA initiator methionine 0 formylated 1 nonformylated
187. G tetra- and pentaphosphates
188. RNAP C
189. 1-methylpseudouridine 0 + 1 -
190. N, N -dimethylguanosine
191.
192. archeosine (in D loop)
193. queuine 0 + 1 -
194. RNA modification levels 0 low 1 high
195. tRNA anticodon loop 0 + 1 -
196. tRNA transamidation of asparagine 0 asparaginyl synthase 1 aspartyl synthase
197. tRNA transamidation of glutamine 0 glutamyl synthase 1 glutaminyl synthase
198. tRNA mischarging
199. tRNA spacers 0 + 1 -
200. tRNA 3' terminal CCA added posttranscriptionally
201. mRNA ends 0 without cap or tail 1 with tail only 2 with cap and tail
202. RNA methionine methyl donor 0 S-adenosylmethionine 1 folate derivative

DNA

203. histones
204. topoisomerase II gyrase activity 0 + 1-
205. 2-stranded DNA repair Ku protein with C-terminal HEH domain
206. type II DNAP VI meiotic protein
207. B DNAPs, no. of subunits per ring 0 1 1 2 2 3 3 4
208. DNA helicase 0 DNAB 1 MCM (minichromosome maintenance)
209. DNA binding protein 10b
210. G-C ratio 0 low 1 medium 2 high
211. sliding clamp

antibio sensitvity RNA polymerase

212. rifampicin
213. streptolydigin
214. actinomycin D
215. novobiocin
216. DNA polymerase aphidicolin butylphenyl-dGTP protein synthesis

ribosome-targeted

217. antibio group A 0 high 1 low
218. antibio group B 0 high 1 low

219. antibio group C 0 low 1 high
220. antibio overall 0 high 1 low
221. antibio very high
222. antibio high
223. antibio low

EF-targeted antibios

224. kirromycin
225. pulvomycin
226. fusidic acid
227. penicillins 0 lo 1 hi

miscillaneous

228. introns 0 - 1 self-splicing 2 protein splicing
229. promoter type 0 eubacterial 1 box A
230. box A in ICR
231. TATA-binding protein (TBP)
232. SECIS (sec insertion sequence) binding protein
233. genome size 0 small 1 large
234. silybin stimulation

<u>addenda</u>

235. flagella 0 external 1 internal
236. gas vacuoles
237. rosettes
238. mycelia
239. wall peptidoglycan with LL-DAP acid
240. wall peptidoglycan with L- and D-lysine
241. wall peptidoglycan with arabose-galactose
242. calmodulin
243. oligosaccaryl transferases
244. proton pumping H-ATPase catalytic subunit insertion 0 F 1 V
245. rhodopsin
246. C40-biphytanyl diol chains 0 acyclic 1 mono and bicyclic 2 tri and tetracyclic
247. S oxidation
248. hyperacidophily
249. corkscrew motion
250. EF-5A
251. DNAP uracil sensitivity
252. DNAP exonuclease function 0 + 1 -
253. tRNA 5' terminal base, molecular stalk, paired
254. (complex) replication factor C
255. HSP (heat shock protein) 10 0 + 1 -
256. plasmologens

257. STK (succinate thiokinase) size 0 small 1 large
258. rotund bodies
259. LSU lobe size 0 small 1 intermediate 2 large
260. epsilon B DNAP
261. division by invagination
262. halophily 0 - 1 moderate 2 extreme

Added characteristics for 2nd analysis

263. reaction center type 0 PS1 1 PS2
264. reaction center and antenna proteins 0 1 1 multiple
265. reaction center and antenna dimers 0 homodimer 1 heterodimer
266. carotenoids
267. chlorophylls 0 bchl a 0 bchl b 0 bchl g 0 chl a and b 1 c 1 d 1 e
268. 0 bchl a0 bchl b0 bchl g 0 chl a and b 0 c 1 d 1 e
269. 0 bchl a 0 bchl b 0 bchl g 0 chl a and b 0 c 0 d 1 e
270. 0 bchl a 0 bchl c 0 bchl d 0 e 1 bchl b 1 bchl g 1 chl a and b
271. 0 bchl a 0 bchl c 0 bchl d 0 e 0 bchl b 1 bchl g 1 chl a and b
272. 0 bchl a 0 bchl c 0 bchl d 0 e 0 bchl b 0 bchl g 1 chl a and b
273. oxidase
274. dissimilatory iron reduction
275. C DNA polymerases 0 Class I 1 Class II 2 Class III
276. H degradation

Deactivated Characteristics for 2nd analysis

5, 24-28, 40, 85-87, 111, 123-129, 140, 145, 154, 169, 191, 235, 249, 257

added for 3rd analysis

277. habitat 0 acquatic 1 terrestrial
278. parasitism
279. ATP coupling factor A
280. methanol

Table A2. Data Matrix for Analysis of Only Prokaryotes Using TNT and Wagner.

0Togabacteria
[01][012]?00-000100000100000000????????000-00?00-----000000000?00000000000000-0-----0---
000000000100000000-0000-00000--0000000000---------------00-0010-
00000000000000000000000000000000-0-000000000000000000000000000000?
00000000000000000000001000000100000-00000 0000000-000---0------?0[01]0
1Thermales
0------0010000000000000000000-101001100?01??[01]0[01]000000000???100000000000-0000000---
0000100000000000000-00011000001-0000010000---------------00-0010-
00000000000000000000000000000000-0-
0000000000000000000000000000001000001000000000000000001000000000000-000000000011-
010---1------[01]000
2Deinobacteria
0------00[01]00000000000000000000-101001000?01?[01][01]0[01]00000000?????10000000000-
0000000---00001000000000001000-0001100011?-00000[01]0000-------------00-0010-
00000000000000000000000000000000-0-00000000000000
000[01]0000000000001000000000000000000000001000000000000-000000000010-000---1------
[01]000
3Rickettsiae
??????0010000100000000000000-110001010001[01][01][01][01][01]00000000??????0000000000-
10000[01]1000[01]0001000000000000000-0001100010?-0000000010---------------00-0010-
00000000000000000000000000000000-0-000000000000000000000000000000?
0000000000000000000000010?0000000000-000000000010-010---0------?000
4Chlamydiae
0------0010000000000000010------------010?01-[01][01]0[01]00000000?????00000000000-0000000---
000010000000001000-0001000010?-0000000010---------------00-0010-
00000000000000000000000000000000-0-0000000000
000000000000000000000000000000000000000000000000-000-000000-000000000010-010---0------?
00[01]
5Planctobacteria
1??????0010000000000010-------------10?01?0?0[01]00000000??????0000000000-0000000---
000010000000001000-0001100010?-0000000000---------------00-011[01]-
00000000000000000000000000000000-0-000000000000000000000000000000?
000000000000000000000001000[01]0-000000-0 00000000010-010---0------[01]000
6Spirochetes
13--1000010110000000000000000-1[01]0001010?01?0[01]0[01]00000000?????00000000000-
00000[01]0---00001000000000000000-0001
[01]000101-0000000100---------------00-0010-00000000000000000000000000000000-0-
000000000000000000000000000000100000000000000000000000101000000000000-000000000[01]10-
010---[01]------[01]000
7Chloroflexi
0----0-?0000000000000000000000-100001000-01?????00000000?????00000000000-
000[01]0[01]100000001000000000001000-00010010[12]0?000000000000100[01][01]00--000000000-
00[01]1000000000000000000000000000000000-0-
00000000000000000000000000000100000000000000000000000010-0000000000-[01]000000000010-
010111[01][01][01]0000?000

8Aquificae
[01][01][01]110?0010000000000000-----110001?10?0?[01][01]00000??000??????[01]?????0?000-
0-----11000?00?????00???0000???01[01]101[01]??-110001000-------------[01]1??0-
00[01]0000000000000000000?0000000000000--------00000?00???0000000000000000
[01]0?????????????????00001000000???0000?100?00000??00??0---0------?000
9Saprospirae
0------0010[01][01][01][01]00000000000000110001?1000?[01]0000-000000?1[01][01]1[01]??0??0?
000-??????[01]100[01]?00???0?00??00?00-???[01][01]0001??-00000[01]000---------------??
[01]100[01][01]-0000000000000000?0000000000000000----00000000000?0000000000000000000101??
10???????????0000010-0000000?00[01]?000000000?1000?[01]---[01]------[01]00[01]
10Chlorobia
0------[01]0[01]000000[01]00000000000010001010?0------0000000?1??????????0?000-0?????0---0?
00???0000??000000?00101111[01]-10000000000[01][01]0[01]0--011[01]010?0-00[01]
[01]000000000000000000000000000000000----000000000000000000000000000?0000????
0???????????0000010-[01]001000?00?100000000??000?01101111000?00
11Acidobacteria
[01][02]??10?00[01]000[01]0000000000000000110001?10?0?000000000000?0??11[01]?????0?000-
[01]00001[01]1000?00?????00??00000-???[01][01][01][01]01??-00[01]00[01]000[01]00[01][01]00--
01[01]00???0-00[01]0000000000000000000[01]000000000000----0000000?00??00000000000?
000000?0????????????????0000010000000000?00?[01]00000000?1000?00??1[01]00000[01][01]00
12Proteobacteria
[01][012]0[01]100[01][01][01]0[01][01][01][01]0[01][01]0000000000[01]100[01]1[01]10[01]0[01]
[01][01][01][01][01]00000000[01]
[01][01]11[01]0000000000-[01][01]0[01]0[01][01][01][01]0[01][01]00100000[01]000[01]
[01]001000[01][01][01][01][01][12][01][01]
[12][01][01][01][01][01][01]000[01]0[01][01][01][01]0--0[01][01][01]10[01]0[01]1[01][01][01][01]
[01][01]000000000000000000000000000000000-0-
00000000000000000[01]000000000000[012]000000000000000000000000
[01]00[01][01]00000000-0000[01]0000[01]10-01[01]111[01]000[01]00[01][01]0[01]
13Cyanobacteria
0------?????10000100000000000-111001010[01]01?000000000000????110000000000-[12]00[01][01]
[01]1[01]000[01]0010110000001000-0001111110[01]2000[01]0000011-----0012021210[01]100[01]
[01]1000000000000000000000000000000000-0-
00000000000000000000000000000000[02]000000000000000000000000010-1000000000-000000000010-
0[01]11100011100[01]0
14Heliobacteria
[01][12]??0-00010000000000000000000-100000000?00?????00000010?????00000000000-0?????0---
00001000000000001000-1001001010??000000000100110 1--0??10??000001[01]-
000000000000000000000000000000000-0-000000000000000000000000000000000?
00000000000000000000000100000 0000000-000000000010-0000001000110?000
15Mollicutes
0------00[01]0[01]0000000-----------------00?01-----000000000-???00000000000-0-----
0000000000000000001000-0001[01]000101-0000000000---------------00-0[01][01]
[01]0000000000000000000000000000000000-0-
0000000100000000000100010000000000001100000000000001000000-00000000000-00000[01]000000-
000---0------[01]010
16Endospora

11--0-000[01]0000000000000000[01]0-[01][01]1100--0?01[01][01][01]00000000000-???
[01]0000100000-0-----11000000100[01]0000001100-100[01][01]101[02]0020[01]0[01][01]
[01]0100---------------0100010-00000000000000000000000000000000-0-
0000000000000000001000100000000000011000000000001000001000001000000-0000[01]
[01]000[01]00-00[02]---0------[01]0[01]0
17Actinobacteria
[01]1--0-0[01]0[01][01][01]00000000000[01][01][01][01]0[01][01][01]11[01]0-00101[01][01]
[01]0[01]0000[01][01]00[01]?001[01]
[01]000[01][01]0000-[02][01]100[01]11000000[01][01]0[01][01]0[01][01][01]010[01]0-[01]00[01]
[01]000[12]01-00000[01]0000---------------0[01]100[01]0-00000000000000000000000000000000-0-
0000000100000000010[01]0100
[01]00002000[01][01]0000000000010000010000[01][01]0[01][01]000-0000[01]0000[01]0 0-000---
[01]------[01]00[01]
18Archeoglobi
????0-?000000000000?0??-----------0--1?00????111???????????0????0?111?0000000---
0011000000000010001-01110[01]0000?-0100001000---------------00-0010-02101???0???0????????
0110111111110????????00???00?00?0110000????????00???0????????0?000??00000---0??0?000?000?
00000000---0------?0?0
19Thermoplasmata
[01]???0-?000000000000---0-----------0--1?00????111???????????0????0100110000000---
0011000000000010001-0111[01]00010?-0000000000---------------00-0010-021010111???0????????
011011111111111111101?100???00100?0110200??????1100?010111111100?00?010000---0??0[01]110?
000?00001000---0------?0?0
20Halobacteria
100?0-100[01]0000100002[01]11-----------0--1?00[01][01][01][01]0111111110???1?0111?0?
001000000[01]1100000000000000000010001-0101[01]01020?30000000000---------------00-0010-
010000[01]1011101111111101101111111110011100110111111110101100001111?
1110011102110110021000 1100000---0111-0001001110000002---1------[01]0?0
21Methanobacteria
[01]0??0-000[01][01][01]00100001[01]11-----------0--1?00?????1111111??????0111?0?
111[01]0000000---0011000000000010001-0111010120?30000001000---------------00-0010-
02[01]0[01]00[01]1???01111[01][01]1101101111111101111001100111 1?
1101011020011111110011101001110021001[01]00000---01100000100011000010[01]---0------?0?0
22Sulfobacteria
[01][02]010-000[01]0000000012011-----------0--1?10?????11211111??????0111?0?
0011000000110000110000000000010001-0111[01]10000?-0000000000---------------10-0010-
021011111???1????1?11111111111111111111
1111011111110111102101111111100111021121110211110100000---0110[12]1101000110002100---
0------?0?0

Table A3. Data Set-Eukaryota-2nd Analysis (the character states are 0 absent and 1 present unless otherwise described).

<u>chloroplasts</u>

general

0. shape 0 spherical 0 cup 1 discoid 2 laminate 3 lenticular 4 stellate
1. 0 spherical 0 discoid 0 laminate 0 lenticular 0 stellate 1 cup
2. position 0 in cytosol 1 in SER 2 in RER
3. membrane envelopes 0 2 1 3 2 4
4. thylakoid no. 0 1 1 2 2 3 3 4+
5. grana
6. girdle lamella
7. phycobilisomes
8. ER continuous with nuclear envelope
9. PR (periplastidial reticulum)
10. nucleomorphs
11. starch position 0 internal 1 external
12. starch type 0 floridean 1 plant
13. DNA arrangement 0 centrally concentrated 1 scattered nodules 2 fine skein of tiny granules 3 ring-shaped nucleoids

stigma

14. position 0 in cytoplasm 1 in plastid
15. photoreceptor type 0 membrane 1 flagellar
16. structure 0 A 0 B 0 D 1 C
17. 0 A 0 C 1 B 1 D
18. photomotive action spectrum type 0 UV-Blue 1 green- yellow 2 red- infrared

pyrenoid type

19. 0 A 0 C 0 D 0 E 1 B
20. 0 A 0 D 0 E 0 B 1 C
21. 0 A 0 B 0 C 0 E 1 D
22. 0 A 0 B 0 C 0 D 1 E

pigments

23. chlorophylls 0 a 0 c 0 b2 (secondary) 1 b1 (primary)
24. 0 a 0 b1 0 b2 1 c
25. 0 a 0 b1 0 c 1 b2
26. xanthophylls 0 zeanthin 0 A 1 B 1 C 1 D
27. 0 zeanthin 0 A 0 C 0 D 1 B
28. 0 zeanthin 0 A 0 B 0 D 1 C
29. 0 zeanthin 0 A 0 B 0 C 1 D

30. 0 zeanthin 0 B 0 C 0 D 1 A
31. reaction center 0 phycobilins 1 chlorophylls

morphology

kinetids

flagella

32. flagella presence
33. relative length 0 isokontic 1 anisokontic
34. position 0 anterior 1 posterior
35. mastigonemes 0 - 1 on 1 flagellum 2 on both flagella
36. mastigoneme structure 0 nontubular 1 tubular
37. mastigoneme composition 0 polysaccharide 1 glycoprotein
38. formation of mastigonemes 0 in GB 1 in perinuclear compartment 2 in ER
39. paraflagellar rod
40. autoflourescence
41. swelling 0 - 1 on short whiplash 2 on long tinsel
42. groove
43. wing
44. trailing flagellum forms undulating membrane
45. collar around base
46. vane
TZ (transition zone)
47. concentric rings
48. helix
49. star
50. constriction
51. striation
52. ED (electron dense) plaque
53. length

basal bodies (kinetosomes)

54. arrangement 0 parallel 1 60 1 90 1 90 with 1 closer to N 1 widely scattered 1 180
55. 0 parallel 0 60 0 90 0 90 with 1 closer to N 0 widely scattered 1 180
56. 0 parallel 0 3 parallel, one 90 0 180 0 4 as cross 0 60 1 90 1 90 with 1 closer to N 1 widely scattered
57. 0 parallel 0 3 parallel, one 90 0 4 as cross 0 60 0 180 0 90 1 90 with 1 closer to the N 1 widely scattered
58. 0 parallel 0 3 parallel, one 90 0 4 as cross 0 60 0 180 0 90 0 90 with 1 closer to the N 1 widely scattered
59. 0 parallel 0 60 0 180 0 90 0 90 with 1 closer to the N 0 widely scattered 1 3 parallel, one 90 1 4 as cross

60. BBs form amorphous EDM (electron dense material)
61. with hairpin
62. with plate
63. with struts
64. MTRs parallel to BBs 0 - 1 terminate at cell surface 2 curve under cell surface
65. MTR forms 3 bands
66. MTR at flagellar base consists of microtubules radiating perp. to BB
67. MTR arise 0 all around from base 1 all around from side(and with EDM rings)
1 on 1 side 1 on 1 side bunched together 1 on 1 side with EDM
68. 0 all around from base 0 all around from side(and with EDM rings)
1 on 1 side 1 on 1 side bunched together 1 on 1 side with EDM
69. 0 all around from base 0 all around from side(and with EDM rings)
0 on 1 side 0 on 1 side with EDM 1 on 1 side bunched together
70. 0 all around from base 0 all around from side(and with EDM rings)
0 on 1 side 0 on 1 side bunched together 1 on 1 side with EDM
71. semicircle of mts around BBs
72. curved arm extends from BB to plasma membrane
73. MTR connection
74. Mtr point of origin lining oral structure 0 BB 0 oral structure lip 1 flagellar inpocketing
75. 0 BB 0 flagellar inpocketing 1 oral structure lip
76. mt bundle surronds cytopharynx
77. mt bundle in cytoplasm
78. mt packing 0 quadratic 1 hexagonal
79. rhizoplast
80. rhizostyle
81. mt ribbons around cytopharynx
82. mts ribbons around cytopharynx as 0 short bands 0 with ED connectors 1 spiralling bands
83. 0 short bands 0 spiralling bands 1 with ED connectors
84. mts perp. to mts
85. mt extends to N
86. N-associated mts as bundle
87. N-associated mts as sheet 0 sheet 0 axostyle 1 spread out as individuals
88. 0 sheet 0 spread out as individuals 1 axostyle
89. cone of mts subtends N
90. linked mts underlie cell membrane 0 as small bands 0 as large bands 1 under entire membrane
91. 0 under entire membrane 0 as small bands 1 as large bands
92. spacing of mt linkage under cell membrane 0 narrow 1 wide
93. mts 0 singlet 1 doublet
94. cytoplasmic mts arise from differentiated pads on NE

rootlets

95. no. 0 2 1 1 2 3
96. single rootlet runs along groove floor
97. split R1

98. ppks (posterior postkinetosomal structure)
99. no. of mts in R4
100. cruciate rootlets 0 + 1 -
101. MLS (multilayered structure)
102. amorphous rootlet with concentric circles
103. striated rootlet
104. rootlet configuration 0 single and extending perp. to BB and parallel to cell surface
1 single and extending into cell 2 between mtr and N membrane
105. crypomonad-type rootlets
106. heterokont-type ribbed rootlets
107. dinoflagellate-type rootlet pattern
108. polymastigote rootlet pattern

extrusomes

109. dinociliate trichocysts
110. spindle trichocysts
111. filamentous trichocysts
112. mucocysts
113. kinetocysts
114. ejectisomes
115. ED bodies with rectangular arrays

pseudopods

116. form 0 lobose 0 reticulate 0 axopodial 1 filose
117. 0 lobose 0 filose 0 axopodial 1 reticulate
118. 0 lobose 0 filose 0 reticulate 1 axopodial
119. 2-way streaming
120. gel tube contraction 0 - 1 unidirectional flow 2 shuttle flow
121. eruptive pseudopodial motion
122. meroplasmodial reticulopod complex
123. pheodarian-type system

reproduction

124. oomycete-type zoosporangial exit canals
125. perforated septa (arising from centripeatal invagination)
126. trichogyne

cell surface

127. cortical alveoli
128. thecal plates
129. peripheral plates
130. dinociliate thecal vesicles
131. mineralized cell body scales

132. prasinophycean-type scales
133. haptomonad scale-forming vesicles

miscellaneous

134. cristal shape 0 lamellar 1 discoid 2 flattened tubular 3 tubular 4 tubular-vesicular 5 ampulliform
135. cristal branching
136. cristal filament
137. dictysomes (Golgi Bodies)
138. dictyosome stacking
139. dictyosome association 0 ER 1 between nucleus and BBs 2 nucleus 3 BBs
140. 0 ER 0 between nucleus and BBs 0 nucleus 0 BBs 1 mitochondrion
141. karyomastigont
142. cytoplasmic organization 0 diffuse 1 transitional 2 zonal
143. tubulemma
144. SDVs
145. mitochondrion
146. ribosome size 0 70S 1 80S
147. ribosomal cap
148. myonemes
149. honeycomb lattice on inner nuclear membrane
150. PNB (paranuclear body)
151. choanocytes
152. tissue
153. hypha/mycelium 0 - 1 + 2 plectenchyma
154. protonemas

addenda

155. BB length 0 short 1 long
156. flagellar direction 0 same 1 different
157. granular pseudopods

<u>chemistry</u>

polysaccharides

158. wall 0 cellulose 1 chitin 2 –
159. chitin
160. storage glycan 0 alpha 1-4 1 beta 1-3

amino acids

161. lysine pathway 0 DAP acid 1 AAA
162. tryptophan pathway with nicotinic acid
163. tryptophan pathway sedimentation patterns 0 0 1 IV 1 I 1 III - VI
164. 0 0 0 IV 1 I 1 III - VI

165. 0 0 0 IV 0 I 1 III – VI

proteins

166. urea assimilation 0 amylase 1 oxidase
167. G3P dehydrogenase cofactors 0 NAD 1 NADP
168. GDH types 0 I 1 II 2 III
169. tannins
170. articulins
171. mycozoan histones
172. protein import mechanism 0 cotranslational 1 posttranslational 2 GB vesicles

lipids

173. carotenoid synthesis pathways 0 a-car 1 C 2 D 3 E-F-G
174. 0 a-car 0 C 0 D 0 E-F-G 1 gm-car 1 b-car 1 D 1 Q 1 R
1 T 1 V-L-J
175. 0 a-car 0 C 0 D 0 E-F-G 0 gm-car 0 V-L-J 1 b-car 2 D
3 Q 4 R 5 T
176. 0 a-car 0 C 0 D 0 E-F-G 0 gm-car 0 b-car 0 D 0 Q 0 R
0 T 1 V-L-J
177. carotenes 0 a 0 e 1 g 1 b
178. 0 a 0 g 0 b 1 e
179. xanthophyll cycle 0 violaxanthin 1 antheraxanthin 2 zeaxanthin
180. sporopollenin
181. gibberellin
182. ophiobolin
183. hopanoids
184. sterol types 0 cycloartenol 1 lanosterol
185. phytosterol side chain alkylation mechanism 0- 1 1 2 2 3 3 4 4 5 5
186. monounsaturated fatty acid synthesis pathway 0 Desaturase I 1 Desaturase II
187. fatty acid omega 9
188. galactosyls 0 - 1 mono 2 di 3 tri
189. sulfonolipids
190. phosphatidic acid
191. sphingolipids
192. phosphonolipids
193. ergolines

<u>physiology</u>

cell division

194. mitosis 0 closed 1 semiopen 2 open
195. spindle position 0 intranuclear 1 extranuclear
196. nonkinetochore mts form central bundle
197. central spindle 0 lateral 1 axil

198. central spindle position 0 in cytoplasm 1 in N
199. TIZ (telophase intranuclear zone) spindle 0 persistent 1 collapsing
200. spindle development/formation 0 intranuclear 1 extranuclear
201. spindle vesicles
202. spindle elongation
203. osmophilic sphere
204. striated bar
205. centriole 0 - 1 V 2 180
206. plaque
207. rings
208. polar structure migration 0 prior to spindle formation 1 during spindle formation 2 mt array that reorients during spindle formation
209. equatorial plate
210. shape of mitotic apparatus 0 bilateral 1 monaxonic
211. position of mitotic apparatus 0 in cytoplsm 1 in N
212. nucleolar behaviour 0 persistant 0 fragmentary 0 associative 0 dispersive 1 discardive
213. 0 persistant 0 discardive 1 fragmentary 2 associative 3 dispersive
214. MTOC position 0 in cytoplasm1 under NE 2 in N
215. kinetochore location 0 on chromatin(in nucleoplasm) 0 collar present 1 on NE
216. 0 on chromatin 0 on NE 1 collar present
217. no. of kinetochore mts 0 1 1 few 2 many
218. close packing of framework (nonkinetochore) mts
219. framework mt organization 0 continuous(pole-to-pole) 1 interdigitating
220. telophase behaviour 0 equatorial constriction of NE 0 septum formed 1 constricted in 2 places 1 new NE formed
221. 0 equatorial constriction of NE 0 constricted in 2 places 0 septum formed 1 new NE formed
222. 0 equatorial constriction of NE 0 constricted in 2 places 0 new NE formed 1 septum formed
223. chromosome pattern 0 uncondensed 1 separate 2 fused
224. chromosomes permanently condensed
225. anaphase chromosome migration (chromosome-pole movements)
226. perinuclear ER cisterna
227. pericentriolar aster

general

228. cell cleavage 0 starts at anterior 1 starts at posterior
229. cell plate
230. cell plate formation 0 furrow 1 phycoplast 2 phragmoplast
231. meiosis(position) 0 zygotic 1 sporic 2 gametic
232. meiosis(no. of steps) 0 1 1 2
233. no. of chromosome sets 0 monoploid 1 diploid 2 mono-diploid
234. type of gametic fusion/production 0 isogamy 1 anisogamy 2 oogamy
235. multiple fission
236. aggregation behaviour
237. conjugation
238. glycolate oxidation 0 oxidase 1 dehydrogenase
239. TCA cycle 0 incomplete 1 complete

240. heterokaryosis
241. gliding motility
242. feeding groove

addenda

243. no. of flagella 0 1 1 2 2 4 3 8 4 8+
244. flagellar transformation 0 anterior to posterior 1 posterior to anterior
245. composite fiber
246. I fiber
247. B fiber
248. C fiber
249. tubulemma
250. testate amebas
251. silica
252. calcium
253. extranuclear spindle development formation 0 – 1 from cytoplasm 2 from between extranuclear structures
254. DHFR-TS gene fusion
255. CAD gene fusion
256. proteinaceous wall 0 + 1 -
257. mastigoneme rows 0 2 on both flagella 0 2 on 1, 1 on the other 0 2 on 1, 0 on the other 0 2 on I, with the other a reduced BB 0 2 on 1 with the other a barren BB 1 1 on both 1 1 on 1, 0 on the other
258. 0 2 on both flagella 0 2 on 1, 1 on the other 0 2 on 1, 0 on the other 0 2 on I, with the other a reduced BB 0 2 on 1 with the other a barren BB 0 1 on both 1 1 on 1, 0 on the other
259. 0 2 on both flagella 0 2 on 1, with the other a reduced BB 0 2 on 1 with the other a barren BB 0 1 on both 0 1on 1, 0 on the other 1 2 on 1, 1 on the other 1 2 on 1, 0 on the other
260. 0 2 on both flagella 0 2 on 1, with the other a reduced BB 0 2 on 1 with the other a barren BB 0 1 on both 0 1 on 1, 0 on the other 0 2 on 1, 1 on the other 1 2 on 1, 0 on the other
261. 0 2 on both flagella 0 1 on both 0 1 on 1, 0 on the other 0 2 on 1, 1 on the other 0 2 on 1, 0 on the other 1 2 on 1, with the other a reduced BB 1 2 on 1 with the other a barren BB
262. 0 2 on both flagella 0 1 on both 0 1 on 1, 0 on the other 0 2 on 1, 1 on the other 0 2 on 1, 0 on the other 0 2 on I, with the other a reduced BB 1 2 on 1 with the other a barren BB
263. Pitelka TZ types 0 I 1 II
264. ribbed R1
265. R3 n R4 support cytostome margins (n R1 mts link into pellicle)
266. cytostome
267. vanes around cytostome
268. vacuoles- position 0 anterior 1 posterior
269. vacuoles- feeding type 0 tubular 1 vesicular
270. no. of pseudopods 0 1 1 mutiple

271. starch type 0 floridean 1 plant 2 animal
272. meiosis
273. sporulation
274. stalked sporophore
275. cell shape 0 spherical 1 non-spherical
276. cellularity 0 single 1 filamentous 2 multicellular
277. nucleation 0 single 1 multiple
278. genome size (in Mbp) 0 under 10 1 between 10 and 20 2 over 20
279. respiration
280. nutritional mode 0 osmotrophic 1 phagotrophic 2 autotrophic
281. thermophily 0 + 1-
282. lobate chloroplasts
283. chloroplast DNA value (in 10, 000 photon units) 0 under 11 1 between 11 and 100 1 over 100
284. phycobilins 0 phycocyanins 1 phycoerythrins
285. peripheral thylakoids 0 + 1 -
286. thylakoid arrangement 0 concentric 1 parallel
287. lamellar interconnections in thylakoids
288. phycobilisome arrangement in zig-zag pattern
289. no. of chloroplasts 0 1 1 2 or more
290. ameboflagellation
291. kineties
292. no. of BBs 0 more than 1 1 1
293. TZ cylinder
294. cartwheel in BB
295. dense transitional column
296. cell wall 0 + 1 –
297. MYTH4FERM myosins
298. TH1 myosins
299. MSD myosins
300. Class 2 myosins
301. chloroplast dividing ring
302. endospores
303. apical complex
304. central capsule 0 - 1 with only 3 apertures 2 with many apertures
305. habitat 0 marine 1 freshwater 2 terrestrial
306. axonemal mt pattern 0 imprecise 1 double helix 2 hexagonal
307. radiating pseudopods
308. hollow siliceous scales
309. ameboid form
310. ameboid test composition 0 organic 1 agglutinate 2 calcareous 3 siliceous
311. 2 parallel kinetosomes interconnected by fibrillar bridges
312. pit plugs
313. sulfated polysacccharides
314. R2 split into inner and outer ribbons
315. R3 associated with array of superficial mts.

316. R4 0 + 1 -
317. R3 originates fromo B2 and curves clockwise from anterior end of cell
318. RM (right mtr)-DL (distal left mtr) association
319. MFS (multilayered fibrillar structure)

Table A4. Data Matrix for 2nd Eukaryotic Analysis.

Hypothetical Ancestor
000000000000000000000000000000000--
-0000000--------000000000000000000000000000000000000000------
000000000000000000000000000000000000000000000000000000000000000000000000000000000000
0000000000000000000000000000000000000000-----0000000000000000000000000
Cyanidioschyzon
000000000000?0?????????
000000000??
0000000????????000000000000000100000011000000000??0?0000??????000012110100?0000??00????
00??????????????????????????????????0??0??0?000??000??????0000010?????????0????000?
1000110000?0?000????0000010002??00?000000000
Cyanidium
000000000000???????????
0001100000???
0000000????????000000000000000100000011000000000??0?0000??????00012110100?0000??00????
00??????????????????????????????????0??0??0?000??000??????00000000?????????0????
0010000110100?11100????0000011002??00?000000000
Galdieria
00000000000003?????????
0001100000???
0000000????????000000000000000100000011000000000??0?0000??????00012110100?0000??00????
00??????????????????????????????????0??0??0?000??000??????00000000?????????0????1010020010?
01????00????0000001002??00?000000000
Glaucobiota
30000000000000?????????00011000011010??0001000000000000101000000000???0000000?
0000000000000000?000001000000?000000????????00010000000001130000011000000000?
1000000??????0001????1000000???01????02?0?0??1?00000?1?????00??????10?00??0??0?100??
0001000000000?100000000?000?01?000?1001111010?0?0?0000000000000001??00?000000000
Stylonemata
10000000000011?????
1100001100000??
00000000????????000000000000001101?0001100000000??020000???0??0001????10000000000??
000011???00?1?000101??03?000?0???00?00?120200001??000??000000001101?????????0????0010?
1011110110-0000????0000100000??00?000000000
Rhodellae
10000000000011?????
1100001100000??
00000000????????000000000000001120?0001100000000??010000???0??0001????10000000000??
000011???00?1?000101??03?000?0???00?00?120200001??000??000000001101?????????0????0000?
0011111111-0000????0000000000??00?000000000
Porphyridia
10000000000011?????
1100001100000??
00000000????????000000000000001101?0001100000000??010000???0??0001????10000000000??
000011???00?1?000001??03?000?0???00?00?120200001??000??000000001101?????????0????0110?

1?11111101-0000????0000000001??00?011000000
Rhodobiota
30000000000001?????
1100001100000??
00000000???????110000000000111 2?000110000011 1??002000???01?
0001312010000000000110000111??00211000001??03?00020?1002?20?120211001??010??
000000011101?????????0????011011211111111111100????0000001001??00?000000000
Plasmodiophorae
?????????????????????????????1100???00000000100000 0??????0000000????000???0?000????
0000??000?00000000?000000000000????????10000000030001??000011000000001??0??0??????
0000???????0000???000000010???0110020001??00?0021?11020010?00??0?000??
0000000000000?0????????000?????010101110 1????????0000010000?0002??00?000000000
Spongomonada
------------------------------1000---000000000000000010000000000000----101001000010010000000000?
000010011000000000000000000000000000000300011000001110000000010-0--0??????0000-------
0000?-?000?0??????????????????????????0???????0?00?00???000??000000000000100?0-------?
001001--?0-1001101--------?000001--------------100000000
Plantae
41003100000111000020011100110001111110000000000000010000??????0???300???000???0?100??
0000000000010000110110000100001000000000000100001101001110000011000001011000000 0???
11??00131111001101051131110121112111110100011 20300021?
10020111112111210111000400000001001010011001000??11111?121111112?011?
11000001001000002??00?000000000
Chlorarachnia
10222000001?2??????????001110001110 10??0000000000000000??????????0?0?????0???00?
000???????000000?0000100??0000100000000010000?000000000000 30001??00001100000000010?
0100??????0002??????000000??0?????021????????????????12??0?0????0000?00?00????000??
000000000000 0?00???????000??????0?1011111 01?010?1100000100000000 0?001?000000000
Cryptomonada
0112100010102100??10100010000010110 12?1000000000000000100000000000000???000??00?01???
0000000000100000001000001000000????????00000000002100110000011000000000010?
0000??????0002301011?00000??11?000021011?001??????111031000?2???20100?00????000?
10000000000000010?1010001000??1?1?0?100111111101001100000100000000 1??00?000000000
Haptomonada
40022000110121?????0000010101111110 10?1010000000001000110000000000000???001??00?00???
00000000000100001000?000000010000100000000000000101300013000001100000000010?
0100??????0000100000100000??0200000 2?011??1100100?112031?0??????10100000?12?000??
000000000000 10001??????1100??1??10?110111112?010?11000001000000001??01?000000000
Heterokonta
2012301011 0??31101?0110010111111110221111101000110001110000010010 00????100??00?110??
110100000111000000001101001001001000100101000000010130101200000110000000011021110 1000
1??00210001111000010140120101121121110110021112032002111002 0101?
2122000110100000001010010100111100010 11??1101211111 2?110?1111000111000000011113?
000110100
Cercomonada
?????????????????????????????11100???
0000000010000000100000000000000000000000000000000001000001001 11001?0000000?

1000000000000000000030011200000110001000001 1?0??0??????0000??????0000???0001?
1021??????????????????????????????0??0?00????000??0110000000000010????????000?1?1??0?
1001101????????10001010000?0002?001?100000000

Apusomonada

-------------------------------1100---00000000000000?110000000000????0010000-0100000000000010?
000?100000000000000??00000000000000000030011--00001100000000?1?-0??00??????000??????-
0000????0????00????????????????????????????????????0---?-000??0112?00000000???-------?000-???-
000?0?1111--------1000101????-0002-001-?00010111

Myxobiota

??????????????????????????????11000??00000000000000??????00000000000000???0?00???????
01000001001110000?0100000000010001010000000000031001??00001100000000001011??0??????
0000010110?0000???0000000211???0110010111??13?001??10010100?00122?000??
00001111100000011???????000?????111111101????????10001111011?0002?001?000111111

Thaumatomonada

-------------------------------1100---00000000100000???????0000?00????0001000?100--??000000-10?
00001001?0000000010010000000000000001002001 1??0001110000000?11-0??0??????000?????---
00??????0????00???????????????????????????????????0-???-000??0112?00000110???-------?000-??1-
000?001111--------10?0001????-0001-0113100010111

Animalia

??????????????????????????????1110???000000111001101101100000000110000000???0?00???
00000000000?00001000?00000000000200000000000000000000011000001100001100000?10?
1??????0010???????00101?000??0?01?0????0?0020??0??03??0??200010??1?002122000??0000?
0000001200 1???????1000????21101201101????????10000011011?0002??00?000001000

Fungi

??????????????????????????????1?10???0000000010011001000001010001111000??00?000??
000000000000100001001000000000000000000000000000000001000301110020020010110111110100
10010110?11101?00201102 11111?1111111211013120011012010111?102121101 01000??
00000000000000011???????000????21101211101????????20001010111?0001??01?000001000

Foraminifera

??????????????????????????????1101???00000000000000???????????0???????0????0??????????
0??000?0000100??0000000000000101100100000000000030011??010011000000000?11?0??0??????
0000???????0000???0?0101000111?10110100210113100???00010110?001130100??1001000000111?
10?001100?000?????10?1101101????????1000001000?0000?0010000000000

Desmothoracida

??????????????????????????????1100???000000000000000?
100000100100000000000000000000000000000?00001001000000011?0010?
000000101010013011120000110100000?10?0??0??????0000???????0000???000001001??11?
111001?0200103?0010?00110101?00111?100??0001000000110?10????????000??11?10?
1001101????????1000001000 0?00010101?000000000

Taxopoda

??????????????????????????????
0???
00000000010000000000000000301110??001101000000?10?0??0??????0000???????0000???
000001001??11?111001?0200103?0010?00110101?00111?100??0001000000110?10????????000??11?
10?1002101????????1000001000 0?000002101?000000000

Centrohelia

??????????????????????????????1100???000000000000000?

1000001001000000000000000000000000000?0000100100000000011?0010?
0000001010100100111201001101000000?10?0??0??????0000???????0000???000001001??11?
111001?0200100?0010?00110101?00111?100??0001000000110?10????????000??11?10?
1001101????????10000010000?00002111?000000000

Celestina

?????????????????????????????????1100???00000000000000?
1000001001000000000000000000000000000?0000100100000000110?0010?
00000001101001101111??0201101000000?10?0??0??????0000???????0000???000001001??11?
111001?0200102?0010?00110101?00111?100??0001000000110?10????????000??11?10?
1001101????????10000010000?00202101?000000000

Pheodarea

?????????????????????????????????1100???00000000000000?
1000001001000000000000000000000000000?00001001000000000000011?
00100000000000000011??0200110000000?10?0??0??????0000???????0000???00000011??1?111001?
0200103?00?0?00110100?00010?100??0001000000110?10????????000??11?10?0001101????????
10000010000?0010?101?000000000

Dinoflagellata

100120000000111?0121110010101111111100??10?11000001000?
1010000000000000001000000000001010000000000010000110100000110000000000000000001111010300
11101000110100000?1000000??????0102015011100000??110001000?00?
0011100102010000111100001101010000?100??0001000000010110?1100000000?11?1110111111112?
010??1000001000000101?001?000010100

Ciliophora

?????????????????????????????????1110???00000100000000???????
00000000000000000100000100000000000000100001001000001101000????????0001001000300111??
00001100000000?00?0??0??????0000???????00011?00001110011002 1?100000?
1110120021211010100?00211?0010110040000000000?10???????001101???10?1001101????????
11000010000?0001??00?000000000

Apicomplexa

??02?000000?1???????????????????11010??0000000000000000??????????0000000?0????0?
00???????????001?00001001?00000000000????????00011000005001 1??00001100000000?0?00?
0??????0000???????00001??0001010111112?10110111101100120 10?0012010 0?000001100?
100010000000000110???????0000????01101001101????????00000010000?010???00?000000000

Nuclearida

?????????????????????????????????
0???
0000000100000000000000000010011200000110000000??0?0??0??????000????????0000????0????
0000????0?0000??0??03??????000???10?0?????010???10??00000000??????????????0?201?000?0?
1111????????00????1?????0001?001?000000000

Vampyrellida

?????????????????????????????????
0???
000000010000000000000000004001???00001100000000??0?0??0??????000????????0000????0????
000?1???????000?11103????????????????0??????00??000??00000000??????????????0????1?
0001011111????????00????1?????0001?101?000000000

Gromida

?????????????????????????????????

0???
0000000100000000000000000003101???0?00110000000??0?0??0?????000???????0000????
00000000?11?0????0???111031??????????????0?????000??0000?00000100?????????????0???1?
0001001111???????00????1?????0000?0010000000000
Euglyphida
??????????????????????????????
0???
0000000100000000000000001003101???0?01110000000??0?0??0?????000???????0000???00?
00000?1??0????0???11?032????????????0?0?????000??0000?00000110?????????????0???1?
0001001111???????00????1?????0002?0113000000000
Jakobida
????????????????????????????????1100???00010001000000??????
00000000000000000000000000000000000102110011000?00000000100????????00000000030011??
1000110000000001010??0??????0000???????0000???000?0?
00000000000000000000000000000000????00????000??0?11011110000?10????????010??????0?
1001101?????????00000100000?0002??00?000100000
Heterolobosa
????????????????????????????????1100???
000000000000000000000000000000000000000000001000000001020100100100010000000000011
000000000000011010??000011000000000010?0?????????0000???????0000???000101001??0??
0100000???00???????????????00??0?010??0002001000000?10???????0110???0?
0111101101????????10000010000?0002?001?000100000
Polymastigota
????????????????????????????????1110???00010101000000??????
0000010000000000000000000010010100102110010010000100000000???????0000000000???10??
100000001000000010?0??0??????0000???????0000???000001000100?0??000001001??
11011000010100?00220?000?000110411100001100???????000100???10?1011001????????
02000010000?000???00?000100000
Euglenaria
200121000000?211102010000111010111000001?
0121010000000000000000110001000000000111010001100000011010200001100?
0000000000000000000000000000300111000001100000000010?0110????1?
0102112010100000410410000001022?10100100001200100??10011100000????
000110101000001000?10?11000011111111??10?200111102?011?11110001000000001??00?000100000
Pelomyxida
????????????????????????????????1?00???0000000000000000??????00000010000000000?
00000000001000000?000010000000000000000000000011000000000000???00??000000000000000?0?0??
0??????0000???????0000???00??0?0?200???1????????1??03?00??????10100?00????100??
0004000000000?10???????0000??11?01?1012001????????10111110000?0001?001?000111100
Eulobosa
??????????????????????????????
0???
000000001001011000000000000310020?11111100000000?0?0??0??????0000???????0000??
1000201011?22??2?0?000?11303100??????10?00?00??1?100??00040000000000010????????000??11?
00?1011101????????1010001101?0001?1011000111100;

Table A5. Data Set – Eubacteria (all of the states are 0 absent and 1 present except where otherwise indicated).

chemistry

1. LPS (lipopolysaccharide)
2. LPS with 0 DAP 1 ornithine
3. wall composition 0 peptidoglycan 1 protein
4. Gram stain 0 + 1 -
5. citrate synthase size
6. SDK (succinate thiokinase) size
7. KDO (keto-deoxyoctanate)
8. NADH (nicotinamide adenine dinucleotide hydride)(coenzyme 1) resistance
9. type of storage product 0 a-1-4 glucans 1 PHB (poly-beta-hydroxybutyrate)
10. carotenoids
11. carotenoid pathway 0 chlorobactene 1 spirilloxanthin 2 okenone
12. carotenoid structure 0 aryl 1 aliphatic 2 alicyclic
13. carotenoid type 0 a-car. 0 A 0 M 0 gm-car. 1 H
14. 0 A 0 M 0 gm-car. 0 H 1 a-car.
15. 0 a-car. 0 M 0 gm-car. 0 H 1 A
16. 0 a-car. 0 gm-car. 0 H 0 A 1 M
17. 0 a-car. 0 M 0 H 0 A 1 gm-car.
18. quinones
19. quinone type 0 menaquinones 0 phylloquinones 1 ubiquinones 1 plastoquinones
20. 0 menaquinones 0 ubiquinones 0 plastoquinones 1 phylloquinones
21. 0 menaquinones 0 phylloquinones 0 plastoquinones 1 ubiquinones
22. 0 menaquinones 0 phylloquinones 0 ubiquinones 1 plastoquinones
23. fatty acids 0 straight chain 1 branched chain
24. cytochrome aa3
25. sphingolipids
26. carboxysomes

physiology

27. metabolism 0 fermentation 1 respiration
28. respiration 0 anaerobic 1 aerobic
29. carbon sources 0 CH O (heterotrophic) 1 CO (autotrophic)
30. energy sources 0 chemical compounds (chemotrophic) 1 light (phototrophic)
31. electron or H donors 0 organic compounds (organotrophic) 1 inorganic compounds and C (lithotrophic)
32. CO fixation (assimilation) pathway 0 hydroxyproprionate 1 reverse TCA 2 Calvin-Benson
33. S and S compound oxidation
34. sulfate reduction
35. S deposition 0 external 1 internal
36. N fixation
37. N oxidation
38. nitrate reduction
39. H oxidation
40. methylotrophy

41. CO oxidation
42. iron oxidation
43. photosynthesis
44. chlorophylls 0 chlorophylls and b 1 bchls a,b,g 1 bchs c,d,e
45. 0 chlorophylls and b 0 bchls a,b,g 1 bchs c,d,e
46. reaction center type 0 PS1 1 PS2
47. reaction center and antenna complexes contained in 0 one protein 1 different proteins
48. reaction center and antenna dimers 0 homodimers 1 heterodimers
49. photosynthetic system 0 phycobilisomes 1 chlorosomes 2 cytoplasmic membrane
50. photochemical lipids 0 MGDs 1 DGDs 2 TGDs 3 ornithine
51. sporulation 0 - 1 exospores 2 endospores
52. multiplication by budding
53. motility 0 absent 1 by flagella 2 gliding
54. corkscrew motion
55. clastic system
56. adenylate ADP-ATP exchange system

mol. biol.
57. G-C ratio 0 low 1 high
58. plasmids 0 circular 1 linear
59. genome size 0 small 1 large

ecology
60. thermophily 0 + 1 -
61. habitat 0 marine 1 freshwater 2 terrestrial

morphology
62. flagellar ring
63. flagellar position
64. outer membrane

addenda
65. murein 0 N-actelylated 1 N-glycolated
66. peptide cross linkage 0 A 1 B
67. peptide bridge for A 0 none 1 all
68. 0 1 monocarb 2 dicarb
69. 0 1 polymerized subunits
70. peptide bridge for B 0 L-amino acid 1 D-amino acid
71. amino acid at position 3 0 lysine 1 ornithine 1 DAP acid
72. 0 lysine 0 ornithine 1 DAP acid
73. teichoic acid
74. mycolic acid
75. acid fastness
76. ATP coupling factor A
77. Q 8
78. Q 10

Table A6. Data Matrix-Eubacteria.

Hypothetical Ancestor
0000000000000000000000000000000000000000000000000000000000000000000000000000
Togabacteria
0-00000000-------0----?00000000000-00000000-------00100000?020000000????00000
Thermi
1100111101???????100001000110000000-00000000-------0000002000110100000000000101
Deinobacteria
1100111101???????110100000110000000-00100000-------00000020111101000000000001?1
Spirochetes
1101111101???????100001000110000000-00000000-------0010101110211100000000000000
Rickettsiae
1001111100?????????????000010000000--0000000-------000000000121010000000000000
Saprospirae
1001111101??000010100010101100000 0-00010000-------10200010212101000000000000000
Chloroflexi
1001111101?1000001000000001011000000000001111112002000101121010000000000000000
Campylobacteria
1001111100-------10000000011000001-00100000-------00?1000001210100000000000010
Enterobactteria
1001111111??00100110100000011000000-00100000-------00100010112101000000000000000
Caulobacteria
1000111110-------110101100110000000-00100000-------010000201111010000000000010
Siderobacteria
1001111100-------1???000011101000-00000010-------001000?0?12101000000000000?0
Myxospora
1001111101?100100100001010110000000-00100000-------10200020212101000000000000000
Pseudomonada
1001111111???????11010110111101000-10010110-------001000101221011000000000000010
Nitrobacteria
1001111100-------?????0101?1101000-01000100-------01100020?12101000000000000?0
Thiobacteria
1001111100-------1????010011101010-00000000-------00210011112101000000000000?0
Methylomonada
1001111110-------????000011101000-00001000--------0100010?12101000000000000?0
Hydrogenobacteria
1001111100-------10000000011101110-00110000-------001000000021010000000000000000
Acidobacteria
1001111101???????10000000011110000-00000001??0???00100010111101000000000000000
Desulfobacteria
1001111110-------10000100010101111-10010000-------0010000011210100000000000000
Chlorobia
1001111101010000110000010010 11?110?10000001110111200?0000001110100000000000000
Rhodobacteria
1001111101220010111010000110111010110100001101002300100000211101000000000000010
Chromatia

1001111101120010111010010110111010100000001101002300100010111101000000000000000
Cyanobacteria
1001111101?00000111101000111111000-10000101001110111200020212101000000000000000
Chlamydiae
1011111100-------????000010000000-0000000?-------00000110012101--------000000
Planctobacteria
1011111100-------10000000011000000000000000-------01200000111101--------000000
Heliobacteria
0-000000011?001001????00010010000-0000000110100 2?20200010?11000000000000000100
Mollicutes
0-00000000-------00000?01000000000-00000000-------000000100120-0------------00
Endospora
0-00000001???????10000100010000001-00000000-------20101010102000000001001100000
Actinobacteria
0-00000001???????10000100011000000-00000000-------1010001011200010111101111000;

.Figure 1. Majority Consensus (>50) Cladogram for 1st Bacterial Analysis (Using TNT and Wagner).

Figure 2. 50% Majority Consensus Cladogram for Only Prokaryotes (Using TNT and Wagner).

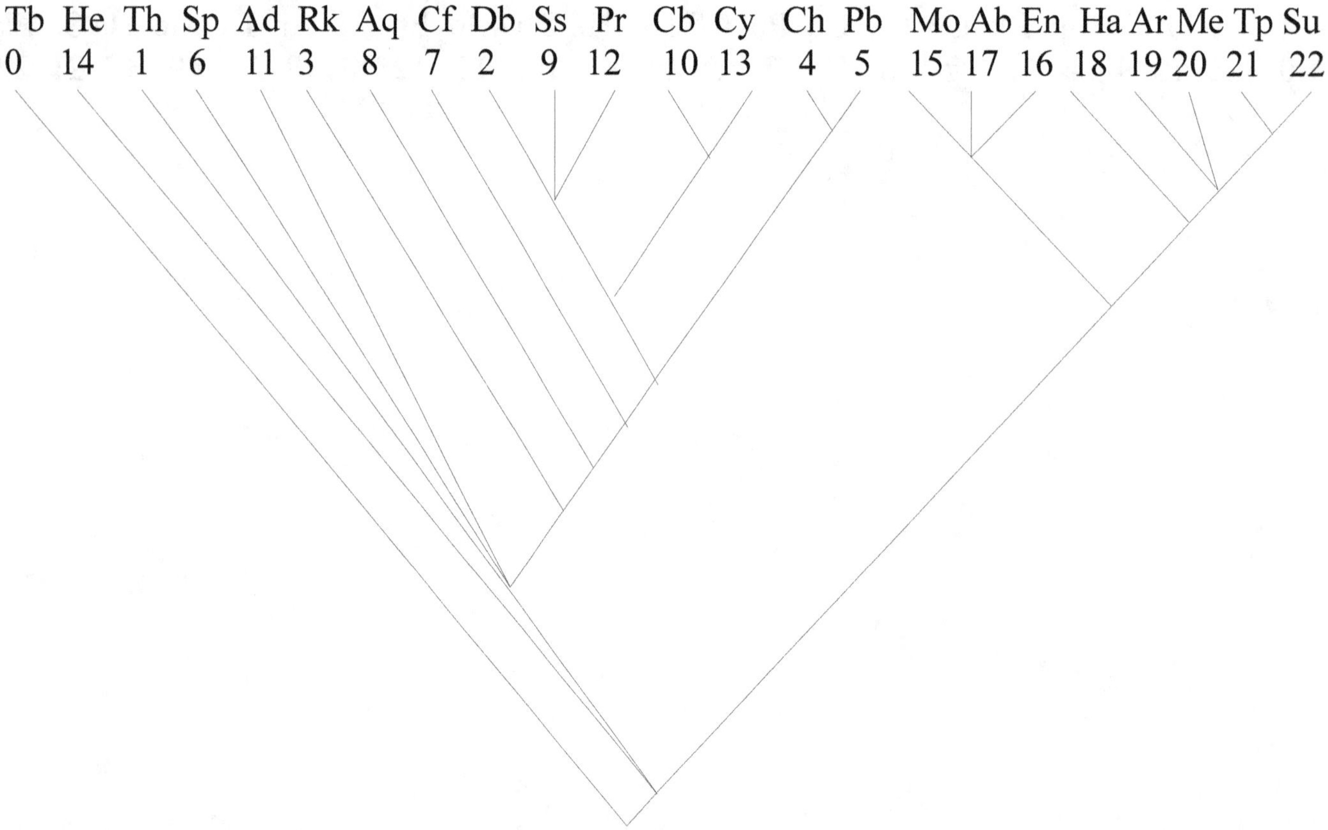

Figure 3. 50% Majority Consensus Cladogram for Prokaryotes+Eukaryota (Using PAUP and Fitch).

Figure 4. 50% Majority Consensus Cladogram for Only Prokaryotes (Using PAUP and Fitch).

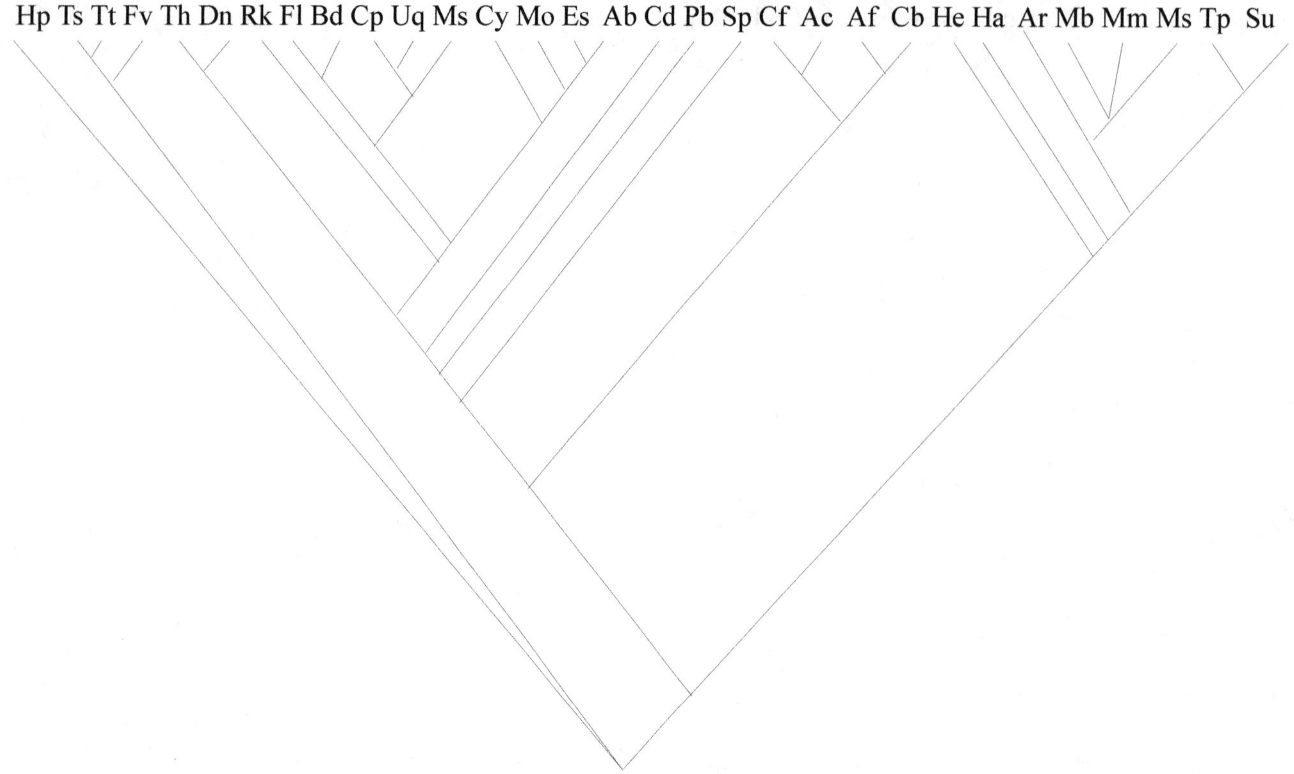

Figure 5. Majority Consensus Tree for 1st Eukaryotic Analysis (Using TNT and Wagner).

Figure 6. Implied Weighting Tree for 1st Eukaryotic Analysis.

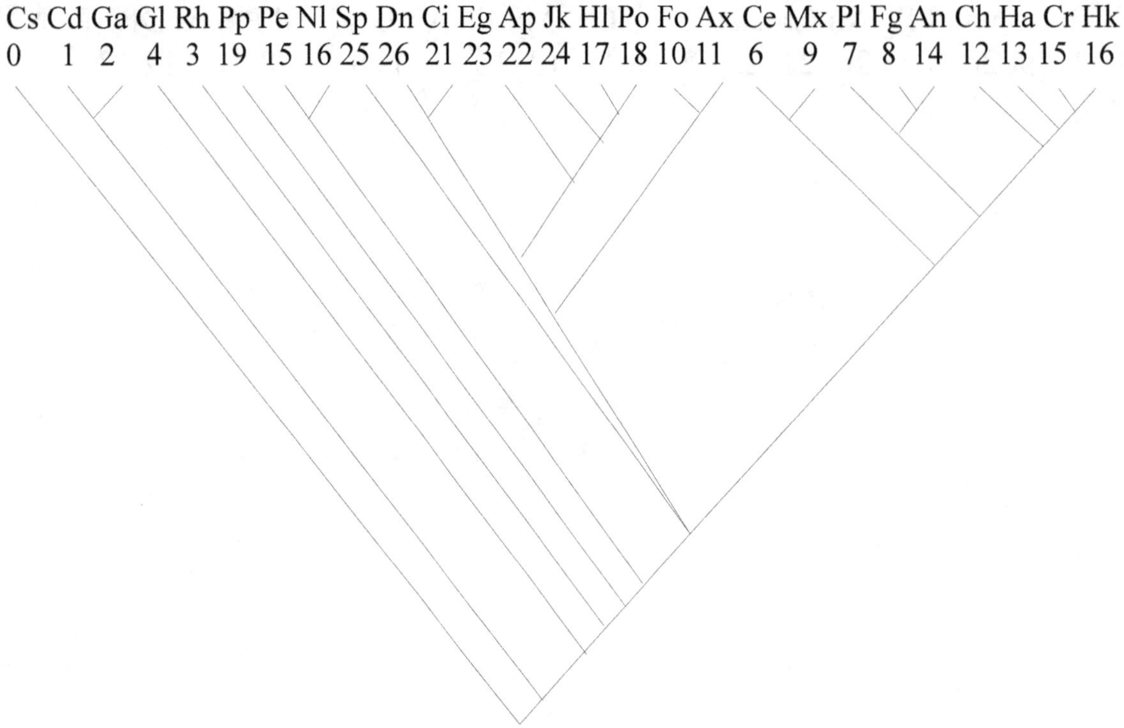

Figure 7. 50 % Majority Consesnsus Cladogram for 2nd Eukaryotic Analysis (Using PAUP and Fitch).

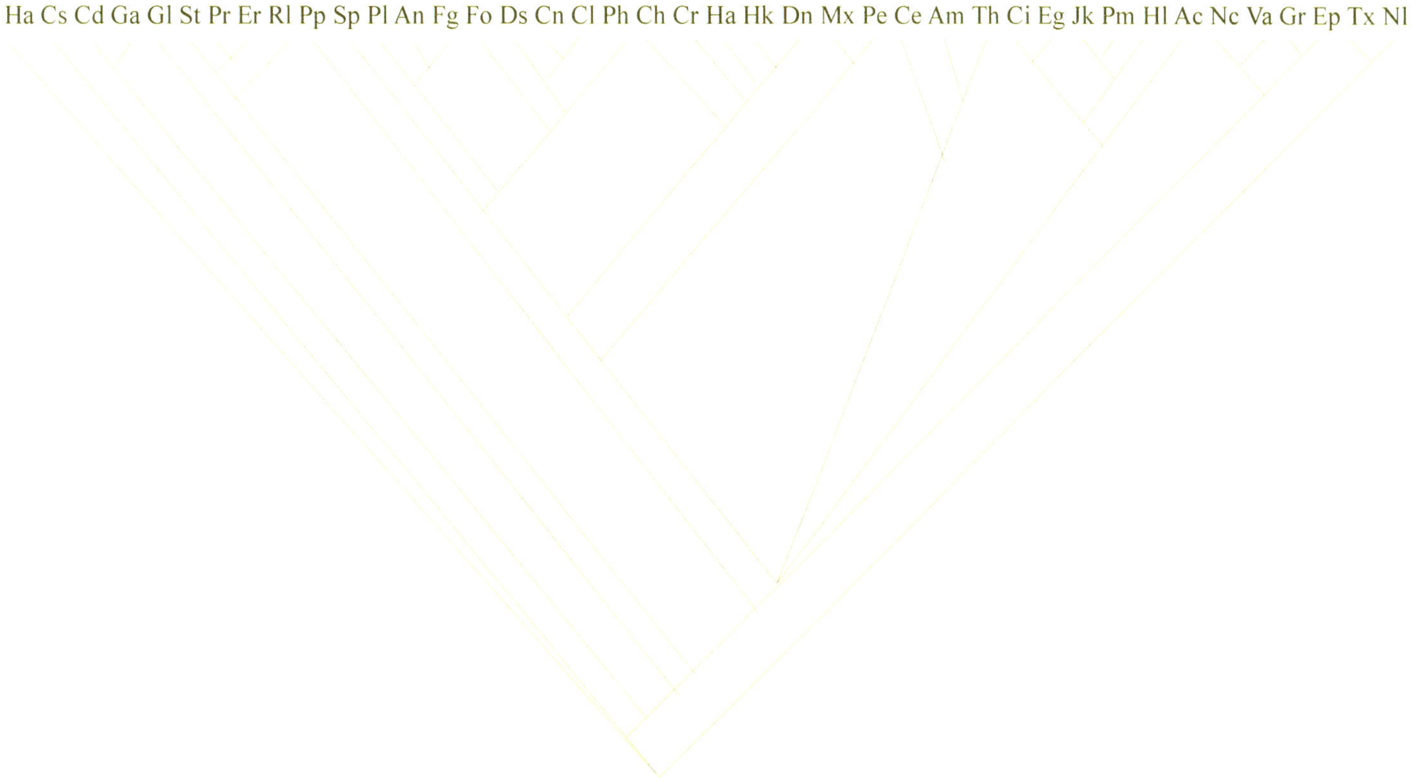

Author Profile

I was born in 1950 in Montreal. I have a Certificate in Metaphysics (ontology and cosmology, 15 lessons) from the Pathways to Philosophy program of the ISP (International Society for Philosophers) (2008), which is affiliated to the University of London; diplomas in adult psychology (2008) and art (2009) at ICS (International Career School, aka International Correspondence Schools); a certificate in Astronomy (2010) and a botany certificate in Plant Biology (2011) at Granton (the Granton Institute of Technology now Granton Tech), and a psychology diploma (3-course program: Human Relations, Psychology and Behavioural Sciences, and Abnormal Psychology) also from Granton (2010); an art certificate from FAS (Famous Artists School) (2011); and a literature certificate from the Saylor Academy (2014). I have taken a graduate course in astronomy at SAO (Swinburne Astronomy On-Line (2008); a linguistics course (Linguistics 101)(2008) and a cultural anthropology course at UNC (Univ. of N. Carolina) (2009); 2 Level 4 courses at Oxford U.: a metaphysics course (Reality, Being, and Existence) (2010) and an epistemology course (the Theory of Knowledge)(2011); and 7 Teaching Company courses: The Theory of Evolution, Origins of Life, Nature of the Earth, Great Ideas of Philosophy, Roots of Human Behaviour (physical anthropology), Masterpieces of the Imaginative Mind (literature), and Understanding the Fundamentals of Music.

I am a member of the ISP (International Society for Philosophers)(lifetime member, since 2007), the International Phycological Society (since 2010), and the Société Française de Psychologie (since 2011), and I was once a member of the Willi Hennig Society and the IAPT (International Association of Plant Taxonomy) (2009-10).

Some of my poems have appeared in 4 anthologies: Quest of a Dream IV ('94, Pacific Rim), River of Dreams ('94, NLP), Illuminating Shadows (2006, SP (Shadow Poetry)), and SP Quill Anthology (2008); and in SP Quill magazine (Autumn, 2006). And I received the SP You've Been Spotted Award (October, 2006).

I have 1045 books, 73 on biology. And I have 5 musical instruments (guitar, recorder, and 3 glockenspiels) but play mostly the glockenspiel (orchestra bells).

www.ingramcontent.com/pod-product-compliance
Lightning Source LLC
Chambersburg PA
CBHW080903170526
45158CB00008B/1979
* 9 7 8 1 3 2 9 8 7 4 0 0 8 *